T0206929

NEUROMETHODS

Series Editor
Wolfgang Walz
University of Saskatchewan
Saskatoon, SK, Canada

For further volumes:
http://www.springer.com/series/7657

Neuromethods publishes cutting-edge methods and protocols in all areas of neuroscience as well as translational neurological and mental research. Each volume in the series offers tested laboratory protocols, step-by-step methods for reproducible lab experiments and addresses methodological controversies and pitfalls in order to aid neuroscientists in experimentation. *Neuromethods* focuses on traditional and emerging topics with wide-ranging implications to brain function, such as electrophysiology, neuroimaging, behavioral analysis, genomics, neurodegeneration, translational research and clinical trials. *Neuromethods* provides investigators and trainees with highly useful compendiums of key strategies and approaches for successful research in animal and human brain function including translational "bench to bedside" approaches to mental and neurological diseases.

Single Molecule Microscopy in Neurobiology

Edited by

Nobuhiko Yamamoto

Laboratory of Cellular and Molecular Neurobiology, Graduate School of Frontier Biosciences, Osaka University, Suita, Osaka, Japan

Yasushi Okada

Laboratory for Cell Polarity Regulation, RIKEN Center for Biosystems Dynamics Research, Suita, Osaka, Japan

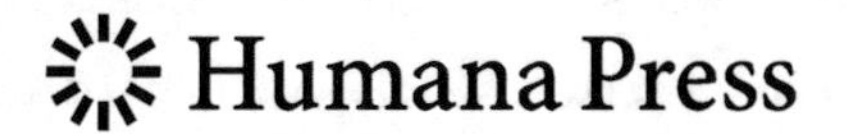

Editors
Nobuhiko Yamamoto
Laboratory of Cellular and Molecular Neurobiology, Graduate School of Frontier Biosciences
Osaka University
Suita, Osaka, Japan

Yasushi Okada
Laboratory for Cell Polarity Regulation
RIKEN Center for Biosystems Dynamics Research
Suita, Osaka, Japan

ISSN 0893-2336 ISSN 1940-6045 (electronic)
Neuromethods
ISBN 978-1-0716-0534-9 ISBN 978-1-0716-0532-5 (eBook)
https://doi.org/10.1007/978-1-0716-0532-5

This Humana imprint is published by the registered company Springer Science+Business Media, LLC, part of Springer Nature.
The registered company address is: 1 New York Plaza, New York, NY 10004, U.S.A.

Preface to the Series

Experimental life sciences have two basic foundations: concepts and tools. The Neuromethods series focuses on the tools and techniques unique to the investigation of the nervous system and excitable cells. It will not, however, shortchange the concept side of things as care has been taken to integrate these tools within the context of the concepts and questions under investigation. In this way, the series is unique in that it not only collects protocols but also includes theoretical background information and critiques which led to the methods and their development. Thus it gives the reader a better understanding of the origin of the techniques and their potential future development. The Neuromethods publishing program strikes a balance between recent and exciting developments like those concerning new animal models of disease, imaging, in vivo methods, and more established techniques, including, for example, immunocytochemistry and electrophysiological technologies. New trainees in neurosciences still need a sound footing in these older methods in order to apply a critical approach to their results.

Under the guidance of its founders, Alan Boulton and Glen Baker, the Neuromethods series has been a success since its first volume published through Humana Press in 1985. The series continues to flourish through many changes over the years. It is now published under the umbrella of Springer Protocols. While methods involving brain research have changed a lot since the series started, the publishing environment and technology have changed even more radically. Neuromethods has the distinct layout and style of the Springer Protocols program, designed specifically for readability and ease of reference in a laboratory setting.

The careful application of methods is potentially the most important step in the process of scientific inquiry. In the past, new methodologies led the way in developing new disciplines in the biological and medical sciences. For example, Physiology emerged out of Anatomy in the nineteenth century by harnessing new methods based on the newly discovered phenomenon of electricity. Nowadays, the relationships between disciplines and methods are more complex. Methods are now widely shared between disciplines and research areas. New developments in electronic publishing make it possible for scientists that encounter new methods to quickly find sources of information electronically. The design of individual volumes and chapters in this series takes this new access technology into account. Springer Protocols makes it possible to download single protocols separately. In addition, Springer makes its print-on-demand technology available globally. A print copy can therefore be acquired quickly and for a competitive price anywhere in the world.

Saskatoon, SK, Canada ***Wolfgang Walz***

Preface

For the past quarter of a century, many gene products, RNAs and proteins, have been identified to operate in the brain for physiological, anatomical, biochemical, and developmental functions. For instance, genetic and molecular biological approaches have identified various synaptic proteins and uncovered their essential roles in synaptic function. Those proteins localize and accumulate in either the pre- or post-synaptic regions, according to their functions. The precise localization of these proteins is inevitable for normal synaptic transmission and is dependent on the intracellular transport systems. Various RNAs and proteins are shown to be transported to their specific destinations in the axons and in the dendrites. The next intriguing issue is how these molecules are expressed and located in specific subcellular regions and function by interacting with other molecules. Conventional biochemical and molecular biological approaches might be insufficient to solve these problems in the highly heterogenous environment of the CNS. There are various cell types. Each cell of the same cell type, or even each synapse in the same neuron, would behave differently according to the neuronal connections and firing and synaptic activities. Therefore, it would be required to examine how the molecules of interest localize and function in specific subcellular locations in a neuron and how it changes dynamically by the physiological or external stimuli or development. Single-molecule imaging techniques enable us to examine the details of the localization and the dynamics of the molecules. There are, however, technical hurdles to apply these techniques to living cells, especially to living neuronal cells. In this book, the experts in this field introduce the methodology and technical issues in detail along with their own results. We believe that it would help readers to start their own single-molecule experiments by themselves.

Suita, Osaka, Japan — *Nobuhiko Yamamoto*
Suita, Osaka, Japan — *Yasushi Okada*

Contents

Preface to the Series v
Preface vii
Contributors xi

1 Single-Molecule Imaging of Intracellular Transport in Neurons and Non-neuronal Cells: From Microscope Optics to Sample Preparations 1
Jay Yuan Jian Wang, Taketoshi Kambara, and Yasushi Okada

2 In Vivo Single-Molecule Tracking of Voltage-Gated Calcium Channels with Split-Fluorescent Proteins in CRISPR-Engineered *C. elegans* 11
Yunke Zhao and Fabien Pinaud

3 Nanocores and Liquid Droplets: Single-Molecule Microscopy of Neuronal Stress Granule Components 39
Benedikt Niewidok, Rainer Kurre, and Roland Brandt

4 Live-Cell Single-Molecule Imaging with Optogenetics Reveals Dynamics of a Neuronal Activity-Dependent Transcription Factor 59
Hironobu Kitagawa, Noriyuki Sugo, and Nobuhiko Yamamoto

5 Single-Molecule Imaging of Recycling Synaptic Vesicles in Live Neurons 81
Merja Joensuu, Ramon Martínez-Mármol, Mahdie Mollazade, Pranesh Padmanabhan, and Frédéric A. Meunier

6 Approaching Protein-Protein Interactions in Membranes Using Single-Particle Tracking and Packing Coefficient Analysis 115
Marianne Renner and Antoine Triller

7 Synaptic Function and Neuropathological Disease Revealed by Quantum Dot-Single-Particle Tracking 131
Hiroko Bannai, Takafumi Inoue, Matsumi Hirose, Fumihiro Niwa, and Katsuhiko Mikoshiba

8 Multipolarization Dark-Field Imaging of Single Endosomes in Microfluidic Neuronal Culture for Simultaneous Orientation and Displacement Tracking 157
Luke Kaplan and Bianxiao Cui

9 Practical Guidelines for Two-Color SMLM of Synaptic Proteins in Cultured Neurons 173
Xiaojuan Yang and Christian G. Specht

10 Single-Molecule Localization Microscopy Propelled by Small Organic Fluorophores with Blinking Properties 203
Akihiko Morozumi, Mako Kamiya, and Yasuteru Urano

11 Highly Biocompatible Super-resolution Imaging: SPoD-OnSPAN 229
Tetsuichi Wazawa, Takashi Washio, and Takeharu Nagai

12 Nanoscale Molecular Imaging of Presynaptic Active Zone Proteins in Cultured Hippocampal Neurons 245
Hirokazu Sakamoto, Shigeyuki Namiki, and Kenzo Hirose

13 Three-Dimensional Super-Resolution Imaging of the Cytoskeleton in Hippocampal Neurons Using Selective Plane Illumination. 261
Frances Camille M. Wu, Feby Wijaya Pratiwi, Chin-Yi Chen, Chieh-Han Lu, Wei-Chun Tang, Yen-Ting Liu, Bi-Chang Chen, and Peilin Chen

14 A Protocol for Single-Molecule Translation Imaging in *Xenopus* Retinal Ganglion Cells. 295
Florian Ströhl, Julie Qiaojin Lin, Francesca W. van Tartwijk, Hovy Ho-Wai Wong, Christine E. Holt, and Clemens F. Kaminski

15 Investigating Molecular Diffusion Inside Small Neuronal Compartments with Two-Photon Fluorescence Correlation Spectroscopy 309
Kazuki Obashi and Shigeo Okabe

Index *329*

Contributors

HIROKO BANNAI • *Japan Science and Technology Agency, PRESTO, ERATO, Kawaguchi, Saitama, Japan; Laboratory for Developmental Neurobiology, RIKEN Center for Brain Science, Wako, Saitama, Japan; School of Medicine, Keio University, Shinjuku-ku, Tokyo, Japan; Department of Electrical Engineering and Bioscience, Waseda University, Shinjuku, Tokyo, Japan*

ROLAND BRANDT • *Department of Neurobiology, University of Osnabrück, Osnabrück, Germany; Center of Cellular Nanoanalytics—Integrated Bioimaging Facility iBiOs, University of Osnabrück, Osnabrück, Germany; Institute of Cognitive Science, University of Osnabrück, Osnabrück, Germany*

BI-CHANG CHEN • *Research Center for Applied Sciences, Academia Sinica, Taipei, Taiwan*

CHIN-YI CHEN • *Research Center for Applied Sciences, Academia Sinica, Taipei, Taiwan*

PEILIN CHEN • *Research Center for Applied Sciences, Academia Sinica, Taipei, Taiwan*

BIANXIAO CUI • *Department of Chemistry, Stanford University, Stanford, CA, USA*

KENZO HIROSE • *Department of Pharmacology, Graduate School of Medicine, The University of Tokyo, Tokyo, Japan*

MATSUMI HIROSE • *Laboratory for Developmental Neurobiology, RIKEN Center for Brain Science, Wako, Saitama, Japan; School of Medicine, Keio University, Shinjuku-ku, Tokyo, Japan*

CHRISTINE E. HOLT • *Department of Physiology, Development and Neuroscience, University of Cambridge, Cambridge, UK*

TAKAFUMI INOUE • *Department of Life Science and Medical Bioscience, Waseda University, Shinjuku, Tokyo, Japan*

MERJA JOENSUU • *Clem Jones Centre for Ageing Dementia Research, Queensland Brain Institute, The University of Queensland, Brisbane, QLD, Australia; Minerva Foundation Institute for Medical Research, Helsinki, Finland*

TAKETOSHI KAMBARA • *Center for Biosystems Dynamics Research (BDR), RIKEN, Osaka, Japan*

CLEMENS F. KAMINSKI • *Department of Chemical Engineering and Biotechnology, University of Cambridge, Cambridge, UK; Department of Clinical Neurosciences, UK Dementia Research Institute at the University of Cambridge, Cambridge, UK*

MAKO KAMIYA • *Graduate School of Medicine, The University of Tokyo, Bunkyo-ku, Tokyo, Japan*

LUKE KAPLAN • *Department of Chemistry, Stanford University, Stanford, CA, USA*

HIRONOBU KITAGAWA • *Laboratory of Cellular and Molecular Neurobiology, Graduate School of Frontier Biosciences, Osaka University, Suita, Osaka, Japan*

RAINER KURRE • *Center of Cellular Nanoanalytics—Integrated Bioimaging Facility iBiOs, University of Osnabrück, Osnabrück, Germany*

JULIE QIAOJIN LIN • *Department of Chemical Engineering and Biotechnology, University of Cambridge, Cambridge, UK; Department of Physiology, Development and Neuroscience, University of Cambridge, Cambridge, UK; Department of Clinical Neurosciences, UK Dementia Research Institute at the University of Cambridge, Cambridge, UK*

YEN-TING LIU • *Research Center for Applied Sciences, Academia Sinica, Taipei, Taiwan*

CHIEH-HAN LU • *Research Center for Applied Sciences, Academia Sinica, Taipei, Taiwan*

RAMON MARTÍNEZ-MÁRMOL • *Clem Jones Centre for Ageing Dementia Research, Queensland Brain Institute, The University of Queensland, Brisbane, QLD, Australia*
FRÉDÉRIC A. MEUNIER • *Clem Jones Centre for Ageing Dementia Research, Queensland Brain Institute, The University of Queensland, Brisbane, QLD, Australia*
KATSUHIKO MIKOSHIBA • *Laboratory for Developmental Neurobiology, RIKEN Center for Brain Science, Wako, Saitama, Japan; Shanghai Institute for Advanced Immunochemical Studies (SIAIS), ShanghaiTech University, Shanghai, China; Department of Biomolecular Science, Faculty of Science, Toho University, Funabashi, Chiba, Japan*
MAHDIE MOLLAZADE • *Clem Jones Centre for Ageing Dementia Research, Queensland Brain Institute, The University of Queensland, Brisbane, QLD, Australia*
AKIHIKO MOROZUMI • *Graduate School of Pharmaceutical Sciences, The University of Tokyo, Bunkyo-ku, Tokyo, Japan*
TAKEHARU NAGAI • *The Institute of Scientific and Industrial Research, Osaka University, Ibaraki, Osaka, Japan*
SHIGEYUKI NAMIKI • *Department of Pharmacology, Graduate School of Medicine, The University of Tokyo, Tokyo, Japan*
BENEDIKT NIEWIDOK • *Department of Neurobiology, University of Osnabrück, Osnabrück, Germany*
FUMIHIRO NIWA • *Laboratory for Developmental Neurobiology, RIKEN Center for Brain Science, Wako, Saitama, Japan; École Normale Supérieure, Institut de Biologie de l'ENS (IBENS), INSERM, CNRS, École Normale Supérieure, PSL Research University, Paris, France*
KAZUKI OBASHI • *Department of Cellular Neurobiology, Graduate School of Medicine, The University of Tokyo, Tokyo, Japan*
SHIGEO OKABE • *Department of Cellular Neurobiology, Graduate School of Medicine, The University of Tokyo, Tokyo, Japan*
YASUSHI OKADA • *Center for Biosystems Dynamics Research (BDR), RIKEN, Osaka, Japan; Department of Physics and Universal Biology Institute (UBI), Graduate School of Science, The University of Tokyo, Tokyo, Japan; International Research Center for Neurointelligence (WPI-IRCN), The University of Tokyo, Tokyo, Japan*
PRANESH PADMANABHAN • *Clem Jones Centre for Ageing Dementia Research, Queensland Brain Institute, The University of Queensland, Brisbane, QLD, Australia*
FABIEN PINAUD • *Department of Biological Sciences, Dana and David Dornsife College of Letters, Arts and Sciences, University of Southern California, Los Angeles, CA, USA; Department of Chemistry, Dana and David Dornsife College of Letters, Arts and Sciences, University of Southern California, Los Angeles, CA, USA; Department of Physics and Astronomy, Dana and David Dornsife College of Letters, Arts and Sciences, University of Southern California, Los Angeles, CA, USA*
FEBY WIJAYA PRATIWI • *Research Center for Applied Sciences, Academia Sinica, Taipei, Taiwan*
MARIANNE RENNER • *Institut du Fer à Moulin (IFM), INSERM, UMR-S 1270, Sorbonne Université, Paris, France*
HIROKAZU SAKAMOTO • *Department of Pharmacology, Graduate School of Medicine, The University of Tokyo, Tokyo, Japan*
CHRISTIAN G. SPECHT • *Institute of Biology of the École Normale Supérieure (IBENS), CNRS, INSERM, PSL Research University, Paris, France*

FLORIAN STRÖHL • *Department of Physics and Technology, UiT The Arctic University of Norway, Tromsø, Norway; Department of Chemical Engineering and Biotechnology, University of Cambridge, Cambridge, UK*

NORIYUKI SUGO • *Laboratory of Cellular and Molecular Neurobiology, Graduate School of Frontier Biosciences, Osaka University, Suita, Osaka, Japan*

WEI-CHUN TANG • *Research Center for Applied Sciences, Academia Sinica, Taipei, Taiwan*

ANTOINE TRILLER • *Institute of Biology of the École Normale Supérieure (IBENS), CNRS, INSERM, PSL Research University, Paris, France*

YASUTERU URANO • *Graduate School of Pharmaceutical Sciences, The University of Tokyo, Bunkyo-ku, Tokyo, Japan; Graduate School of Medicine, The University of Tokyo, Bunkyo-ku, Tokyo, Japan*

FRANCESCA W. VAN TARTWIJK • *Department of Chemical Engineering and Biotechnology, University of Cambridge, Cambridge, UK*

TAKASHI WASHIO • *The Institute of Scientific and Industrial Research, Osaka University, Ibaraki, Osaka, Japan*

TETSUICHI WAZAWA • *The Institute of Scientific and Industrial Research, Osaka University, Ibaraki, Osaka, Japan*

HOVY HO-WAI WONG • *Department of Physiology, Development and Neuroscience, University of Cambridge, Cambridge, UK; Centre of Research in Neuroscience, Brain Repair and Integrative Neuroscience Programme, Department of Neurology and Neurosurgery, The Research Institute of the McGill University Health Centre, Montreal General Hospital, Montréal, QC, Canada*

FRANCES CAMILLE M. WU • *Research Center for Applied Sciences, Academia Sinica, Taipei, Taiwan*

NOBUHIKO YAMAMOTO • *Laboratory of Cellular and Molecular Neurobiology, Graduate School of Frontier Biosciences, Osaka University, Suita, Osaka, Japan*

XIAOJUAN YANG • *Institute of Biology of the École Normale Supérieure (IBENS), CNRS, INSERM, PSL Research University, Paris, France; East China Normal University, Shanghai, China*

JAY YUAN JIAN WANG • *Center for Biosystems Dynamics Research (BDR), RIKEN, Osaka, Japan*

YUNKE ZHAO • *Department of Biological Sciences, Dana and David Dornsife College of Letters, Arts and Sciences, University of Southern California, Los Angeles, CA, USA*

Chapter 1

Single-Molecule Imaging of Intracellular Transport in Neurons and Non-neuronal Cells: From Microscope Optics to Sample Preparations

Jay Yuan Jian Wang, Taketoshi Kambara, and Yasushi Okada

Abstract

Intracellular transport is crucial for maintaining normal function and morphogenesis of cells and mediated by motor proteins. These motor proteins transport a variety of cellular components. Translocation of motor proteins would be highly regulated, since various cargos are transported to their appropriate destinations. Single-molecule investigation of dynamic movement of motor proteins in living cells has enabled us to explore these regulatory mechanisms. Total internal reflection fluorescent microscopy (TIRFM) was a milestone development in microscopy, allowing spatiotemporal investigation of kinetics and dynamics of single molecules in living cells. However, sample preparation is equally important for the successful single-molecule imaging. Here, we describe the equipment of TIRFM as well as the procedures for sample preparations for the single-molecule imaging in living cultured cells.

Key words TIRFM, Intracellular transport, Kinesin, Microtubule, Single-molecule imaging, Transfection

1 Introduction

Proteins, RNAs, organelles, or other cellular components are often localized to specific regions in the cell to do their functions. Such localization is mainly achieved by the directional intracellular transport, especially in polarized cells such as neurons and epithelial cells. Cellular components are transported as various cargoes such as membrane vesicles, protein complexes, or ribonucleoprotein complexes. They are transported along cytoskeletal filaments such as microtubules, by using molecular motors like kinesin superfamily proteins and cytoplasmic dynein [1].

Intracellular transport of cargoes is essential to maintain functions of cells, especially in neurons that have extremely polarized structures with dendrites and an axon. It serves as an essential logistics for the morphogenesis, function, and survival of neurons.

Nobuhiko Yamamoto and Yasushi Okada (eds.), *Single Molecule Microscopy in Neurobiology*, Neuromethods, vol. 154,
https://doi.org/10.1007/978-1-0716-0532-5_1,

Various neurodegenerative diseases are reported to be related to the defects in the intracellular transport systems in neuron [2].

Since the identification of kinesin as the motor proteins that convey the transport, many studies have clarified the molecular mechanisms as well as their biological functions of the transport [3]. However, many fundamental questions are remaining. For example, it has been unclear until recently how many motor proteins or how much force are required to transport cargoes of various sizes through crowded cytoplasm [4]. The navigation mechanism that enables efficient transport of various materials to their cognate destinations is still controversial. KIF5 (a member of kinesin-1) transports various membrane proteins into the axon toward its terminals [5–7], while same KIF5 transports some other membrane proteins such as GABA receptors or RNA granules into the dendrites [8, 9]. Single-molecule imaging of KIF5 in vitro and in living cells has suggested that KIF5 preferentially moves along a small subset of microtubules in the cell, and they serve as the directional cue for the KIF5-based transport [5, 6]. Single-molecule imaging further demonstrated that binding of KIF5 to a microtubule triggers a conformational change in the microtubule to increase its affinity to the next binding of KIF5, which would serve as a self-organizing mechanism for the directional cue [10].

As exemplified in this latest example, single-molecule imaging of motor proteins in vitro and in living cells is a powerful and indispensable method for the study of the mechanisms of intracellular transport. In this chapter, we describe the detailed protocols from microscope optics, and sample preparations to image analysis.

2 Microscope Optics

Most of the widely used fluorescent proteins like GFP or other fluorescent dyes such as rhodamine, Alexa Fluor (Thermo Fisher), Cy-dyes (Thermo Fisher), STELLA Fluor dyes (Goryo Chemical), or ATTO dyes (ATTO-TEC) are bright enough to see single molecules with your naked eyes. If they are very sparsely dispersed in the sample, for example, less than 0.1 nM, each fluorescent molecule is well separated. Modern fluorescence microscope is powerful enough to see each molecule even with the conventional epifluorescence illumination. However, in living cells, the concentration of the target molecule is usually higher. More importantly, culture medium, cytoplasm, and other cellular components contain fluorescent molecules that cause background fluorescence or autofluorescence. Epifluorescence microscope can be used for the imaging of single GFP molecules in living cells, only when the background fluorescence is minimized by optimizing the sample preparations. Instead, total internal reflection fluorescence microscopy (TIRFM) has been widely used. Evanescent field of light is

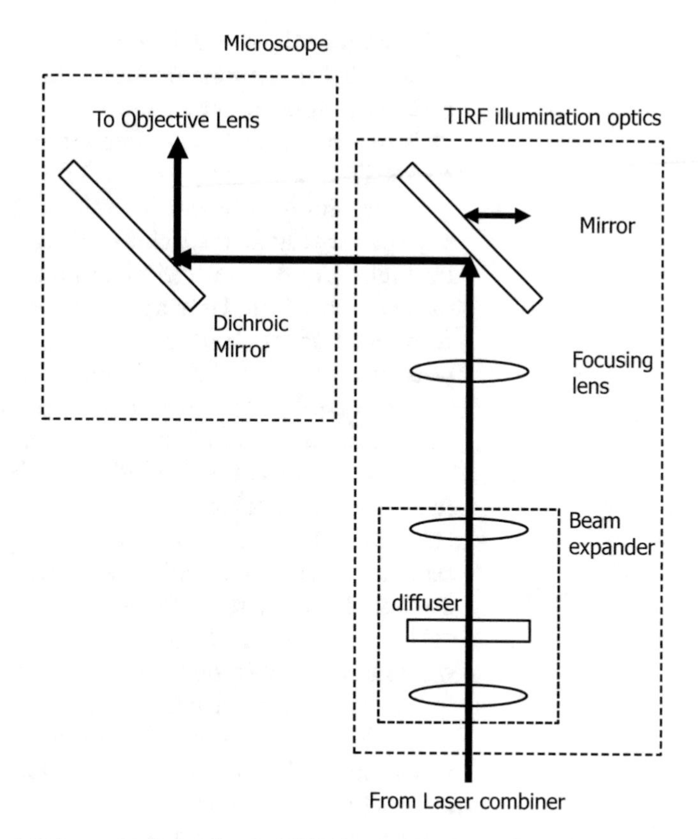

Fig. 1 Schema for the optics for TIRF illumination. It consists of four parts: laser beam expander, focusing lens, diffuser, and mirror. The diffuser is used to reduce speckle noise, and is placed at the focus of the Kepler-type beam expander. A kinematic mirror mount is used for the mirror, which is adjusted to achieve TIRF illumination

generated by total internal reflection and only the basal surface regions of the cell near the cover glass are illuminated, leading to higher contrast of single-molecule images without the background fluorescence from the cellular components and medium above the illuminated basal surface region.

Illumination optics for TIRFM can be purchased from the microscope manufacturers, or can be made by placing several lenses and mirrors (Fig. 1). Multiple laser lines can be introduced into this optics through a laser combiner, which can be purchased from several manufacturers or custom-made. We have been using a TIRFM system based on the Olympus inverted microscope (IX81). Five laser lines of 445, 488, 514, 552, and 640 nm (Sapphire or OBIS, Coherent) are introduced through a custom-made laser combiner (similar combiner is commercialized as LightHUB from Omicron). An acousto-optic tunable filter (AOTF) is used to switch between the laser lines and to control the laser power. 445 nm laser is used for CFP variants such as mTurquoise2;

488 nm for GFPs; and 514 nm for YFPs. Orange-red dyes such as tetramethyl rhodamine or Alexa Fluor 568 and RFPs are excited at 552 nm, while other near-IR dyes at 640 nm. Flatness of the dichroic mirror strongly affects the image quality. We use Semrock's TIRF dichroic mirrors (Di03-Rxxx-t1-25x36).

The highly coherent laser illumination can produce interference patterns such as speckle or fringes. These artifacts are reduced by high-frequency electrical modulation of laser [11] or optical modulation of the laser beam by vibrating diffuser (Optotune or Edmund optics). In our hand, the better suppression of speckle noise was achieved with a rotating diffuser (light stirrer, NANDN, Tokyo, Japan) [12]. The diffuser is placed at the focus of the Kepler laser beam expander (5–20×, depending on the laser power).

The laser beam is focused to the back focal plane of the objective lens. The position of the beam focus on BFP is adjusted by translating the mirror to achieve TIRF. Objective lenses for fluorescence microscope with a numerical aperture (NA) larger than 1.4 can be used for TIRFM, and we routinely use UPlanSApo ×100/1.40, UApoN ×100/1.49, or Plan ApoN ×60/1.42 oil-immersion objective lens (Olympus).

Electron-multiplying charge-coupled device (EMCCD) camera such as Andor iXon3 DU897E-CSO has been widely used for single-molecule imaging, because of its low readout noise and high sensitivity. Recent scientific complementary metal oxide semiconductor (sCMOS) cameras are cheaper alternatives, with enough sensitivity and noise level for single-molecule imaging. We currently use ORCA FLASH 4.0 v3 (Hamamatsu), Zyla 4.0 (Andor), or Prime 95B (Photometrics) at 2 × 2 binning mode to reduce noises. The microscope image is magnified before projection to the camera to the image pixel size of 60–100 nm so that a diffraction-limited image of a single fluorescent molecule is projected to a few tens of pixels.

3 Preparation of the Probe

3.1 Chemically Labeled Protein

Fluorescently labeled target molecule should be introduced into the cell. The most classical and straightforward method is to prepare the target molecule (e.g., protein) and label it in vitro before introducing into the cell. For example, cytoskeletal proteins such as tubulin and actin can be easily purified biochemically. Small and photostable fluorescent dyes can be used for labeling. The labeled protein can be introduced into the cell by several methods, such as microinjection, scrape loading, or electroporation. A detailed protocol for the preparation and the electroporation of fluorescently labeled actin for single-molecule imaging can be found in the literature [13].

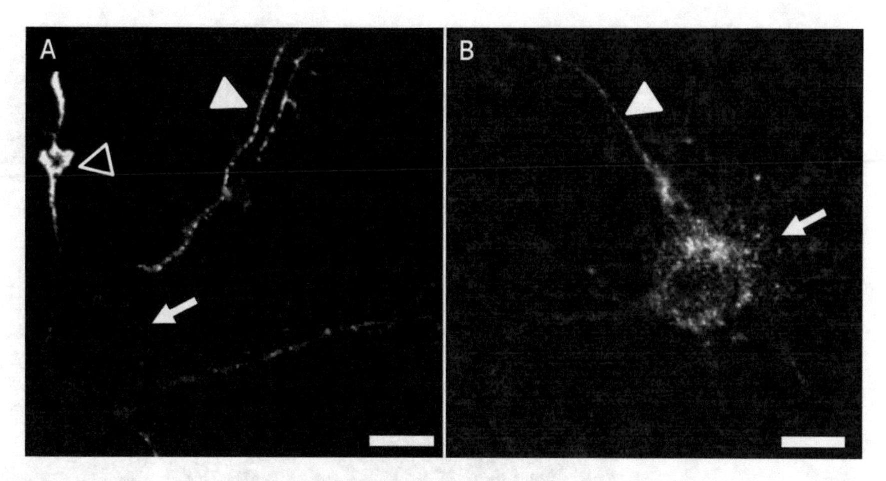

Fig. 2 Mouse hippocampal neuron (DIV5) expressing constitutively active, truncated kinesin (KIF5C 1–560 aa) tagged with EGFP. The plasmid DNA was transfected by calcium phosphate method. (**a**) Neurons were transfected with 2 μg of kinesin plasmid. The expression level was too high, and overexpressed kinesin is mislocalized into dendrites (black arrowhead). (**b**) Lower expression level was achieved by diluting the kinesin plasmid with empty plasmid at a 1:4 ratio. Note that the image contrast is enhanced in this panel to show the much lower fluorescence signals. Each single molecule of kinesin can be traced in the axon (white arrowhead). The arrows show the cell body, and the scale bar is 10 μm

Some expertise in biochemistry such as protein purification and in vitro labeling is required. However, fluorescent dyes are generally much more photostable than fluorescent proteins. More importantly, the protein function is generally less affected by the labeling with small dye molecules than fusion with larger fluorescent proteins. Another practical merit of this method is that the amount of the labeled target protein in the cell is easily and precisely controlled. This is practically important for the single-molecule imaging. If the target protein is expressed too much, it is difficult to observe single molecules in the cell. If the expression level is too low, the target events (binding, movement, ...) would occur too rarely to observe before bleaching. Examples of overexpressed cells are shown in Figs. 2a and 3a.

3.2 Expression as a Fusion Protein

Although the fusion protein can affect the dynamics or the function of the target protein, it is much easier to express fusion protein than purify and label protein in vitro. If bright and photostable fluorescent dye is required to label, Halo-tag (Promega) or SNAP-tag (New England Biolabs) can be used. Therefore, we usually express the target protein as a fusion protein by transfecting the cell with the plasmid DNA.

Since neurons are sensitive to the contamination of endotoxin, we usually purify the plasmid DNA by using EndoFree Plasmid Maxi Kit (QIAGEN). However, for most non-neuronal culture cell lines, the quality of the plasmid DNA is enough, if purified with popular kits such as Plasmid Midi kit (QIAGEN) for transfection of non-neuronal cells. Purified plasmid DNA was diluted to 1 μg/μL

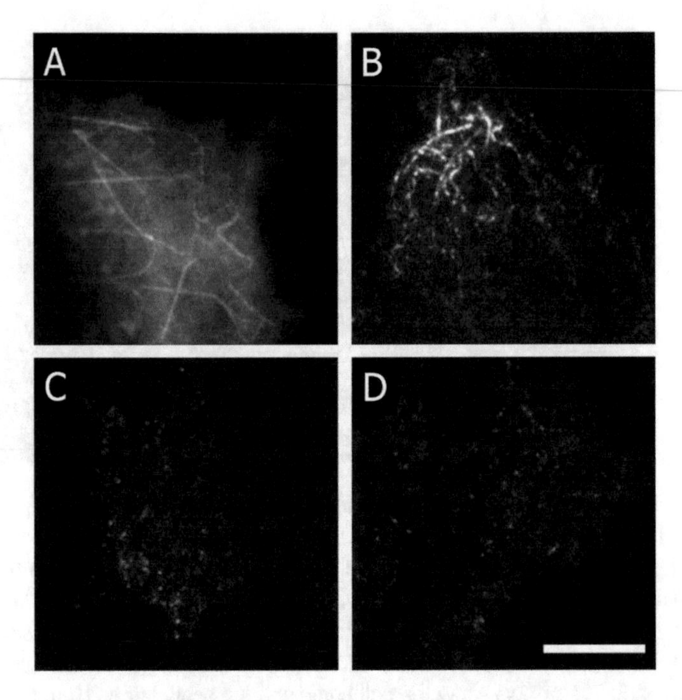

Fig. 3 Attenuation of the expression level for single-molecule imaging. (**a–c**) Mouse embryonic fibroblast (MEF) transfected with the same expression plasmid in Fig. 2. (**a**) The expression level was too high, when the kinesin plasmid was transfected without dilution. (**b**, **c**) The kinesin plasmid was diluted 1:4 with empty plasmid. The expression level was still too high to track single molecules in some cells (**b**), but lower expression cells can be found in the same dish (**c**), which is suitable for the single-molecule tracking. (**d**) The expression level is more uniform and controllable with mRNA transfection. Here, COS7 cells are transfected with 0.5 μg of mRNA coding kinesin-EGFP by electroporation. Scale bar = 10 μm

in endotoxin-free buffer TE for neuron. Sterile TE can be used for non-neuronal cells for plasmid dilution. Generally, the target protein would be expressed too much for single-molecule imaging, if it is expressed from a widely used protein expression plasmid such as pCMV (Stratagene), pcDNA (Thermo Fisher), or pEGFP (Clontech). The expression level can be attenuated by diluting the expression plasmid with noncoding plasmid DNA (e.g., empty vector), or by truncating the promoter region (e.g., CMVd3 vector, Promega).

In many cases, the expression level of the protein does not change linearly to the amount of the plasmid DNA or the promoter strength (Fig. 3a–c). A better control of the expression level can be achieved by transfecting mRNA rather than plasmid DNA (Fig. 3d). A high-quality mRNA can be easily prepared by using the commercial kit such as mMESSAGE mMACHINE High Yield Capped RNA Transcription Kit (Thermo Fisher). The reaction mixture was purified using RNeasy Mini Kit (QIAGEN), according to the manufacturer's instruction. Obtained mRNA was, then, treated with Antarctic phosphatase (New England Biolabs) to prevent the immune activation [14], followed by purification using RNeasy Mini Kit. Purified mRNA was diluted to 1 μg/mL with RNase-free water. The samples (2 μL aliquot) were snap frozen in liquid N_2 and stored at −80 °C.

4 Cells, Transfection, or Electroporation

4.1 Cells

Most of the cultured cell lines such as COS7, MDCK, Vero, and HeLa can be used for the single-molecule imaging. However, pigment granules in melanocytes have strong autofluorescence, which can be reduced by the inhibition of biosynthesis of melanin by phenylthiourea. We also use primary mouse cells, such as mouse embryonic fibroblast (MEF) or primary neurons. Hippocampal neurons are maintained as a mixed culture with glial cells in NbActiv4 medium (BrainBits). The cells dissociated from E16 mouse embryo are plated at a density of 1×10^4 cells/cm^2 to the poly-D-lysine-coated glass (No. 1.5 coverslip, 14 mm glass diameter) in the glass-bottom dish (P35G-1.5-14-C, MatTeK).

4.2 Transfection to Primary Neurons

We use calcium phosphate transfection [15] for neurons by using CalPhos Mammalian Transfection Kit (631312, Clontech). The expression plasmid DNA is mixed with an empty vector DNA at a 1:3–1:9 ratio. Then, 2 μL DNA mixture (concentration of 1 μg/μL) is diluted with sterile water (41.8 μL). 6.2 μL of the calcium chloride solution of the kit is added to make 50 μL DNA-$CaCl_2$ solution. Finally, 2× HEPES buffered saline solution (50 μL) of the kit is added to the DNA-$CaCl_2$ solution in eight installments with 6.2 μL each time. While waiting for the growth of the Ca precipitates for 15–20 min at room temperature, the culture media is withdrawn and set aside in 5% CO_2 incubator at 37 °C, and fresh and pre-warmed NbActiv4 medium (500 μL) is added to the neuron plates. The transfection solution is added dropwise to the culture. After incubating for 3 h in 5% CO_2 incubator at 37 °C, the transfection solution is removed and replaced with 1 mL of NbActiv4 medium preincubated in 10% CO_2 incubator (intentionally acidified to remove Ca precipitates). The culture should be returned to the 5% CO_2 incubator (not to 10% CO_2 incubator). This wash step can be repeated several times at 10-min interval. After the final wash, the original culture medium is returned to the culture, and wait for the protein expression. We usually start observation 4–6 h after transfection, and usually finish observation by ~12 h of transfection.

4.3 Transfection to Non-neuronal Cells

For the transfection of plasmid DNAs to non-neuronal cells, we usually use X-tremeGENE HP transfection reagent (Roche). Briefly, 1 μL of DNA mixture (1 μg/μL, 1:3–1:9 mixture of the target protein expression plasmid and noncoding plasmid) is mixed with 2 μL X-tremeGENE HP transfection reagent in 150 μL Opti-MEM. Then, keep the transfection mixture at room temperature, while trypsinizing the cell to prepare a cell suspension. The cell suspension (0.3–0.6 × 10^5 cells) is mixed with the transfection mixture and plated onto the collagen-coated 35 mm glass-bottom dish (MatTek). Then, the culture medium (DMEM with FBS) is gently added and incubated for 10–12 h in a 37 °C 5% CO_2 incubator before observation.

4.4 Electroporation

Electroporation in micropipette enables us to transfect small volume of cells with high efficiency, and we preferentially use Neon Transfection System (Invitrogen) for lipofection-resistant cells such as primary non-neuronal cells, neuronal cell lines, or primary neurons. Either plasmid DNA, mRNA, or protein can be introduced into the cells. We routinely use a custom-made electroporation buffer (5 mM potassium phosphate, 25 mM HEPES, 2 mM EGTA, 5 mM $MgCl_2$, 150 mM trehalose, 1% DMSO, 2 mM ATP, 2 mM reduced form of glutathione, pH 7.4). The cells are suspended in this buffer at a final density of 2.0 × 10^7 cells/mL, and 0.2–1 μL mRNA (1 μg/μL), 0.5–1 μL plasmid DNA (1 μg/μL), or 0.1–1 μL labeled protein (10 μM) was added to 10 μL of cell suspension. The electroporation condition should be optimized for each cell type. The electroporated cells are mixed with 150 μL culture medium (DMEM with FBS). For plasmid DNA or mRNA, the cells can be plated directly to the glass-bottom dish. However, cells should be washed by centrifugation to remove labeled protein in solution before plating. After 30-min incubation in a 37 °C 5% CO_2 incubator, 0.5–1.5 mL of fresh culture medium is added to the dish. We typically wait 4 h before imaging.

5 Observation and Image Acquisition

Riboflavin in the medium is the major cause of autofluorescence and accelerates photobleaching [16]. Contrastingly, another vitamin, tocopherol, is known to prevent photodamages. We, therefore, use Hibernate E low fluorescence medium (BrainBits) with B27 supplement (MACS NeuroBrew-21, Miltenyi Biotec).

Excitation laser power should be carefully tuned to balance the signal-to-noise ratio of the image and the photobleaching/photodamages. We typically use 1–3 W/cm^2 laser power density (at the sample plane).

Our typical imaging conditions for the EMCCD camera (Andor iXon3 DU897E-CSO) are as follows: readout rate = 10 MHz, frame rate = 30 frames/s, EM gain = 1000, and no binning. Similar imaging conditions are used for sCMOS cameras, but without EM gain and with 2 × 2 binning.

6 Analysis

The image data should be analyzed according to the aims of the experiment. However, it is important to evaluate the image before starting in-depth analyses. Firstly, the size and the brightness of the image of each fluorescence spot should be examined and compared with the control image of the same fluorophore molecule under the same imaging conditions. For example, purified recombinant GFP can be used as a control for GFP-tagged proteins. If the target protein is monomeric in the cell, the fluorescence signal should distribute around a single peak that corresponds to the single GFP signal. If dimers or oligomers are formed in the cell, the signal would show multiple peaks (Fig. 4). Secondly, the time course of photobleaching should be examined. The fluorescence signal

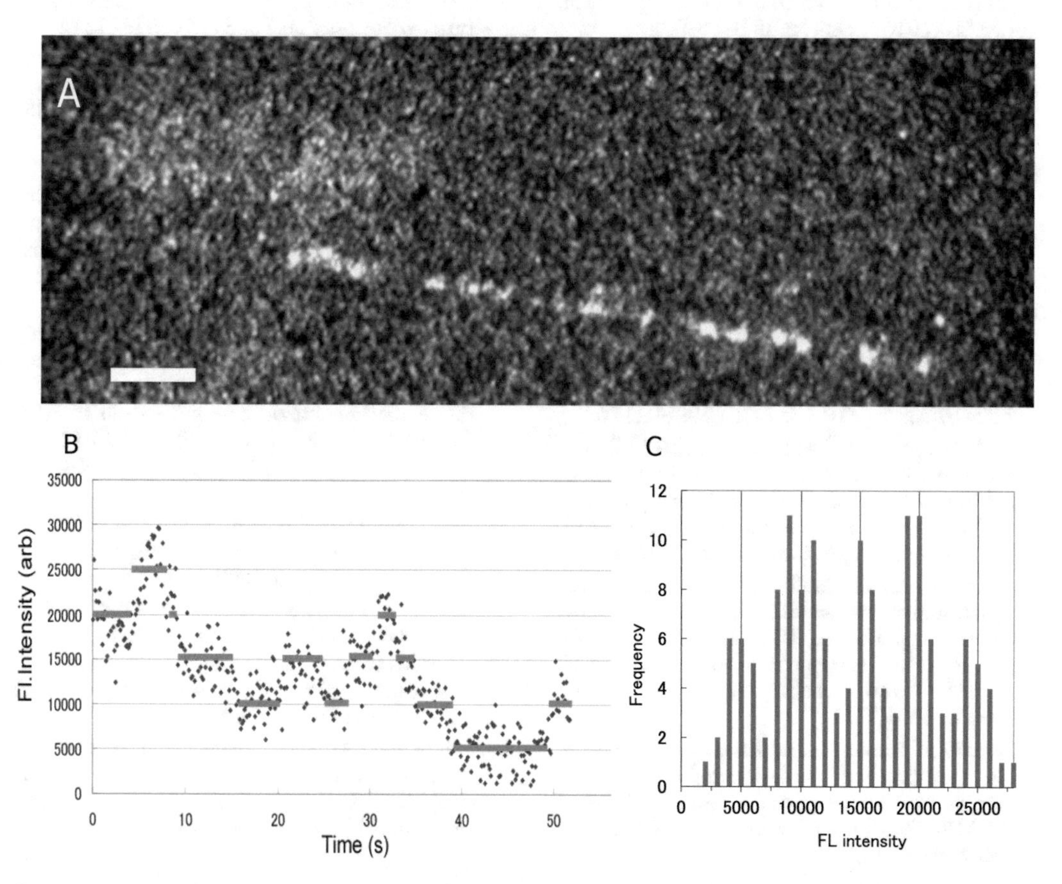

Fig. 4 Example of the analysis of kinesin-EGFP in the neuronal axon. (**a**) A full-length kinesin construct was expressed, and they bound to axonally transported vesicles. Bar, 2 μm. (**b**) A time course or the fluorescence intensity change of a single vesicle. The fluorescence intensity of a single kinesin-EGFP molecule was ~5000 in this imaging condition. (**c**) The distribution of fluorescence intensity of 95 vesicles, which showed multiple peaks at the multiples of 5000, the intensity of single kinesin-EGFP molecule. These results collectively suggest that each axonally transported vesicle can bind 1–5 kinesin-EGFP molecules

should change in a stepwise manner, with the step size of single fluorescence signal. The half-life or the bleaching kinetics should be dependent on the laser illumination intensity. These are the classical but still effective criteria to confirm that a single fluorescent molecule is actually imaged.

References

1. Hirokawa N, Noda Y, Tanaka Y, Niwa S (2009) Kinesin superfamily motor proteins and intracellular transport. Nat Rev Mol Cell Biol 10:682–696. https://doi.org/10.1038/nrm2774
2. Hirokawa N, Niwa S, Tanaka Y (2010) Molecular motors in neurons: transport mechanisms and roles in brain function, development, and disease. Neuron 68:610–638. https://doi.org/10.1016/j.neuron.2010.09.039
3. Hirokawa N, Tanaka Y (2015) Kinesin superfamily proteins (KIFs): various functions and their relevance for important phenomena in life and diseases. Exp Cell Res 334:16–25
4. Hayashi K, Tsuchizawa Y, Iwaki M, Okada Y (2018) Application of the fluctuation theorem for noninvasive force measurement in living neuronal axons. Mol Biol Cell 29:3017–3025. https://doi.org/10.1091/mbc.e18-01-0022
5. Nakata T, Hirokawa N (2003) Microtubules provide directional cues for polarized axonal transport through interaction with kinesin motor head. J Cell Biol 162:1045–1055. https://doi.org/10.1083/jcb.200302175
6. Nakata T, Niwa S, Okada Y et al (2011) Preferential binding of a kinesin-1 motor to GTP-tubulin-rich microtubules underlies polarized vesicle transport. J Cell Biol 194:245–255. https://doi.org/10.1083/jcb.201104034
7. Chiba K, Araseki M, Nozawa K et al (2014) Quantitative analysis of APP axonal transport in neurons: role of JIP1 in enhanced APP anterograde transport. Mol Biol Cell 25:3569–3580. https://doi.org/10.1091/mbc.E14-06-1111
8. Kanai Y, Dohmae N, Hirokawa N (2004) Kinesin transports RNA: isolation and characterization of an RNA-transporting granule. Neuron 43:513–525. https://doi.org/10.1016/j.neuron.2004.07.022
9. Nakajima K, Yin X, Takei Y et al (2012) Molecular motor KIF5A is essential for GABAA receptor transport, and KIF5A deletion causes epilepsy. Neuron 76:945–961. https://doi.org/10.1016/j.neuron.2012.10.012
10. Shima T, Morikawa M, Kaneshiro J et al (2018) Kinesin-binding–triggered conformation switching of microtubules contributes to polarized transport. J Cell Biol 217:4164–4183. https://doi.org/10.1083/jcb.201711178
11. Dulin D, Barland S, Hachair X, Pedaci F (2014) Efficient illumination for microsecond tracking microscopy. PLoS One 9:e107335. https://doi.org/10.1371/journal.pone.0107335
12. Hayashi S, Okada Y (2015) Ultrafast superresolution fluorescence imaging with spinning disk confocal microscope optics. Mol Biol Cell 26:1743–1751. https://doi.org/10.1091/mbc.E14-08-1287
13. Yamashiro S, Mizuno H, Watanabe N (2015) An easy-to-use single-molecule speckle microscopy enabling nanometer-scale flow and wide-range lifetime measurement of cellular actin filaments. Methods Cell Biol 125:43–59. https://doi.org/10.1016/BS.MCB.2014.10.013
14. Avci-Adali M, Behring A, Keller T et al (2014) Optimized conditions for successful transfection of human endothelial cells with in vitro synthesized and modified mRNA for induction of protein expression. J Biol Eng 8:8
15. Jiang M, Chen G (2006) High Ca^{2+}-phosphate transfection efficiency in low-density neuronal cultures. Nat Protoc 1:695–700
16. Bogdanov AM, Kudryavtseva EI, Lukyanov KA (2012) Anti-fading media for live cell GFP imaging. PLoS One 7:e53004. https://doi.org/10.1371/journal.pone.0053004

Chapter 2

In Vivo Single-Molecule Tracking of Voltage-Gated Calcium Channels with Split-Fluorescent Proteins in CRISPR-Engineered *C. elegans*

Yunke Zhao and Fabien Pinaud

Abstract

How voltage-gated calcium channels (VGCCs) mediate signal transductions in response to changes in membrane potential is primarily studied in in vitro and ex vivo systems via biochemical and electrophysiological methods. With the emergence of single-molecule (SM) fluorescence microscopy techniques, it is now possible to characterize the molecular organization and the biophysical dynamics of ion channels in cells with precisions on the order of a few nanometers. However, performing such SM measurements within excitable tissues in intact animals is challenging. Here we describe protocols for an in vivo and tissue-specific SM imaging technique called complementation-activated light microscopy (CALM). By combining native expression of CRISPR-engineered split-fluorescent protein (split-FP) fusions and controlled fluorescence activation of split-FPs in vivo, CALM enables researchers to study the dynamics of individual calcium channels with a precision better than 30 nm directly within neuromuscular synapses of adult *Caenorhabditis elegans* (*C. elegans*) nematodes or at the sarcolemma of their body-wall muscle cells. With the availability of various split-FP spectral variants and of tissue-specific fluorescent markers, CALM can be extended to multicolor and nanoscale dynamic studies of virtually any membrane proteins and channels expressed at physiological levels in live animals.

Key words Voltage-gated calcium channel, Single-molecule microscopy, Split-fluorescent proteins, Membrane diffusion, Quantitative imaging

1 Introduction

Monitoring the motion of individual biomolecules by single-molecule tracking is a powerful approach to quantify their diffuse behaviors at the nanoscale within cells and to determine how their spatial distribution and dynamic interactions with a variety of nanodomains govern cellular functions [1–3]. For instance, key neuronal responses are often driven by a limited number of membrane channels and neuronal receptors whose molecular functions hinge on their subunit compositions, their dynamic partitioning in distinct membrane nanostructures, or their local clustering within pre-

Nobuhiko Yamamoto and Yasushi Okada (eds.), *Single Molecule Microscopy in Neurobiology*, Neuromethods, vol. 154, https://doi.org/10.1007/978-1-0716-0532-5_2,

or postsynaptic membrane areas [4, 5]. These critical molecular events occur on length scales of just a few nanometers and cannot be easily probed using traditional ensemble and diffraction-limited optical imaging methods. Advances made in SM microscopy techniques over the past two decades are now allowing such biomolecular interactions and heterogeneous cellular events to be investigated at the nanoscale.

One exciting avenue in SM research is to extent the exquisite sensitivity of SM microscopy to live-animal imaging, for instance to quantitatively study the nanoscale dynamics, the stoichiometry, or the spatial distribution of membrane channels and receptors under controlled cellular homeostasis and at endogenous expression levels within intact tissues in vivo [6]. Indeed, recent advances in gene editing technologies by clustered regularly interspaced short palindromic repeats (CRISPR-Cas9) [7, 8] now provide efficient means to insert fluorescent protein (FP) tags at given chromosomal loci for protein tagging and for SM imaging in genetically engineered cell lines and animal models. A key requirement to achieve nanometer precision imaging in a live animal is to maintain a sub-nanomolar SM detection regimen within specific cellular areas of the 3D tissues and to ensure high-contrast detection by limiting out-of-focus and background signals. Genomic tagging with photoactivable FPs does allow for sparse detections by light-induced activation of a controlled number of molecules [8]. Yet, the ability to turn on photoactivatable FP fusions and image them at specific subcellular structures (e.g., only at the plasma membrane) is complicated by the unknown orientation of tissue-embedded cells with respect to the optical axis of the microscope. Hence, high signal-to-noise ratio cell membrane imaging such as that obtained in vitro by total internal reflection fluorescence (TIRF) microscopy [9, 10] is difficult to achieve when attempting to track the dynamics of individual ion channels in vivo.

Recently, we developed an in vivo SM imaging technique called complementation-activated light microscopy (CALM) [6, 11] where the controlled activation of FP fusions is accomplished by physical contact of small complementary peptides with FPs that have been asymmetrically split into fragments (Fig. 1). These split-FP (sFP) fragments are obtained by the strategic removal of one or more β-strands from the β-barrel structure of a FP [12, 13]. For instance, complementary split-green fluorescent protein (GFP) fragments are obtained by splitting a super-folder GFP into a nonfluorescent GFP 1–10 domain (sGFP, amino acids 1–214) and a small peptide domain corresponding to the terminal 11th β-strand of the super-folder GFP β-barrel (M3, amino acids 214–230) [12]. Both fragments, including synthetic versions of the M3 peptide, spontaneously and irreversibly self-assemble in solution to form a fully folded GFP with a mature chromophore [13]. In CALM, the large and nonfluorescent sGFP fragment is

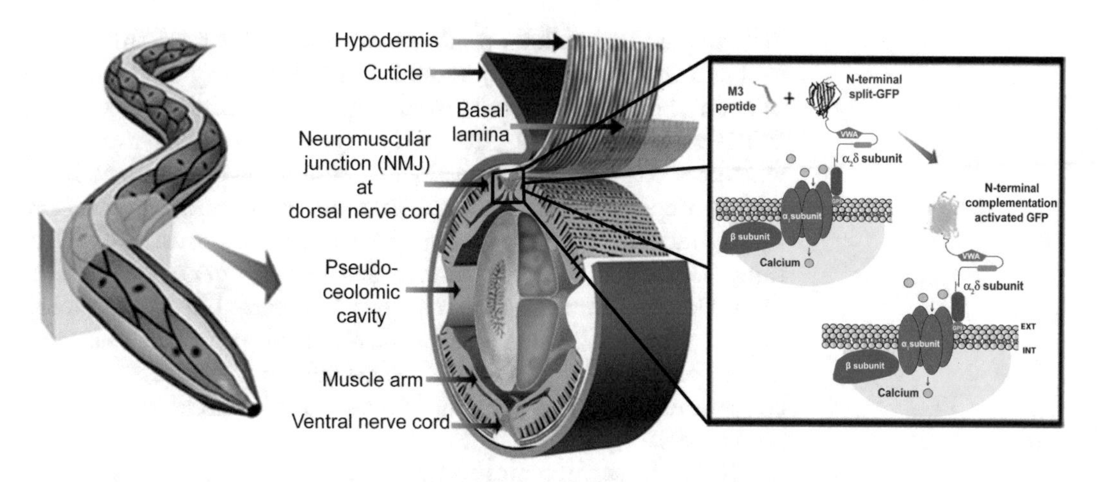

Fig. 1 CALM imaging of individual VGCCs in *C. elegans* using CRISPR-Cas9-mediated tagging of the $\alpha_2\delta$ subunit with split-GFP fusions. Schematic of *C. elegans* tissues and of functional VGCCs, which are comprised of a α_1-subunit, a β-subunit, and a $\alpha_2\delta$ subunit fused to a nonfluorescent split-GFP. In CALM, synthetic M3 peptides are microinjected in the pseudocoelomic cavity of worms to control the fluorescence activation of GFP signals by complementation of the split-GFP fusion and to track individual VGCCs at neuromuscular junctions or at the membrane of muscle cells with nanometer precision. (Adapted from Zhan et al. [6])

fused to a protein of interest and expressed in cells or tissues as traditionally done with a protein fused to a full-length FP. Controlled detection of the fusion proteins is then achieved by reconstitution of the β-barrel, chromophore maturation, and fluorescence activation of the sGFP fragment upon binding synthetic peptide sequences that encode the missing 11th β-strand [6, 11, 13]. By adjusting the concentration of exogenous complementary peptides and tuning incubation times, a controlled subset of split-FP fusions are switched on, thus allowing SM detections at specific subcellular locations and with minimal background interference both in cells and animals. Indeed, switching sGFP fusions from dark to bright depends on their physical interaction with the exogenous complementary M3 peptides, such that fluorescence detection by CALM is spatially limited to the sites of complementation. This is particularly advantageous for SM imaging of membrane channels in vivo because the complementation of split-FP fusions only takes place at the plasma membrane of expressing cells and high-contrast SM detections are not degraded by out-of-focus intracellular GFP fluorescence. This allows for high-resolution localization of individual channel positions by Gaussian fitting and nanometer precision tracking of their diffusion dynamics at the plasma membrane.

We have used CALM imaging to study the nanoscale membrane dynamics of individual voltage-gated calcium channels (VGCCs) within excitable tissues of live *C. elegans* adult worms [6] (Fig. 1). VGCCs are multi-subunit channels composed of a main pore-forming α_1 subunit and up to three auxiliary subunits, including an intracellular β subunit, an extracellular $\alpha_2\delta$ subunit, as well as a transmembrane γ subunit [14]. According to the diverse calcium currents detected in mammals, five different types of VGCCs (L-, N-, P/Q-, R-, and T-types) have been described, which are encoded by ten distinct α_1 subunit genes [14]. VGCCs initiate depolarization-activated calcium influx and serve as key signal transducers to trigger neurotransmission, muscle contraction, secretion, and calcium-dependent gene regulation [15]. Using SM tracking microscopy and diverse bio-tagging techniques, various diffusion studies have started to reveal the importance of VGCC mobility in modulating calcium domain cooperativity during synaptic release and muscle contraction [6, 16–19].

In *C. elegans*, genetics and molecular studies have revealed that the N/P/Q-type and L-type VGCC α_1 subunits are, respectively, encoded by the genes *unc-2* and *egl-19*, compared to their homologs in mammal [20]. As shown by calcium imaging and electrophysiological recordings, these two types of VGCCs play essential roles in electrochemical coupling at presynaptic terminals [21] and in excitation-contraction coupling at the sarcolemma of body-wall muscle cells [6, 22, 23]. The N-type (including P/Q types) α_1 subunit UNC-2 is known to be localized in motor neurons, in several sensory neurons, as well as in hermaphrodite-specific neurons, and its proper function requires its association with a UNC-36 $\alpha_2\delta$ subunit [20, 24–26]. A putative β subunit CCB-1 also co-localizes with several presynaptic markers but the functional interaction between CCB-1 and UNC-2 remains unclear [27]. In striated muscles, the L-type α_1 subunit EGL-19 associates with the $\alpha_2\delta$ subunit UNC-36 and the putative β subunit CCB-1 to form the only functional VGCC for conducting voltage-dependent calcium currents during contraction [28, 29].

This chapter describes (1) methods to establish a CRISPR-based transgenic worm strain where the VGCC $\alpha_2\delta$ subunit UNC-36 is N-terminally fused to sGFP and expressed from its original genomic locus and (2) methods for SM imaging of neuronal and muscular VGCC dynamics by CALM at physiological states in live *C. elegans*.

2 Materials and Equipment

2.1 *C. elegans* Strains

C. elegans strains are grown on NGM agar plates at 20 °C on *Escherichia coli* OP50 lawns. N2 Bristol wild-type worms, UNC-29-tagRFP (kr208[unc-29::tagRFP]) worms, a kind gift of Dr. Jean-Louis Bessereau, and UNC-36-split-GFP/UNC-29-tagRFP (yk25135[UNC-36::sGFP];kr208[unc-29::tagRFP]) CRISPR-engineered worms are used.

2.2 DNA and RNA

The targeting vector pHZ043 described in a previous study [6] is used as a PCR template to generate homology-dependent repair templates (HDRT) in CRISPR-Cas9 genome editing experiments. All DNA oligos are ordered from Integrated DNA Technology (IDT) through standard desalting. Phusion High-fidelity DNA Polymerase (NEB, M0530S) is used for the amplification of HDRT, which is later purified using MinElute columns (Qiagen, 28006). To release genomic DNA from worms for genotyping, we use a lysis buffer (50 mM KCl, 10 mM Tris pH 8.3, 2.5 mM $MgCl_2$, 0.45% NP-40, 0.45% Tween-20, 0.01% gelatin) with 6% of 10 mg/mL proteinase K (Sigma, P2308). The single chimeric guide RNA (sgRNA) for CRISPR experiments is transcribed in vitro using a MEGAshortscript T7 kit (Thermo Fisher, AM1354). The reaction is quenched with DNase I from the kit, after which the sgRNA is purified using Ultrapure phenol:chloroform:isoamyl alcohol (Thermo Fisher, 15593031) at 25:24:1 v/v/v ratio. The final sgRNA product is dissolved in RNase-free water and its concentration is adjusted to 2 μg/μL. The quality of sgRNA can be verified on a Nanodrop UV/Vis spectrophotometer ($\lambda_{260}/\lambda_{280} \sim 2.0$) and by running it on a 6% urea polyacrylamide electrophoresis gel (PAGE). Alternatively, the sgRNA can be directly purchased from commercial vendors.

2.3 Worm Microinjection and Imaging Reagents

Injection Halocarbon oil 700 (Sigma, H8898), purified synthetic complementary peptides such as M3-biotin (biotin-acp-GSGGGSTSRDHMVLHEYVNAAGIT) or M3-cysteine (C-acp-GSGGGSTSRDHMVLHEYVNAAGIT) (LifeTein, LLC. 75% purity), and a peptide injection buffer (recipe modified from Gottschalk and Schafer [30]: 20 mM K_2HPO_4, 3 mM potassium citrate, 2% PEG 3000, pH = 7.4) are used. For CRISPR-Cas9 injection mix, we use a 10 mg/mL Cas9 endonuclease solution, which is a kind gift from Dr. Thomas Duchaine. Cas9 endonucleases can be purified from a bacterial expression system following a protocol available on the laboratory website of Dr. Geraldine Seydoux [33]. Alternatively, the Cas9 protein can be purchased from commercial vendors. This chapter does not include a protocol for Cas9 production. Injected worms are recovered in standard M9 buffer (42 mM Na_2HPO_4, 22 mM KH_2PO_4, 86 mM NaCl, 1 mM MgSO4, pH 7.2) and recovery buffer (5 mM HEPES pH 7.2, 3 mM $CaCl_2$, 3 mM $MgCl_2$, 66 mM NaCl, 2.4 mM KCl, 4% w/v glucose). The injection and imaging pads are both made of 2% low-melting-point agarose (Invitrogen) in M9 buffer which is dried on 24 × 60 mm micro cover glass (injection pads, VWR, 48393-106) or on micro slides (imaging pads, VWR, 48300-048). Live worms are anesthetized in 1 mM levamisole (Sigma) or 30 mg/mL 2,3-butanedione monoxime (Sigma) together with 0.1 μm polybeads (Polyscience, 16586-5). We use No. 1.5H circular 25 mm high-precision Deckglaser microscope cover glasses (Marienfeld,

0117650) for CALM imaging. The alignment of two fluorescence emission channels is achieved by imaging 40 nm diameter TransFluoSphere beads (488/685, Invitrogen).

2.4 Equipment

Worms are handled using a Nikon dissecting microscope (SM2745). For worm microinjection of CRISPR mix solutions or M3 peptide solutions, we use an inverted IX50 Olympus microscope, equipped with an Eppendorf FemtoJet micro-injector connected to FemtotipII microinjection capillaries that are held on a Leitz manual micromanipulator. The capillaries are backloaded with the solutions to be injected using Microloader tips (Eppendorf, 930001007). For single-molecule CALM imaging of VGCC, we used an inverted Nikon Eclipse Ti-E microscope equipped with TIR and HILO optics, a ×100 1.49 NA objective (Nikon), 488 and 561 nm fiber-coupled excitation lasers (Agilent), as well as a two-camera imaging splitter (Andor) and two iXon EMCCD cameras (Andor). A quad-band ZT405/488/561/647 dichroic mirror (Chroma), a multiband pass ZET405/488/561/647× excitation filter (Chroma), an emission-splitting FF560-FDi01 dichroic mirror (Semrock), and two emission filters at 525/50 nm (Semrock) and 600/50 nm (Chroma) for GFP and tagRFP, respectively, are used for simultaneous dual-color imaging of VGCCs with the marker for cholinergic synapses UNC-29-tagRFP.

3 Methods

3.1 Genomic Insertion of sGFP Using CRISPR-Cas9 Ribonucleoprotein Complexes in *C. elegans*

The highly optimized CRISPR-Cas9 editing systems in *C. elegans* allow convenient genetic modifications to be made at innate loci without interference with gene copy number or chromosomal regulation [31–33]. The genome editing starts with a sequence-specific double-strand break (DSB) which is introduced by CRISPR-Cas9 endonucleases that interact with custom-designed short guide RNAs (gRNAs). The gRNAs generally include two small RNAs: a CRISPR RNA (crRNA) that guides the Cas9 protein to its complementary sequence in the targeted genome, and the trans-activating crRNA (tracrRNA) that interacts with both the Cas9 protein and the crRNA to form the ribonucleoprotein (RNP) complex [34]. The two small RNAs can be separately synthesized from commercial sources [33] or combined into a single chimeric guide RNA (sgRNA) that is synthesized by in vitro transcription [35–37]. Both approaches are easy and rapid. After a DSB is introduced, the target modification is completed by the endogenous homology-dependent repair (HDR) mechanism with a custom-made DNA repair template. Previous CRISPR methods required cloning HDR templates into plasmids with 1Kb or longer homology arms [31, 38–40]. However, in

C. elegans, it was found that linear DNAs having about 35 nucleotides as flanking homology arms provide robust HDR repair templates for gene conversion [41]. Such linear DNAs can be directly ordered as synthetic single-stranded oligodeoxynucleotides (ssODNs) if they are shorter than 200 nucleotides (nt), or they can simply be PCR-amplified if they are shorter than 1.6 kb [33]. Since most gene sequences coding for FP tags fall within this 0.2–1.6 kb range, PCR amplification of HDR repair templates is an efficient strategy for CRISPR engineering of FPs in *C. elegans*.

To engineer our CRISPR transgenic *C. elegans* strains, we adopted a slightly modified protocol from the Seydoux lab [33]. Specifically, we produced single chimeric sgRNAs from T7 in vitro transcription and we injected worms with an in vitro-assembled Cas9 RNP complex mixed with linear HDR templates. As a co-CRISPR screening marker, we used the locus *dpy-10*, where the dominant *cn64* mutation causes easily identifiable phenotypes such as Roller (twisted body, heterozygous mutation) or Dpy (dumpy body, homozygous mutations or null mutant) in the F1 offspring [42]. The experiment is generally separated into three parts: (1) Design and preparation of the CRISPR mix, which includes Cas9 endonucleases, sgRNAs, and HDR templates; (2) *C. elegans* microinjection of the CRISPR mix; and (3) screening and establishment of the strain. In this protocol, we describe how to use CRISPR-Cas9 to introduce a fusion of sGFP to VGCCs in the genome of *C. elegans*. For more detailed and general CRISPR editing protocols, the reader is referred to additional method descriptions available from Paix et al. [33].

3.1.1 Preparation of the CRIPSR Mix Components

Design and Preparation of sgRNAs

The fusion of sGFP to the N-terminus of UNC-36 (extracellular $\alpha_2\delta$ subunit for both muscular and neuronal VGCCs) does not interfere with the function of VGCCs [6]. Thus, we screened for several Cas9 recognition protospacer adjacent motifs (PAM, canonical sequence 5′-NGG-3′) around the N-terminus region and scored for different crRNA sequences using an online CRISPR design platform (http://crispr.mit.edu/). It is important to examine the specificity and the predicted efficiency of each potential crRNA, which is usually around 16–20 bases complementary to the genome region immediately preceding the PAM. Sequences with high GC content and ending with G or GG are believed to give more efficient gRNAs [33]. Our screen resulted in the selection of a PAM positioned at the end of the first exon of gene *unc-36*. This exon encodes for the membrane-targeting signal peptide of UNC-36, which is cleaved after translation, effectively leaving sGFP at the N-terminus. We selected a 20 nts crRNA sequence immediately preceding this PAM and two 35 nts homology arm sequences that flank the Cas9 cleavage site, which is positioned 3 nts before the PAM (Fig. 2). The selected crRNA sequence was then combined with a common tracrRNA sequence to form a

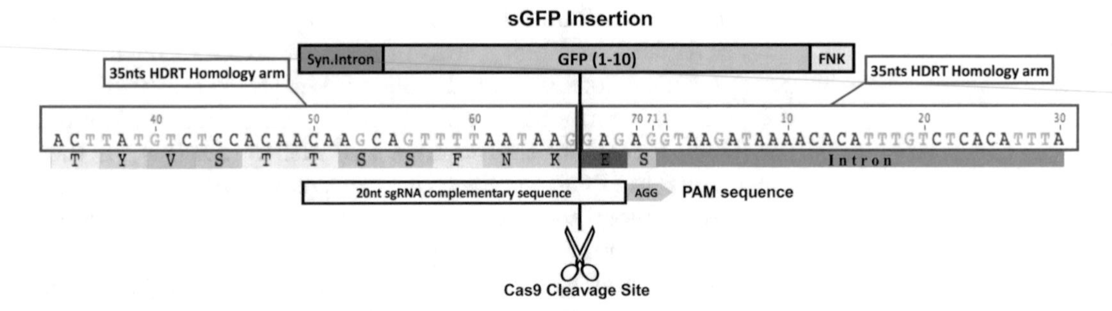

Fig. 2 CRISPR-based HDR template design for the genomic insertion of split-GFP. The letter sequences correspond to the encoding first exon and the first intron of *C. elegans* $\alpha_2\delta$ subunit *unc-36*. The red boxes highlight the two 35 nt-HDRT homology arms flanking the Cas9 cleavage site. The black box underlies the 20 nt sgRNA complementary sequence which is immediately preceding the PAM sequence at the end of the first exon (AGG). The insertion contains a synthetic intron, sGFP sequence (GFP 1–10), and three amino acids (FNK) from *unc-36* coding sequencing (20th–22th). These amino acids are the beginning of UNC-36 protein after the signal peptide (1st–19th) is cleaved post-translationally

sgRNA by in vitro transcription. The DNA template for the transcription was produced by PCR-based fusion of two overlapping DNA oligos coding for the crRNA and the tracrRNA, respectively, and containing a T7 promoter sequence. Oligo sequences employed to make the *unc-36* and the *dpy-10* sgRNA transcription templates are provided below. After PCR, the double-strand 120 bp DNA templates were purified on a 2% agarose gel and eluted with DNase/RNase-free water. The PCR products were then transcribed into sgRNA at 30 °C overnight using a T7 RNA polymerase (Ambion MEGAshortscript T7 kit). The reaction was then quenched and the sgRNA was purified with phenol:chloroform:isoamyl alcohol (25:24:1) and precipitated in pure ethanol. The quality of sgRNA was verified by running a 6% urea PAGE to make sure that no degradation or nonspecific transcripts were present. One transcription reaction generally produces enough sgRNA for multiple edits. sgRNAs were kept at −80 °C until further use.

Common tracrRNA oligo sequence:
5′AAAAGCACCGACTCGGTGCCACTTTTTCAAGTTGATAA CGGACTAGCCTTATTTAAACTTGCTATGCTGTTTCCAGCA TAGCTCTTAAAC3′.

unc-36-specific crRNA oligo sequence:
5′GCAGCTAATACGACTCACTATAGGCAAGCAGTTTTAA TAAGGAGGTTTAAGAGCTATGCT3′.

dpy-10 crRNA oligo sequence:
5′GCAGCTAATACGACTCACTATAGGCTACCATAGGCACC ACGAGGTTTAAGAGCTATGCT3′.

50 μL PCR reaction mix for sgRNA template:			
H_2O: 33 μL	98 °C	2 min	
5× Phusion buffer HF: 10 μL 10 mM dNTP mix: 2 μL 10 μM Gene-specific sgRNA oligo: 2.5 μL	98 °C 50 °C 72 °C	10 s 15 s 15 s	×15 cycles
10 μM Common sgRNA oligo: 2.5 μL	72 °C	3 min	
Phusion DNA polymerase: 0.5 μL	10 °C	hold	
In vitro transcription (Ambion MEGAshortscript T7 kit) NTP buffer mix (ATP + UTP + GTP + CTP + 10× buffer): 10 μL DNA template: 75 ng T7 RNA pol mix: 2 μL H_2O to 20 μL Incubate at 30 °C overnight			

Purification of sgRNA

(a) Add 1 μL of TurboDNase (or DNase I) to digest the DNA template.

(b) Incubate at 37 °C for 20 min.

(c) Add 115 μL of H_2O and 15 μL stop solution (provided in the kit).

(d) Add 1 volume (150 μL) of phenol:chloroform:isoamyl alcohol (25:24:1, v/v/v), vortex for 15 s, and let stand for 2 min.

(e) Centrifuge at 17,000 × *g* at room temperature for 5 min.

(f) Take aqueous phase (upper) into a new tube.

(g) Add 10 μL of 5 M ammonium sulfate.

(h) Add 300 μL (2 volumes) of 100% ethanol.

(i) Precipitate for 1 h at −20 °C.

(j) Centrifuge at 17,000 × *g* at 4 °C for 5 min (a white and dense precipitation should be visible after centrifugation).

(k) Carefully remove supernatant and add 1 mL of pre-cold 70% ethanol.

(l) Centrifuge at 17,000 × *g* at 4 °C for 2 min.

(m) Carefully remove supernatant and air-dry for 15 min or until the pellet is dry.

(n) Resuspend in 20 μL RNase-free H_2O, quantify using a UV/Vis spectrometer (e.g., Nanodrop), and adjust concentration to 2 μg/μL with RNase-free H_2O.

(o) Verify the sgRNA purity on a 6% urea PAGE.

(p) Keep the sgRNA at −80 °C for long-term storage.

Amplification of the HDR Templates

The homology-dependent repair template (HDRT) for inserting sGFP into the *unc-36* gene is PCR-amplified from the targeting vector pHZ043 [6] using a pair of primers that include the 35 nts homology arms (Fig. 2). It is important to verify that there is no PAM recognition site in the repair template, as it will otherwise be cleaved by the Cas9 RNP complex. If a PAM is present, silent mutations can be introduced to either disrupt the PAM sequence or prevent gRNA complementation at the 3′ end [33]. In order to specifically and efficiently amplify the HDRT, we recommend optimizing the thermal-cycle steps by first running a gradient PCR with annealing temperature between 60 and 72 °C. If the efficiency remains low, for instance when long primers (>60 nts) are used, a secondary nested PCR can improve the amplification. In our hands, an efficient PCR amplification using nine tubes of 50 μL reactions will generate a 10 μL final solution of DNA at 1 μg/μL after cleanup with a MinElute kit (Qiagen, 28006). The primer sequences for the amplification of the sGFP coding sequence in pHZ043 and the PCR reactions are described below. To generate a co-CRISPR *cn64* point mutation in the locus of the *dpy-10* gene (screening marker), a short ssODN is used as HDRT.

Primers for sGFP HDRT:

Forward primer: 5′CGCCACTTATGTCTCCACAACAAGCAG3′.

Reverse primer: 5′ TGTAAATGTGAGACAAATGTGTTTTAT CTTACCTCTCCTTATTAAATCCTTTTTCATTTGG3′.

ssODN for Dpy-10 HDRT:

5′CACTTGAACTTCAATACGGCAAGATGAGAATGACTGGA AACCGTACCGCATGCGGTGCCTATGGTAGCGGAGC TTCACATGGCTTCAGACCAACAGCCTAT3′.

50 μL PCR reaction mix for HDR template:			
H_2O: 30 μL	98 °C	2 min	
5× Phusion buffer HF: 10 μL 10 mM dNTP mix: 2 μL 10 μM Forward primer: 2.5 μL	98 °C 61.5 °C 72 °C	30 s 30 s 45 s	×35 cycles
10 μM Reverse primer: 2.5 μL	72 °C	5 min	
Plasmid template (50 ng/μL): 2.5 μL	10 °C	hold	
Phusion DNA polymerase: 0.5 μL			

Before pooling together all the PCR products in a single tube, the quality of the PCR-amplified HDRT is determined by electrophoresis analysis of 5 μL from each tube on an ethidium bromide agarose gel. Make sure that there is only a single band present at the expected DNA size for each PCR product, before combining and purifying the rest of the PCR products using a MinElute kit (Qiagen, 28006). Finally, elute the HDRT with 10 μL of pre-warmed nuclease-free water. The PCR-amplified HDRT is kept at −20 °C until further use.

3.1.2 Microinjection of the CRISPR Mix

The evening before microinjection, prepare about 30 N2 wild-type worms at the L4 larval stage on a NGM plate with a thin layer of OP50 bacteria at the center. Fifteen worms will be sufficient if both arms of the gonad are successfully injected. For detailed procedures describing microinjection of *C. elegans*, a protocol is available in Wormbook [43]. Assemble 15 μL of CRISPR microinjection mix as follows:

15 μL CRISPR mix solution	
250 mM HEPES	0.6 μL
3 M KCl	0.75 μL
Cas9 (at 10 mg/mL)	3 μL
sgRNAs (at 2 μg/μL)	
unc-36	4.225 μL
dpy-10	1.375 μL
0.5 μg/μL *dpy-10* ssODN	0.55 μL
1 μg/μL *unc-36* HDRT	4.5 μL

Mix well and incubate at 37 °C for 10 min. Spin down the mix at high speed (e.g., 18,000 × *g*) for 5 min and back-load a microinjection needle with 3 μL of solution. After positioning worms into a drop of halocarbon oil 700 on an injection pad, inject both gonads with the CRISPR mix solution as described [43]. For an intact needle, 2–3 injections lasting for about 1 s are generally performed for each gonad, or until an obvious flow of solution into the cytoplasm of the distal germline is visible. Conditions might vary between microinjection systems. The key steps are (1) to prevent the needle from clogging, (2) to recover the injected worms in a big drop of recovery buffer (5 mM HEPES pH 7.2, 3 mM $CaCl_2$, 3 mM $MgCl_2$, 66 mM NaCl, 2.4 mM KCl, 4% w/v glucose), and (3) to transfer them into M9 buffer once the worms start moving a bit. Once the injected worms have recovered, gently transfer each worm onto separate NGM plates with a thin layer of OP50 bacteria at the center. This is the generation P0. After 24 h at 20 °C, move P0 onto a second plate to continue laying eggs.

3.1.3 Screening and Establishment of the Strain

After 4 days, examine the F1 generation from the 15–30 injected P0 worms. If CRIPSR-Cas9 was efficient, the majority of F1 worms will display Roller or Dpy phenotypes. Select the top three plates that produce the most of Roller and Dpy phenotypes in F1. These three plates are the "jackpot broods" [33] and are the most likely to contain worms where *unc-36* has been successfully edited by the insertion of the sGFP coding sequence. Pick around 20 Roller and all the Dpy F1 worms, each onto separate plates. We often prioritize the screening of Dpy worms because in our hands (1) Dpy worms carry successful insertion more often than Rollers, (2) the amount of homozygous Dpy animals is about five times less than Rollers which accelerates the screening process, and (3) *unc-36* and *dpy-10* are on different chromosomes which allows easy segregation of the Dpy phenotype. 24 h after cloning the Dpy F1, lyse the F1 worms to release the genome using lysis buffer (50 mM KCl, 10 mM Tris pH 8.3, 2.5 mM MgCl2, 0.45% NP-40, 0.45% Tween-20, 0.01% gelatin) with 6% of 10 mg/mL proteinase K, at 65 °C for 1 h, followed by 95 °C deactivation for 15 min. To detect the insertion of the sGFP sequence, we designed a forward PCR screening primer that overlaps with the promoter sequence of *unc-36* (~200 bp upstream of the sGFP insertion) and a reverse primer that overlaps with the 3′ homology arm in the first intron, after the sGFP sequence (Fig. 2). These primers and the PCR reaction are described below. After running the PCR screening product on an agarose gel, successfully edited heterozygous worms display a 1 kb band corresponding to the sGFP insert and a 250 bp band corresponding to the original length between the promoter and the first intron. Once a successful edition is confirmed in F1, individual worms from the F2 generation are screened to identify *unc-36::sGFP* homozygous worms, which are then sequenced for the entire insertion. Backcross the sequence-confirmed worms with N2 wild-type worms at least twice to eliminate the Dpy phenotype and to dilute any unknown off-target background. The final transgenic strain should behave like wild-type worms. The expression pattern of the sGFP-tagged protein is closely examined by CALM imaging. In our case, we inject a high concentration of complementary peptide to confirm the proper plasma membrane targeting of UNC-36-sGFP and we analyze the diffusing coefficient of single calcium channels, as described in the next section. The locomotion of transgenic worms was also scored by thrashing assay as previously described [6]. The final transgenic strain is established once the correct phenotypes and genotypes are confirmed.

Primers to screen for sGFP insertion:

Forward primer: 5′TCCACCAAAAAAATCTTAATTTTCAG ATCTCTTTTTTGAG3′.

Reverse primer: 5′TGTAAATGTGAGACAAATGTGTTTTA TCTTACCTCTCC3′.

25 μL PCR reaction mix for genotype screening:			
H_2O: 17.875 μL	95 °C	2 min	
10× Taq standard buffer: 2.5 μL 10 mM dNTP mix: 1 μL 10 μM Forward primer: 1 μL	95 °C 55 °C 68 °C	20 s 20 s 1 min	×30 cycles

(continued)

10 μM Reverse primer: 1 μL	68 °C	5 min
Worm lysis (10 μL/worm): 1.5 μL	10 °C	hold
Taq DNA polymerase: 0.125 μL		

The UNC-36-sGFP transgenic strain was further crossed with the UNC-29-tagRFP knock-in strain kr208 to introduce a marker for the cholinergic synapse locations of N/P/Q-type Ca^{2+} channels (Fig. 4c, d). Other transgenic fluorescent markers that do not spectrally overlap with sGFP fluorescence are available for CALM imaging of VGCCs in different types of synapses (e.g., GABAergic motor neurons) or in specific cell types, such as sensory neurons or interneurons.

3.2 In Vivo Single-Molecule Microscopy Imaging

For in vivo SM imaging of N-type VGCCs in neurons and L-type VGCCs at the muscle surface of live *C. elegans* by CALM, we use the established CRISPR-worm strain where the $\alpha_2\delta$ subunit UNC-36-sGFP and UNC-29-tagRFP are co-expressed. We microinject M3 peptides that encode the missing 11th β-strand of sGFP [6, 11, 13] into the pseudocoelomic cavity at the head region of the animals (Fig. 3). The activation of GFP fluorescence is only enabled when the complementary M3 peptide cycling in the pseudocoelomic cavity physically interacts with the sGFP fragment, reconstitutes the β-barrel, and triggers the GFP chromophore maturation in tissues where UNC-36-sGFP is expressed (Fig. 1). By adjusting the concentration of complementary peptides and incubation times, a controlled subset of GFPs are switched on for SM detections of diffusing VGCCs at specific subcellular locations. This section describes how to perform CALM imaging, nanoscale localization of individual VGCC positions by Gaussian fitting, and tracking of their lateral diffusion at the muscle surface as well as at cholinergic synapses using UNC-29-tagRFP as marker.

3.2.1 Microinjection of M3 Peptides

Lyophilized synthetic M3 peptides can be ordered from commercial sources in the form of M3-biotin or M3-cysteine peptides having an aminocaproic acid (acp) linker and a "GSGGGSTS" spacer (>75% purity, biotin-acp-GSGGGSTSRDHMVLHEYVNAAGIT or C-acp-GSGGGSTSRDHMVLHEYVNAAGIT). Smaller peptide sequences encoding only the 11th β-strand can also be used (e.g., RDHMVLHEYVNAAGIT), but we have found that M3-biotin or M3-cysteine has prolonged circulation time in the pseudocoelomic cavity and might be less easily filtrated and degraded in coelomocyte scavenger cells. M3 peptides are reconstituted in DMSO at 100 mM (e.g., 2.8 mg/10 μL for

M3-biotin) and diluted at 1 mM in microinjection buffer (20 mM K_2HPO_4, 3 mM potassium citrate, 2% PEG 3000, pH = 7.4). This 1 mM peptide solution is stable for a week without obvious degradation and decrease in complementation when conserved at 4 °C. Higher peptide concentrations can be prepared for increased activation of sGFP, but the peptides are relatively hydrophobic and precipitation might occur over time above 5 mM.

Twelve hours before microinjection, place L4 worms on a NGM plate with a thin layer of bacteria. M3 peptides are injected in the pseudocoelomic space, which is accessible at the head region between the two pharyngeal bulbs (Fig. 3a, b). Use narrow microinjection capillaries (e.g., Eppendorf FemtotipII, 0.5 μm inner diameter and 0.7 μm outer diameter) to minimize perturbation of the nerve ring and nearby ganglions. Backfill the capillary with 5 μL of peptide solution, connect the capillary, and set the backpressure to 500 hPa. Put a drop of mineral oil on a 2% agarose pad (dried overnight), set the pad on the Olympus microscope, move the capillary in the drop, focus on the capillary tip, and test the flow of liquid using the "Clean" function on the FemtoJet. Set the injection pressure to 1200 hPa and retest the flow using the "Inject" function. Lift the tip, transfer the pad to a dissecting microscope, position a worm in the oil drop, and gently push it to swim in oil to get rid of bacteria around its body. Then, gently push the worm on the agarose pad until the head region of the worm becomes immobilized by self-motion. Back on the Olympus microscope, position the head at a ~80–90° angle to the needle as shown in Fig. 3a, change into high magnification to visualize the nerve ring, and position the needle in between the nerve ring and the first pharyngeal bulb (Fig. 3). Push the needle against the worm until the cuticle is slightly depressed and gently tap the micromanipulator once (with one finger) to induce insertion in the pseudocoelomic cavity. A mild flow of solution inside the cavity can be seen once the pseudocoelomic space is reached. Inject the peptide solution in 3–5 steps of 0.5 s until the worm body is swelling up and expanding in length (Fig. 3c, black arrow). The backpressure and injection pressure can be modified according to the condition of the injection tip. After injection, release the worm from the pad and isolate it from oil using a drop of recovery buffer (5 mM HEPES pH 7.2, 3 mM $CaCl_2$, 3 mM $MgCl_2$, 66 mM NaCl, 2.4 mM KCl, 4% w/v glucose). Transfer the worm onto a clean area on an NGM plate and let it recover in a new drop of recovery buffer. Successful injections are often characterized by one or two eggs being pushed out of the vulva due to the body expansion (Fig. 3c, red arrow). Substitute recovery buffer with M9 buffer once the worm starts moving. If the injection has not damaged nerves in the head region, a worm will recover within 3–10 min (Fig. 3c, blue arrow). Worms injected with 1 mM of M3 peptides are incubated for at least 3 h at

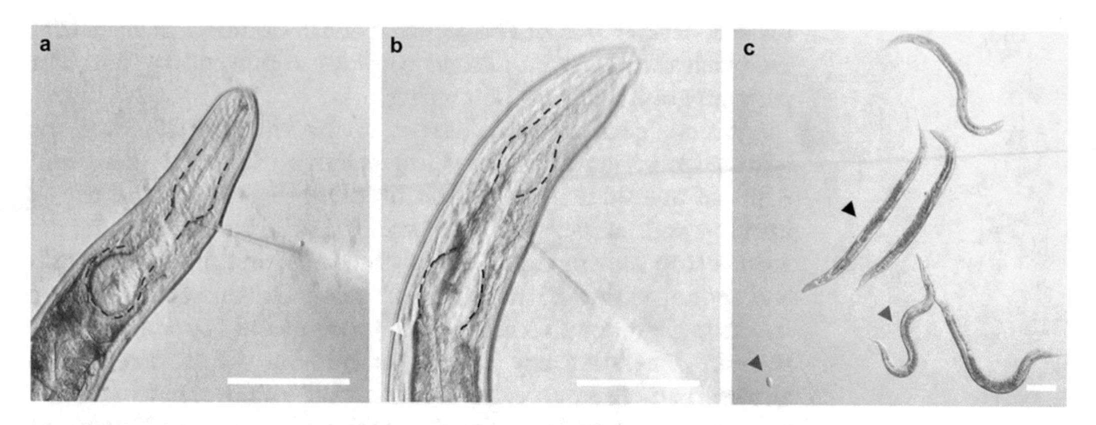

Fig. 3 Microinjection of complementary peptides into the pseudocoelomic cavity of *C. elegans*. (**a**) A *C. elegans* worm immobilized on a 2% agarose pad in a drop of mineral oil is positioned with its head region at an ~80–90° angle to a Femtotipll microinjection capillary to inject a solution of M3 peptide in the pseudocoelomic space, between the two pharyngeal bulbs (dashes). (**b**) After penetration of the capillary, the worm is injected with M3 peptides, which results in the body swelling up and expanding in length. A flow of transparent solution inside the pseudocoelomic cavity can be seen (white arrow). (**c**) Several injected worms are shown recovering on a plate in a drop of recovery buffer. One worm is elongated and lying still after microinjection (black arrow). One worm already recovered and shows body motions (blue arrow). At the corner, one egg has been pushed out of one injected worm, an indication of a successful microinjection (red arrow). All scale bars: 50 μm

20 °C before CALM imaging of muscle VGCCs and for 12 h before imaging neuronal VGCCs, because of lower channel numbers at presynaptic terminals.

3.2.2 Preparation for CALM Imaging

Prepare the following solutions:

1. 2% Low-melting-point agarose in M9 buffer for imaging pads.
2. 2.5% 0.1 μm Polybeads as immobilizing solution.
3. Anesthetics: 1 mM Levamisole (contracted muscles) or 30 mg/mL 2,3-butanedione monoxime (BDM, relaxed muscle) in M9 buffer. Store the 10 mM levamisole solution as a 10× stock at −20 °C and dilute into 1 mM with M9 buffer just before use.
4. 40 nm diameter TransFluoSphere beads (488/685 nm, 2% solids, Invitrogen) diluted at 0.01% per volume in PBS buffer for dual-channel alignment.

Fresh 2% agarose pad is prepared on the day of imaging. Clean micro slides (VWR) with a Kimwipe soaked in 100% ethanol and air-dry before use. Tape one layer of labeling tape (VWR) on both sides of micro slides (VWR), leaving the middle area blank to deposit one drop of melted 2% agarose (70 °C) using a 1 mL pipette. Immediately put another non-tapped micro slide on top of the agarose drop to create an even agarose pad surface with the thickness of tape. Avoid introducing bubbles on the agarose surface. Several pads can be prepared ahead of time if they are kept sandwiched in between two micro slides to prevent dry-out. The pads are good for 1 day of imaging.

When injected worms are ready for imaging, flip over the agarose pad sandwich and gently remove the taped slide. Add 5 μL of anesthetics (levamisole or BDM) in the middle of the imaging pad, and deposit 1–3 worms in the solution. Once the worms stop moving add 2 μL of immobilizing polystyrene beads and mount with a 25 mm high-precision Deckglaser microscope coverslip right away. Gently position the coverslip onto the worms without generating any bubbles or pushing the worms off the agarose pad. Aspirate extra solution that might run to the side and seal the coverslip on the slide with nail polish. It is important to make sure that the worms are immersed in the anesthetic/ immobilizing solution on the agarose pad after mounting and are away from the nail polish. For dual-color-channel alignment, a few μL of TransFluoSphere bead (488/685 nm) solution is added before mounting and sealing the coverslip. To image N-type VGCCs at the nerve cord, worms need to be orientated with the vulva on top or the bottom at around 40° relative to the slide. The orientation is done after the polybeads are added, by gently turning the worms using two eyelashes glued to the end of two pipette tips.

3.2.3 CALM Imaging

The optical setup consists of a Nikon Eclipse Ti-E microscope, equipped with an ×100 and 1.49 NA oil objective, a motorized stage, perfect focus drift compensation optics, and optics for total internal reflection (TIR) and highly inclined and laminated optical sheet (HILO) fluorescence imaging. The microscope is fiber-coupled to a set of laser lines at 488 and 561 nm controlled via an acousto-optical modulator (Agilent). Fluorescence detection is done using a multiband-pass ZET405/488/561/647× excitation filter (Chroma), a quad-band ZT405/488/561/647 dichroic mirror (Chroma), two EMCCD cameras (Andor iXon), and a two-camera imaging adapter (TuCam, Andor) that contains additional magnification lenses, an emission-splitting FF560-FDi01 dichroic mirror (Semrock), and emission filters at 525/50 nm (Semrock) for GFP and 600/50 nm (Chroma) for tagRFP imaging. Imaging of individual VGCCs in microinjected worms is done at room temperature by continuous laser excitation and using acquisition frame rates of 80 ms/frame, with an EMCCD gain of 300 and with a 1 × 1 binning of the image pixel size, which is 107 nm on each camera. Note that this imaging pixel size is adapted to the estimated resolution of the microscope, which is mathematically defined by the Rayleigh equation: $d = \frac{1.22 \times \lambda}{2 \times \mathrm{NA}}$, where λ is the emission wavelength and NA is the numerical aperture of the microscope objective. Indeed, in order to work at the maximum possible spatial resolution allowed by an optical system it is the convention to adapt the size of the pixels in the final image such that they are equal to or less than one-half the resolution of the optical system [44]. This fulfills the Nyquist criterion [45], which requires a spatial sampling at least equal to twice the optical resolution in order to preserve the spatial resolution in the acquired digital images. These conditions for image acquisition may be varied depending on the optical setup and/or the type of ion channels being studied.

Place the mounted worms on the microscope stage, with the coverslip facing the oil on the objective. Focus on the coverslip surface, and use the TIRF optics (*x*, *y* micromanipulator and laser focusing lens) to bring and focus the circularly polarized 488 nm laser excitation beam to the center of the objective back focal plane such that it emerges collimated from the objective. In order to get a HILO illumination, move the laser beam away from the optical axis at the back-focal plane to a position slightly off from the location normally required to obtain a critical TIRF angle. HILO illumination angles are generally between 50° and 40° but will vary depending on the worm area being imaged. Align the GFP and tagRFP emission channels by acquiring images of the TransFluoSphere beads (488/685 nm) using both 488 and 561 nm excitation lasers. Then, locate the worm head region and, using the 488 nm laser, excite VGCCs at the surface of body-wall muscles, which are in close proximity to the coverslip surface, passing the worm cuticle (Fig. 4a, b). To image VGCCs at the nerve cord, use the 561 nm excitation and the tagRFP signal to first focus on acetylcholine receptor puncta and then open the 488 nm excitation laser to excite VGCCs at this focal plane (Fig. 4c, d). Starting at the head, image different areas along the worm body by moving the microscope stage. With an ×100 objective, the entire imaging of a single worm typically takes around 10–15 min, which includes 1–2-min imaging periods in a single two-dimensional plane at each region along the worm body and slight optical readjustments of the microscope focus and HILO angles after moving the stage from one region to the other in order to achieve GFP and tagRFP excitation at different focal planes and different body areas. Individual muscular or neuronal VGCCs should display single-step photobleaching, blinking events, and sudden fluorescence appearance typical of SM detections by CALM [6, 11]. A unimodal fluorescence intensity distribution should also be observed from fluorescence time trace analyses of multiple VGCCs. After imaging, worms can be carefully recovered by unsealing the coverslip and transfer on NGM plates with OP50 bacteria lawns.

3.2.4 Tracking and Diffusion Analyses

While there are many software available for single-particle detection and tracking [46], we perform tracking and diffusion analysis of individual VGCCs using SlimFast, a single-particle detection and

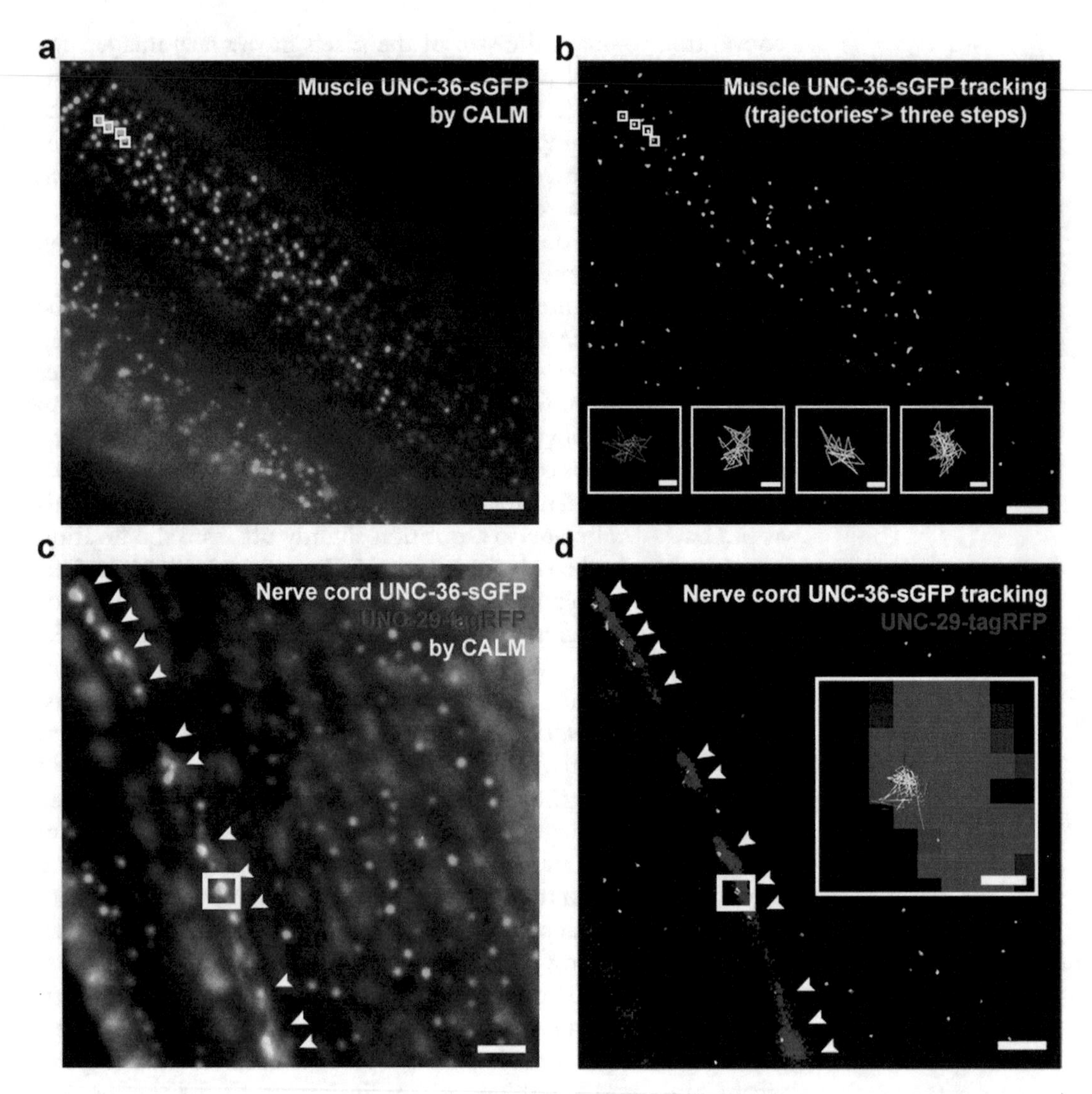

Fig. 4 CALM imaging and single-molecule tracking of neuronal and muscular VGCCs. (**a**) Sum intensity image of a CALM movie showing VGCCs on body-wall muscle cells. (**b**) Corresponding trajectory map of individual diffusing VGCCs from the movie in (**a**) and enlarged representative trajectories (white squares). (**c**) Sum intensity image of a dual-color CALM movie showing neuronal VGCCs along the nerve cord (arrows) as determined by the RFP-tagged acetylcholine receptor UNC-29 marker (red). (**d**) Corresponding trajectory map of VGCCs from the movie in (**c**). Several neuronal VGCC trajectories are seen at each synapse of neuromuscular junction (arrows), while some muscular VGCCs are also visible away from the nerve cord. Representative VGCC trajectories from one synapse are enlarged (white square). Scale bars: 2 μm for all images; 35 nm for insets of (**b**) and 200 nm for inset of (**d**)

tracking software written in Matlab and based on multiple-target tracing algorithms [47] (Fig. 4a, c). If necessary, and depending on the mechanical stability of the microscope stage, images acquired during the 1–2-min acquisition periods can be corrected for lateral drifts using the fluorescent beads as fiducial markers. We have found, however, that lateral drift is so small that corrections are not necessary. Import CALM movies as .tiff stacks in the tracking software, and define the imaging conditions (e.g., acquisition frame rate, pixel size, estimated size of the point spread function, one or two fluorescence channel imaging). Run the routine for sub-diffraction limit localization of VGCCs in each frame of the stack by 2D Gaussian fitting of the point spread function for each complemented green fluorescent VGCC. Then, build diffusion trajectories by linking individual localized positions from one frame to the other, taking into account blinking statistics and local particle densities to avoid error during linkage (Fig. 5a). Only keep trajectories with at least three steps for further analyses of diffusion from individual mean square displacements (MSD, Fig. 5b, d), from the ensemble distribution of diffusion coefficients of all VGCCs (Fig. 5e) or from the probability distribution of their squared displacements (PDSD, Fig. 6). Determine the localization precisions using the method of Thompson et al. [48]. Precisions are reported as the mean ($\pm$ standard deviation) of the distribution of localization precisions for an individual VGCC (Fig. 5c) or for all the localized VGCCs across multiple worms. Under appropriate imaging conditions, localization precisions for VGCCs are expected to be 20–30 nm or better. For each VGCC trajectory, compute an individual MSD curve [49] (Fig. 5b, d), extract a diffusion coefficient, and assess the size and the shape of visited membrane nanodomains by least square fitting the MSD with an appropriate 2D diffusion model such as:

1. A free Brownian diffusion model with measurement error:

$$< r^2 > = 4Dt + 4\sigma^2 \tag{1}$$

where σ is the position error and D is the diffusion coefficient.

2. A circularly confined diffusion model with measurement error:

$$< r^2 > = R^2\left(1 - A_1 e^{-\frac{4A_2Dt}{R^2}}\right) + 4\sigma^2 \tag{2}$$

where R is the confinement radius, σ is the position error, D is the diffusion coefficient, $A_1 = 0.99$, and $A_2 = 0.85$ as previously described [50]. Alternative 2D models such as anomalous diffusion or directed diffusion can also be tested. Note that, in these fits, σ is a free parameter that is expected to return a value similar to the localization precision determined independently.

In our experience, a reliable fit of the circularly confined diffusion model requires a VGGC trajectory with reasonable averaging of the squared displacements over at least the first ten time steps (Fig. 5d). Similarly, a reliable fit of the free Brownian diffusion model can be performed on the MSD values of the first three or four time steps (Fig. 5b, d). Use this last approach to build a distribution of diffusion coefficients for thousands of VGCC

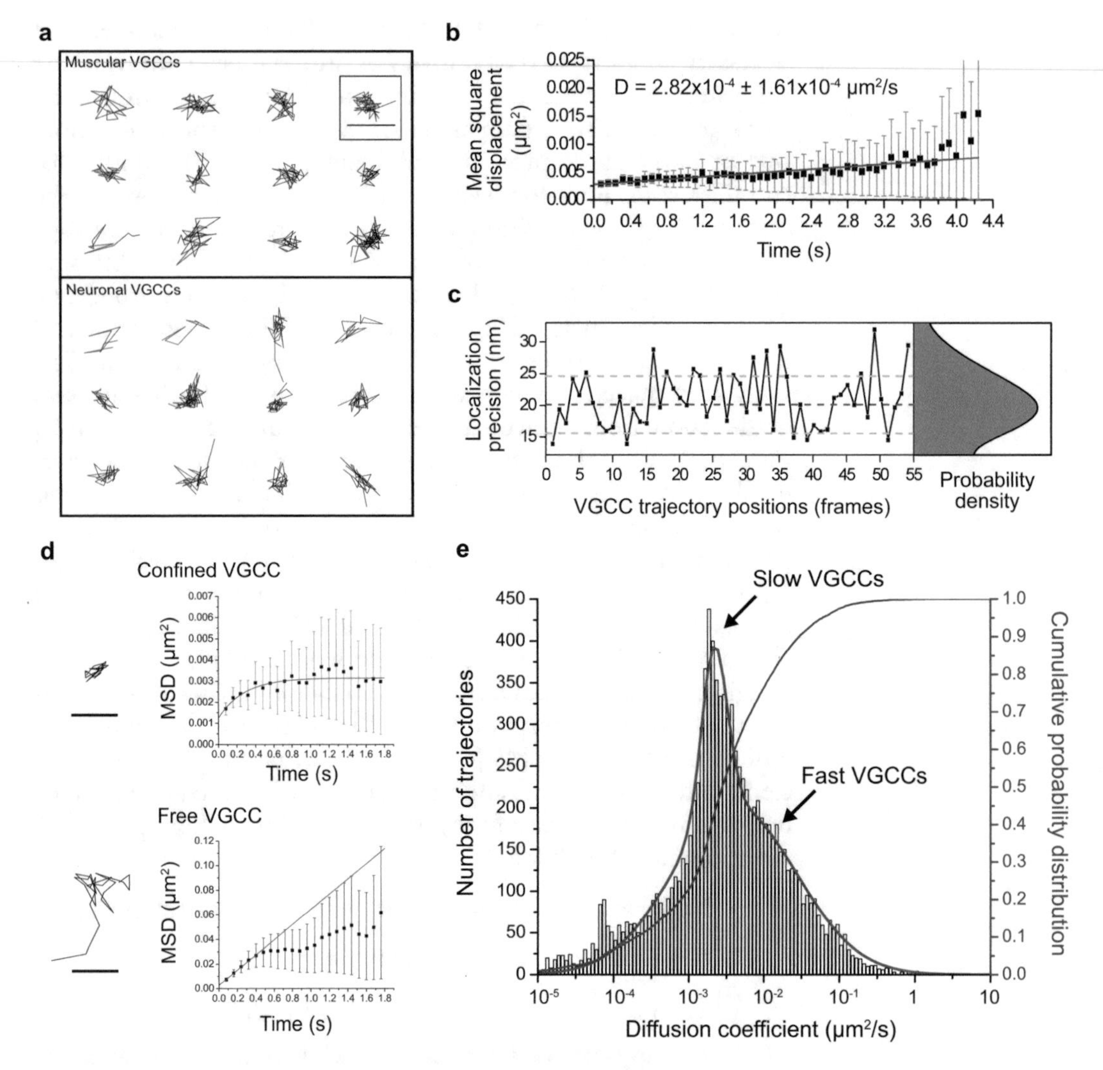

Fig. 5 Diffusion analysis of individual VGCCs from MSD measurements. (**a**) Typical examples of nanoscale diffusion trajectories for muscular (top) and neuronal (bottom) VGCCs in live, CRIPR-engineered *C. elegans* anesthetized with levamisole. Scale bar: 200 nm. (**b**) MSD curve for an individual muscular VGCC (square in (**a**)), corresponding fit of the free Brownian diffusion model on the first three time lags (red) and corresponding diffusion coefficient value *D*. Error bars in the MSD curve correspond to the standard error of the mean at each time lag. (**c**) Distribution of localization precisions for consecutive frames along the trajectory of an individual muscular VGCC (square in (**a**)). The red dotted line is the mean localization precision from the distribution (20.2 nm) and the two green dotted lines correspond to the standard deviation of the mean ($\pm$4.5 nm). (**d**) Examples of confined and Brownian diffusion models fitted on the MSD curve of two individual muscular VGCCs and corresponding diffusion trajectories. Scale bars: 200 nm. (**e**) Distribution of diffusion coefficients and cumulative probability distribution for 10,481 VGCC trajectories (39 muscle cells, 6 worms) determined from individual MSD analysis. Two populations of slow- and fast-diffusing channels coexist at the sarcolemma of resting muscles. ((**d**) and (**e**) are reproduced from Zhan et al. [6])

trajectories and to assess if different and independent diffusing populations of channels coexist in the cell membrane (Fig. 5e). As shown in Fig. 5e, VGCCs diffuse in two independent populations of slow- and fast-diffusing channels at the sarcolemma of muscle cells in live adult *C. elegans*. When the distribution of diffusion coefficients is wide, it can also be rendered as a cumulative distribution, which facilitates comparison between conditions via statistical tests (Fig. 5e).

Alternatively, different populations of diffusing channels can be detected and quantified from a PDSD analysis of all the tracked VGCCs [6]. For PDSD, perform computations on the first ten observation time lags (Fig. 6a). This number of time lags is sufficient to detect the general diffusive behavior (Brownian, confined, anomalous . . .) of distinct VGCC populations on *C. elegans* muscles under the imaging conditions described above. In brief, for the first ten time lags t, fit each Pr^2 curve with the general model:

$$P\left(\vec{r}^{\,2}, t\right) = 1 - \sum_{i=1}^{n} \alpha_i(t) e^{\frac{-r^2}{r_i^2(t)}} \tag{3}$$

$$\sum_{i=1}^{n} \alpha_i(t) = 1$$

where the fitting coefficients $r_i^2(t)$ and $\alpha_i(t)$ are the square displacement and the fraction corresponding to i numbers of diffusion regimes at each time lag t, respectively. Determine the appropriate number of diffusion regimes by analyzing the residuals of different fits with $i = 1, 2, 3$, or more on the Pr^2 curve. In the case of individual VGCCs diffusing on *C. elegans* muscles, Pr^2 distributions are well fitted with $i = 2$ regimes, similar to the two populations observed by individual MSD analyses. Use the fit values at each time lag to define the fraction of VGCC belonging to each population and to build $r_i^2(t)$ curves corresponding to i numbers of diffusing populations (Fig. 6b). Determine the error bars for each r_i^2 in $r_i^2(t)$ curves using $\frac{r_i^2}{\sqrt{N}}$, where N is the number of independent data points used to build each probability distribution function. Fit each $r_i^2(t)$ curve with an appropriate 2D diffusion model to extract the typical diffusion coefficient of each VGCC population and to assess their respective diffusive behavior. Report the measured diffusion coefficients D in micrometer squared per second ± standard deviation of the fit value. Determine significant differences in diffusion coefficients between conditions using F-tests. PDSD analyses are generally performed for thousands of individual VGCCs that are pooled together after single-molecule tracking on tens of muscle cells from multiple worms.

Using a combination of individual MSD and ensemble PDSD analyses, we have shown that a large majority of muscular VGCCs (91%) are nearly immobile ($D_{\text{Leva-slow}} = 9.4 \times 10^{-5} \pm 4 \times$

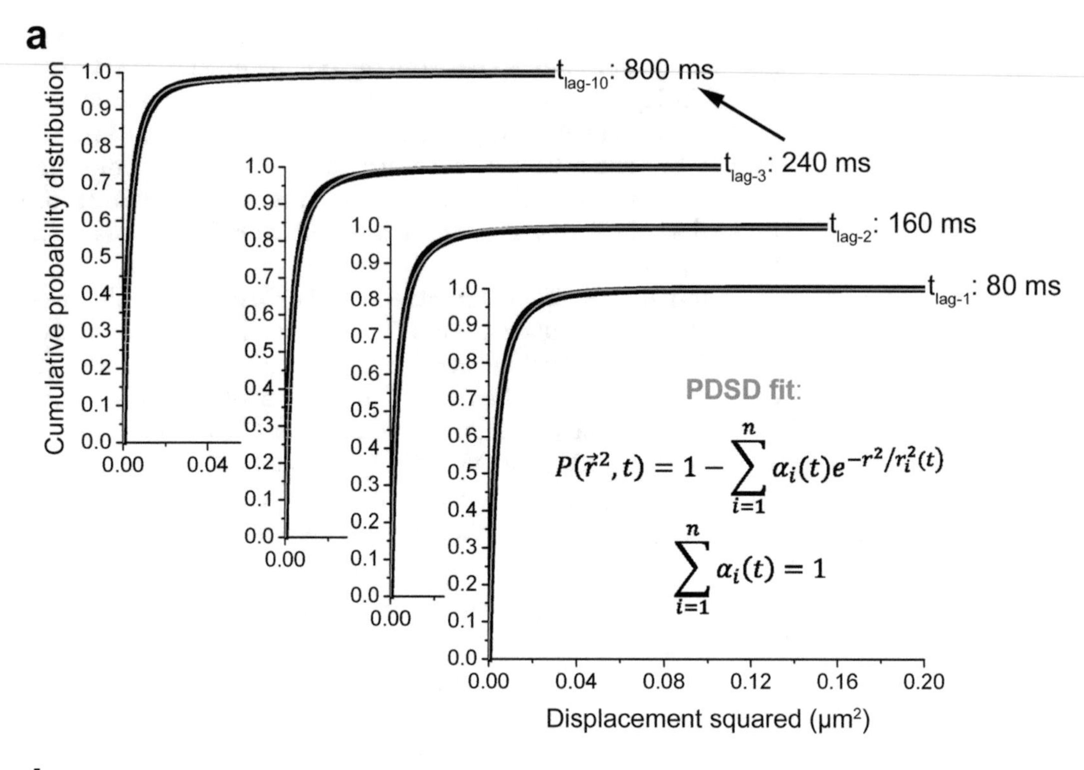

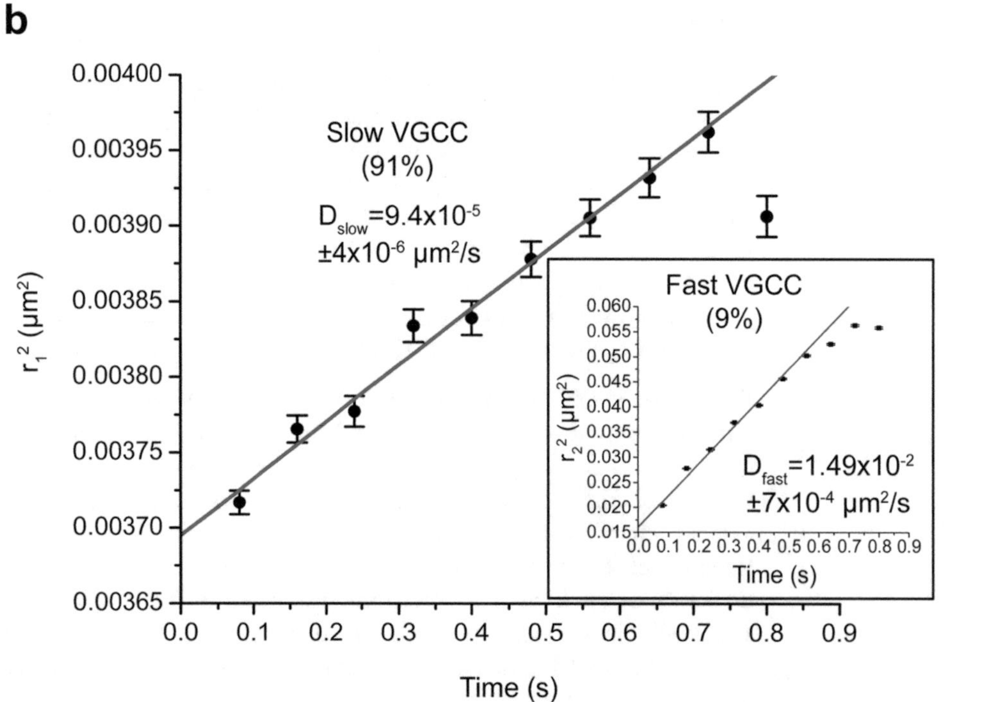

Fig. 6 Diffusion coefficient measurements from PDSD analysis of VGCC behaviors. (**a**) Example of fitted PDSD curves at different time lags (t_{lag}) for a pool of 36,325 individual VGCCs tracked in 62 muscle cells from 18 levamisole-treated worms. (**b**) Two r_i^2 displacement squared parameters (r_1^2 and r_2^2 in inset) extracted

10^{-6} μm^2/s) on muscle cells undergoing sustained contraction (levamisole-treated worms, Fig. 6b), while a minority of VGCCs (9%) diffuse significantly faster (Fig. 6b, $D_{\text{Leva-fast}} = 1.49 \times 10^{-2} \pm 7 \times 10_{-4}$ μm^2/s) [6]. This is in contrast with the nanoscale behavior of VGGCs on resting muscle cells (untreated worms), where the majority of calcium channels (81%) are confined in 80 nm nanodomains and diffuse slowly ($D_{\text{slow}} = 1.22 \times 10^{-3} \pm 1.2 \times 10^{-4}$ μm^2/s) while a minority of VGCCs (19%) are not confined and diffuse faster ($D_{\text{fast}} = 2.22 \times 10^{-2} \pm 1 \times 10^{-3}$ μm^2/s) [6].

To assess the diffuse behavior of neuronal VGCCs, perform the diffusion analyses only on trajectories that overlap with the fluorescent neuronal marker (e.g., cholinergic synapse marker UNC-29-tagRFP, Fig. 4c, d). In this case, we recommend that additional alignment between the CALM channel (green) and the neuronal marker channel (red) be performed digitally. This involves using the TransFluoSphere bead images to generate a color-alignment correction file from an ImageJ plug-in such as UnwarpJ, and applying the correction to the data files. Note that for multicolor imaging, split-FP spectral variants other than sGFP are available [13]. These variants can facilitate CALM imaging in combination with a variety of neuronal or other tissue markers. They additionally provide possibilities for multicolor single-molecule tracking of different types of ion channels within synapses or on muscle tissues in vivo.

4 Conclusion

CALM permits tissue-specific SM imaging in live *C. elegans* and allows quantitative studies of the nanoscale diffusive behaviors of VGCCs under homeostatic control in vivo. Using CRISPR engineered of a split-FP fusion to the extracellular $\alpha_2\delta$ subunit, complementation-activated GFP signals are correlated to natively expressed VGCCs at 1:1 stoichiometric ratios. With a controlled delivery of complementary synthetic peptides via microinjection, a subset of membrane-specific GFP signals are activated by CALM, which allows for sub-resolution localization and temporal tracking of individual calcium channels in intact adult animals. These SM studies on thousands of physiologically functioning VGCCs provide unprecedented insights into the nanoscale dynamics of ion channels and enable the characterization of subpopulation behaviors, including differential diffusions in plasma membrane

Fig. 6 (continued) from PDSD curve fitting in (**a**) are plotted as a function of time lag. Both r_i^2 curves are fitted with a free Brownian diffusion model (red) to extract the D_{slow} and D_{fast} diffusion coefficients characterizing the two populations of VGCCs detected at the sarcolemma of contracted muscles. The respective fractions of both populations (α_1 and α_2) are provided as percentages. ((**b**) is reproduced from Zhan et al. [6])

nanodomains. Combined with tissue-specific markers in *C. elegans*, CALM also allows the study of different types of VGCCs at various cellular locations. In future applications, multicolor in vivo CALM imaging can be realized with additional split-FP fusions to study transient interactions or clustering among membrane channels and proteins. Compared with other SM imaging techniques that use conventional FP, CALM ensures a high-contrast detection of target biomolecules at specific subcellular areas of 3D tissues by limiting out-of-focus and background signals. In principle, CALM can also be applied to other animal species suitable for fluorescence imaging or to thick acute/cultured slice preparations of mammalian brains. In these cases, combining CALM imaging with light-sheet excitation formats where illumination and detection paths are decoupled [51, 52], and using adaptive optics to correct for aberrations of the point spread function due to refractive index variations at high depth [53–56], might be required to optimize deep-tissue SM detection and tracking. Performing in vivo SM studies in *C. elegans* is, however, highly advantageous given its small size, optical transparency, simple anatomy, easy handling, powerful genomic editing systems, and availability of many *C. elegans* mutants to study biomolecules during various developmental or disease stages. In conclusion, CALM and CRISPR-Cas9 techniques open new avenues to explore the nanoscale dynamics of biomolecules in both physiological and pathological states via sub-resolution localization and tracking of individual membrane protein in live animals.

Acknowledgments

We would like to thank Dr. Jean-Louis Bessereau and members from his laboratory for kindly providing the transgenic strains with neuromuscular junction markers, Dr. Hong Zhan for sharing the pHZ043 plasmid and alternative genome editing strategies, and Dr. Thomas Duchaine and Vinay Mayya for many critical advices and reagents for CRISPR-based genome engineering.

References

1. Liu Z, Lavis LD, Betzig E (2015) Imaging live-cell dynamics and structure at the single-molecule level. Mol Cell 58(4):644–659. https://doi.org/10.1016/j.molcel.2015.02.033
2. Lord SJ, Lee HL, Moerner WE (2010) Single-molecule spectroscopy and imaging of biomolecules in living cells. Anal Chem 82 (6):2192–2203. https://doi.org/10.1021/ac9024889
3. Shashkova S, Leake MC (2017) Single-molecule fluorescence microscopy review: shedding new light on old problems. Biosci Rep 37(4):BSR20170031. https://doi.org/10.1042/bsr20170031
4. Triller A, Choquet D (2008) New concepts in synaptic biology derived from single-molecule imaging. Neuron 59(3):359–374. https://doi.org/10.1016/j.neuron.2008.06.022

5. Willig KI, Barrantes FJ (2014) Recent applications of superresolution microscopy in neurobiology. Curr Opin Chem Biol 20:16–21. https://doi.org/10.1016/j.cbpa.2014.03.021
6. Zhan H, Stanciauskas R, Stigloher C, Dizon KK, Jospin M, Bessereau JL, Pinaud F (2014) In vivo single-molecule imaging identifies altered dynamics of calcium channels in dystrophin-mutant C. elegans. Nat Commun 5:4974. https://doi.org/10.1038/ncomms5974
7. Doudna JA, Charpentier E (2014) Genome editing. The new frontier of genome engineering with CRISPR-Cas9. Science 346 (6213):1258096. https://doi.org/10.1126/science.1258096
8. Khan AO, Simms VA, Pike JA, Thomas SG, Morgan NV (2017) CRISPR-Cas9 mediated labelling allows for single molecule imaging and resolution. Sci Rep 7(1):8450. https://doi.org/10.1038/s41598-017-08493-x
9. Axelrod D (1981) Cell-substrate contacts illuminated by total internal reflection fluorescence. J Cell Biol 89(1):141–145
10. Reck-Peterson SL, Derr ND, Stuurman N (2010) Imaging single molecules using total internal reflection fluorescence microscopy (TIRFM). Cold Spring Harb Protoc 2010(3): pdb.top73. https://doi.org/10.1101/pdb.top73
11. Pinaud F, Dahan M (2011) Targeting and imaging single biomolecules in living cells by complementation-activated light microscopy with split-fluorescent proteins. Proc Natl Acad Sci U S A 108(24):E201–E210. https://doi.org/10.1073/pnas.1101929108
12. Cabantous S, Terwilliger TC, Waldo GS (2005) Protein tagging and detection with engineered self-assembling fragments of green fluorescent protein. Nat Biotechnol 23 (1):102–107. https://doi.org/10.1038/nbt1044
13. Koker T, Fernandez A, Pinaud F (2018) Characterization of split fluorescent protein variants and quantitative analyses of their self-assembly process. Sci Rep 8(1):5344. https://doi.org/10.1038/s41598-018-23625-7
14. Catterall WA (2011) Voltage-gated calcium channels. Cold Spring Harb Perspect Biol 3 (8):a003947. https://doi.org/10.1101/cshperspect.a003947
15. Nanou E, Catterall WA (2018) Calcium channels, synaptic plasticity, and neuropsychiatric disease. Neuron 98(3):466–481. https://doi.org/10.1016/j.neuron.2018.03.017
16. Mercer AJ, Chen M, Thoreson WB (2011) Lateral mobility of presynaptic L-type calcium channels at photoreceptor ribbon synapses. J Neurosci 31(12):4397–4406. https://doi.org/10.1523/jneurosci.5921-10.2011
17. Schneider R, Hosy E, Kohl J, Klueva J, Choquet D, Thomas U, Voigt A, Heine M (2015) Mobility of calcium channels in the presynaptic membrane. Neuron 86 (3):672–679. https://doi.org/10.1016/j.neuron.2015.03.050
18. Dixon RE, Moreno CM, Yuan C, Opitz-Araya-X, Binder MD, Navedo MF, Santana LF (2015) Graded Ca(2)(+)/calmodulin-dependent coupling of voltage-gated CaV1.2 channels. Elife 4. https://doi.org/10.7554/eLife.05608
19. Heine M, Ciuraszkiewicz A, Voigt A, Heck J, Bikbaev A (2016) Surface dynamics of voltage-gated ion channels. Channels (Austin) 10 (4):267–281. https://doi.org/10.1080/19336950.2016.1153210
20. Bargmann C (1998) Neurobiology of the Caenorhabditis elegans Genome. Science 282 (5396):2028–2033
21. Mathews EA, Garcia E, Santi CM, Mullen GP, Thacker C, Moerman DG, Snutch TP (2003) Critical residues of the Caenorhabditis elegans unc-2 voltage-gated calcium channel that affect behavioral and physiological properties. J Neurosci 23(16):6537–6545
22. Arellano-Carbajal F, Briseno-Roa L, Couto A, Cheung BH, Labouesse M, de Bono M (2011) Macoilin, a conserved nervous system-specific ER membrane protein that regulates neuronal excitability. PLoS Genet 7(3):e1001341. https://doi.org/10.1371/journal.pgen.1001341
23. Gao S, Zhen M (2011) Action potentials drive body wall muscle contractions in Caenorhabditis elegans. Proc Natl Acad Sci U S A 108 (6):2557–2562. https://doi.org/10.1073/pnas.1012346108
24. Schafer WR, Kenyon CJ (1995) A calcium-channel homologue required for adaptation to dopamine and serotonin in Caenorhabditis elegans. Nature 375(6526):73–78. https://doi.org/10.1038/375073a0
25. Schafer WR, Sanchez BM, Kenyon CJ (1996) Genes affecting sensitivity to serotonin in Caenorhabditis elegans. Genetics 143 (3):1219–1230
26. Saheki Y, Bargmann C (2009) Presynaptic CaV2 calcium channel traffic requires CALF-1 and the α2δ subunit UNC-36. Nature Neuroscience 12, 1257–1265

27. Klassen MP, Shen K (2007) Wnt signaling positions neuromuscular connectivity by inhibiting synapse formation in C. elegans. Cell 130(4):704–716. https://doi.org/10.1016/j.cell.2007.06.046
28. Frokjaer-Jensen C, Kindt KS, Kerr RA, Suzuki H, Melnik-Martinez K, Gerstbreih B, Driscol M, Schafer WR (2006) Effects of voltage-gated calcium channel subunit genes on calcium influx in cultured C. elegans mechanosensory neurons. J Neurobiol 66 (10):1125–1139. https://doi.org/10.1002/neu.20261
29. Laine V, Frokjaer-Jensen C, Couchoux H, Jospin M (2011) The alpha1 subunit EGL-19, the alpha2/delta subunit UNC-36, and the beta subunit CCB-1 underlie voltage-dependent calcium currents in Caenorhabditis elegans striated muscle. J Biol Chem 286 (42):36180–36187. https://doi.org/10.1074/jbc.M111.256149
30. Gottschalk A, Schafer WR (2006) Visualization of integral and peripheral cell surface proteins in live Caenorhabditis elegans. J Neurosci Methods 154(1–2):68–79. https://doi.org/10.1016/j.jneumeth.2005.11.016
31. Dickinson DJ, Ward JD, Reiner DJ, Goldstein B (2013) Engineering the Caenorhabditis elegans genome using Cas9-triggered homologous recombination. Nat Methods 10 (10):1028–1034. https://doi.org/10.1038/nmeth.2641
32. Waaijers S, Boxem M (2014) Engineering the Caenorhabditis elegans genome with CRISPR/Cas9. Methods 68(3):381–388. https://doi.org/10.1016/j.ymeth.2014.03.024
33. Paix A, Folkmann A, Seydoux G (2017) Precision genome editing using CRISPR-Cas9 and linear repair templates in C. elegans. Methods 121-122:86–93. https://doi.org/10.1016/j.ymeth.2017.03.023
34. Deltcheva E, Chylinski K, Sharma CM, Gonzales K, Chao Y, Pirzada ZA, Eckert MR, Vogel J, Charpentier E (2011) CRISPR RNA maturation by trans-encoded small RNA and host factor RNase III. Nature 471 (7340):602–607. https://doi.org/10.1038/nature09886
35. Chiu IM, Morimoto ET, Goodarzi H, Liao JT, O'Keeffe S, Phatnani HP, Muratet M, Carroll MC, Levy S, Tavazoie S, Myers RM, Maniatis T (2013) A neurodegeneration-specific gene-expression signature of acutely isolated microglia from an amyotrophic lateral sclerosis mouse model. Cell Rep 4(2):385–401. https://doi.org/10.1016/j.celrep.2013.06.018
36. Cho SW, Kim S, Kim JM, Kim JS (2013) Targeted genome engineering in human cells with the Cas9 RNA-guided endonuclease. Nat Biotechnol 31(3):230–232. https://doi.org/10.1038/nbt.2507
37. Lo TW, Pickle CS, Lin S, Ralston EJ, Gurling M, Schartner CM, Bian Q, Doudna JA, Meyer BJ (2013) Precise and heritable genome editing in evolutionarily diverse nematodes using TALENs and CRISPR/Cas9 to engineer insertions and deletions. Genetics 195(2):331–348. https://doi.org/10.1534/genetics.113.155382
38. Chen B, Gilbert LA, Cimini BA, Schnitzbauer J, Zhang W, Li GW, Park J, Blackburn EH, Weissman JS, Qi LS, Huang B (2013) Dynamic imaging of genomic loci in living human cells by an optimized CRISPR/Cas system. Cell 155(7):1479–1491. https://doi.org/10.1016/j.cell.2013.12.001
39. Tzur YB, Friedland AE, Nadarajan S, Church GM, Calarco JA, Colaiacovo MP (2013) Heritable custom genomic modifications in Caenorhabditis elegans via a CRISPR-Cas9 system. Genetics 195(3):1181–1185. https://doi.org/10.1534/genetics.113.156075
40. Kim S, Kim D, Cho SW, Kim J, Kim JS (2014) Highly efficient RNA-guided genome editing in human cells via delivery of purified Cas9 ribonucleoproteins. Genome Res 24 (6):1012–1019. https://doi.org/10.1101/gr.171322.113
41. Paix A, Wang Y, Smith HE, Lee CY, Calidas D, Lu T, Smith J, Schmidt H, Krause MW, Seydoux G (2014) Scalable and versatile genome editing using linear DNAs with microhomology to Cas9 sites in Caenorhabditis elegans. Genetics 198(4):1347–1356. https://doi.org/10.1534/genetics.114.170423
42. Arribere JA, Bell RT, Fu BX, Artiles KL, Hartman PS, Fire AZ (2014) Efficient marker-free recovery of custom genetic modifications with CRISPR/Cas9 in Caenorhabditis elegans. Genetics 198(3):837–846. https://doi.org/10.1534/genetics.114.169730
43. Evans TC (2006) Transformation and microinjection. In: Community TCeR (ed) WormBook. https://doi.org/10.1895/wormbook.1.108.1
44. Pawley JB (2006) Points, pixels, and gray levels: digitizing image data. In: Pawley JB (ed) Handbook of biological confocal microscopy. Springer US, Boston, MA, pp 59–79. https://doi.org/10.1007/978-0-387-45524-2_4
45. Nyquist H (1928) Certain topics in telegraph transmission theory. Trans Am Inst Electr Eng

47(2):617–644. https://doi.org/10.1109/t-aiee.1928.5055024

46. Chenouard N, Smal I, de Chaumont F, Maska M, Sbalzarini IF, Gong Y, Cardinale J, Carthel C, Coraluppi S, Winter M, Cohen AR, Godinez WJ, Rohr K, Kalaidzidis Y, Liang L, Duncan J, Shen H, Xu Y, Magnusson KE, Jalden J, Blau HM, Paul-Gilloteaux P, Roudot P, Kervrann C, Waharte F, Tinevez JY, Shorte SL, Willemse J, Celler K, van Wezel GP, Dan HW, Tsai YS, Ortiz de Solorzano C, Olivo-Marin JC, Meijering E (2014) Objective comparison of particle tracking methods. Nat Methods 11(3):281–289. https://doi.org/10.1038/nmeth.2808
47. Serge A, Bertaux N, Rigneault H, Marguet D (2008) Dynamic multiple-target tracing to probe spatiotemporal cartography of cell membranes. Nat Methods 5(8):687–694. https://doi.org/10.1038/nmeth.1233
48. Thompson RE, Larson DR, Webb WW (2002) Precise nanometer localization analysis for individual fluorescent probes. Biophys J 82 (5):2775–2783. https://doi.org/10.1016/s0006-3495(02)75618-x
49. Schutz GJ, Schindler H, Schmidt T (1997) Single-molecule microscopy on model membranes reveals anomalous diffusion. Biophys J 73(2):1073–1080. https://doi.org/10.1016/s0006-3495(97)78139-6
50. Pinaud F, Michalet X, Iyer G, Margeat E, Moore HP, Weiss S (2009) Dynamic partitioning of a glycosyl-phosphatidylinositol-anchored protein in glycosphingolipid-rich microdomains imaged by single-quantum dot tracking. Traffic (Copenhagen, Denmark) 10 (6):691–712. https://doi.org/10.1111/j.1600-0854.2009.00902.x
51. Chen BC, Legant WR, Wang K, Shao L, Milkie DE, Davidson MW, Janetopoulos C, Wu XS, Hammer JA 3rd, Liu Z, English BP, Mimori-Kiyosue Y, Romero DP, Ritter AT, Lippincott-Schwartz J, Fritz-Laylin L, Mullins RD, Mitchell DM, Bembenek JN, Reymann AC, Bohme R, Grill SW, Wang JT, Seydoux G, Tulu US, Kiehart DP, Betzig E (2014) Lattice light-sheet microscopy: imaging molecules to embryos at high spatiotemporal resolution. Science 346(6208):23
52. Power RM, Huisken J (2017) A guide to light-sheet fluorescence microscopy for multiscale imaging. Nat Methods 14(4):360–373. https://doi.org/10.1038/nmeth.4224
53. Izeddin I, El Beheiry M, Andilla J, Ciepielewski D, Darzacq X, Dahan M (2012) PSF shaping using adaptive optics for three-dimensional single-molecule super-resolution imaging and tracking. Opt Express 20 (5):4957–4967. https://doi.org/10.1364/oe.20.004957
54. Burke D, Patton B, Huang F, Bewersdorf J, Booth MJ (2015) Adaptive optics correction of specimen-induced aberrations in single-molecule switching microscopy. Optica 2 (2):177–185. https://doi.org/10.1364/optica.2.000177
55. Tehrani KF, Zhang Y, Shen P, Kner P (2017) Adaptive optics stochastic optical reconstruction microscopy (AO-STORM) by particle swarm optimization. Biomed Opt Express 8 (11):5087–5097. https://doi.org/10.1364/boe.8.005087
56. Booth M, Andrade D, Burke D, Patton B, Zurauskas M (2015) Aberrations and adaptive optics in super-resolution microscopy. Microscopy 64(4):251–261. https://doi.org/10.1093/jmicro/dfv033

Chapter 3

Nanocores and Liquid Droplets: Single-Molecule Microscopy of Neuronal Stress Granule Components

Benedikt Niewidok, Rainer Kurre, and Roland Brandt

Abstract

Stress granules (SGs) are the result of phase separation of different mRNAs and multivalent RNA-binding proteins. Their main function appears to adapt the translatome of a cell to adverse environmental conditions in a fast, adjustable, and reversible manner. While being highly dynamic during physiological conditions, SGs may also be precursors of more rigid aggregates that form during neuropathological processes. Thus, analysis of the localization and mobility of key stress granule components in neural cells is an important aspect to scrutinize the material state and dynamics of SGs that could also be of pathologic relevance. Here we describe an experimental approach to follow the distribution and dynamics of paradigmatic RNA-binding proteins (RBPs) by single-molecule imaging in chemically induced SGs of model neurons. Specifically, we provide detailed information about the preparation, differentiation, and labeling of the cells; image acquisition with a TIRF microscope in the highly-inclined laminar optical sheet (HILO) mode; and image processing for single-molecule localization and tracking. We describe an approach for quantitative determination of the fraction of bound and mobile molecules, determination of the lifetime of RBP binding in nanocores, and determination of the diffusion behavior of the respective proteins to provide information about the biophysical properties of the liquid phase of SGs. Our goal is to present to the reader guidelines on how to apply single-molecule microscopy and quantitative data analysis to determine the behavior of SG components in model neurons. Moreover, the approach should also be easily adjustable for the analysis of other biomolecular condensates with liquid-like properties and for the use of other cell types.

Key words Stress granules, Liquid-liquid phase separation, Single-molecule localization microscopy, RNA-binding proteins, Model neurons

1 Introduction: Single-Molecule Imaging of Stress Granule Components

Stress granules (SGs) are a class of RNA-protein (RNP) complexes, which are induced by environmental stressors such as heat or oxidative stress. They are thought to be the result of a liquid-liquid phase separation (LLPS) leading to the formation of subcellular microcompartments, where RNA and proteins are concentrated in droplet-like structures [1]. RNP complexes are devoid of a lipid membrane and macromolecules are kept together solely by weak intermolecular interactions organized and regulated by multivalent

Nobuhiko Yamamoto and Yasushi Okada (eds.), *Single Molecule Microscopy in Neurobiology*, Neuromethods, vol. 154, https://doi.org/10.1007/978-1-0716-0532-5_3,

RNA-binding proteins (RBPs). During acute stress conditions, SGs appear to have an adaptive, survival-promoting role and they quickly disassemble when stress is released. However, inappropriate formation of SGs or changes from a more liquid-like, dynamic phase to higher insolubility of their components have been implicated in aging and pathological processes, where more stable RNP complexes form [2].

In yeast, SGs appear to behave like unstructured, solid storage depots for mRNAs and proteins. In contrast, mammalian SGs have a more liquid-like consistency [3], although they appear to be less uniform than expected. In particular, substructures, which have been referred to as "cores" with higher concentrations of proteins and mRNA, were reported [4]. In agreement, we have observed the presence of relatively immobile nanometer-sized nanocores in experimentally induced SGs of living model neurons [5]. Nanocores had a diameter of 150–200 nm and reflected regions with multiple binding sites for dynamic interaction of RNA-binding proteins (Fig. 1).

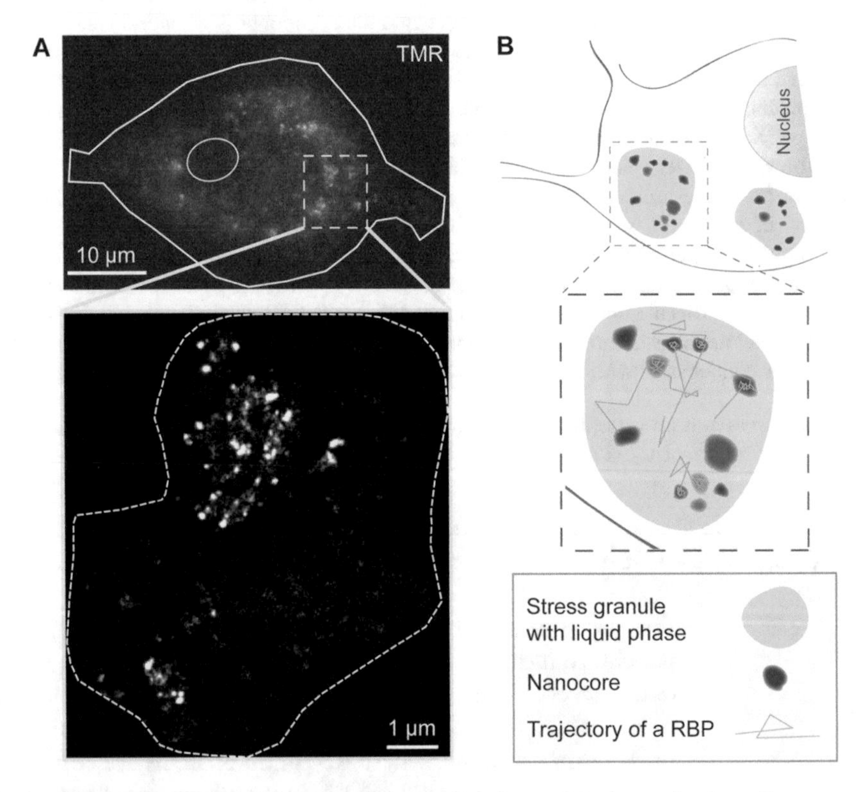

Fig. 1 Single-molecule localization microscopy of neuronal stress granules and schematic representation of stress granules with nanocores. (**a**) The top image shows a single frame of TMR-stained G3BP1 with a stress granule highlighted with a yellow box. Single-molecule localizations of 8000 frames are shown in the indicated stress granule, revealing hot spots of G3BP1 in the lower image. Outline and nucleus of the cell are indicated with a solid white line (top), while the outline of the stress granule is highlighted with a dashed line based on IMP1 SiR labeling (bottom). (**b**) Schematic representation of stress granules with a liquid phase containing distributed nanocores where binding of RBPs occurs as evidenced from the trajectories of single molecules

Ras-GAP SH3-domain-binding protein 1 (G3BP1) appears to be a key organizer of SGs and is constitutively present in SGs of many cell types [6]. We and others have shown that overexpression of G3BP1 nucleates SG formation in the absence of stress or drugs [7, 8]. The activity of G3BP1 to nucleate SGs is influenced by phosphorylation at specific residues [9] suggesting a regulatory role of G3BP1 phosphorylation in SG assembly. G3BP1 contains an RNA recognition motif and four low-complexity (LC) regions, which are thought to mediate low-affinity protein-protein interactions [10, 11]. Thus G3BP1 represents a paradigmatic multivalent RBP of liquid droplets. Another typical multivalent RNA-binding protein is insulin-like growth factor II mRNA-binding protein 1 (IMP1), which contains two RNA recognition motifs and four K homology (KH) domains, which can function in RNA recognition [12]. Both proteins induce the formation of SG-like structures in neural cells, but differ in their exchange kinetics between SGs [8].

Rat PC12 cells are a well-established model system for neural cells with key properties of PNS neurons [13]. As a cell line they are susceptible for gene transfer by lipofection and can be induced to develop axon-like processes following treatment with NGF. They express typical neuronal marker proteins and respond to chemically induced stress (treatment with arsenite) with SG formation [5]. PC12 cells express endogenous G3BP1 and IMP1, and SG formation can be induced by overexpression of exogenous human G3BP1 or IMP1 [8].

Here we describe an experimental approach to follow the distribution and dynamics of SG proteins in living neuronal cells (for a timeline of the workflow *see* Fig. 2a). We provide information about

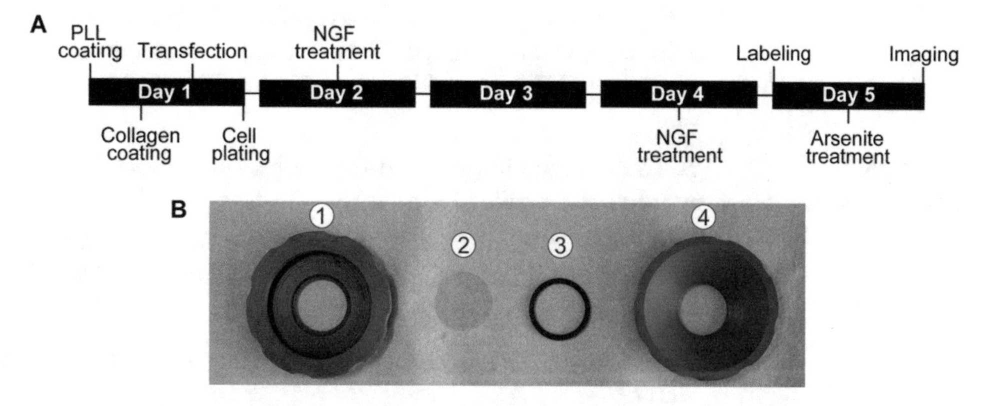

Fig. 2 Workflow of the experiment and components of the imaging chamber. (**a**) Timeline showing the typical workflow of an experiment over a time course of 5 days. (**b**) For imaging, a 24 mm glass coverslip (2) was mounted in a microscopy chamber consisting of a bottom part (1) with an inner thread and a top part (4) with an outer thread. In order to ensure proper sealing a rubber ring (3) was placed between the coverslip and the top part of the chamber. Medium was added after proper mounting of the coverslip

the expression of HaloTag- or SNAP-tagged SG proteins by lipofection in PC12 cells, which allows for substoichiometric, covalent labeling for live-cell single-molecule imaging. We describe how to differentiate the cells to a neuronal phenotype and to experimentally induce SG formation by arsenite treatment. We provide detailed information about labeling of the cells with HaloTag Ligand-TMR (HTL-TMR) and SNAP-Cell 647-SiR (SNAP-SiR), respectively, and image acquisition with a total internal reflection fluorescence (TIRF) microscope in the highly-inclined laminar optical sheet (HILO) mode [14] to excite fluorophores with minimal background. We describe how to perform high-speed single-molecule tracking. Finally, we describe an approach for quantitative determination of the fraction of bound and mobile molecules, determination of the lifetime of RBP binding in nanocores, and determination of their diffusion behavior in the liquid phase of SGs from the mean squared displacement (MSD) with time. A similar approach should also be possible in primary neurons, but has not yet been tested.

2 Materials

2.1 Materials for Glass Coverslip Coating

- 24 mm Glass coverslips (e.g., 24 mm, No. 1.5, 1001 "Assistent" Microscope Cover Glasses—Glaswarenfabrik Karl Hecht GmbH & Co KG or preferentially Roth PK26.1) (*see* **Note 1**).
- Poly-L-lysine (PLL): 100 μg/ml PLL in borate buffer, pH 8.5. Store at 4 °C (*see* **Note 2**).
- Collagen solution: Prepared from rat tails with 20 mM acetic acid diluted to 50 μg/ml, sterile filtered. Store at 4 °C.
- Phosphate-buffered saline (PBS): Dissolve 8 g NaCl, 0.2 g KCl, 0.2 g KH_2PO_4, and 1.15 g $Na_2HPO_4{\cdot}2H_2O$ in 800 ml of ddH_2O; stir on magnetic stirrer until completely dissolved; adjust to pH 7.4 and add up to 1 L with ddH_2O.

2.2 Materials for PC12 Cell Culture

- PC12 cells (cell line derived from a pheochromocytoma of the rat adrenal medulla [13]).
- Dulbecco's modified Eagle medium (DMEM) prepared from powder (containing 4.5 g/L D-glucose) (Biochrom/Merck).
- Phenol red-free DMEM (PAN Biotech).
- 10 mg/ml Penicillin/streptomycin (Pen/Strep) (Biochrom/Merck).
- 15% Serum/DMEM: DMEM with 10% (v/v) fetal calf serum, 5% (v/v) horse serum, 1% (v/v) 0.2 M L-glutamine, and 1% (v/v) Pen/Strep, store at 4 °C.

- 1% Serum/DMEM: DMEM with 0.67% (v/v) fetal calf serum, 0.33% (v/v) horse serum, 1% (v/v) 0.2 M L-glutamine, and 1% (v/v) Pen/Strep, store at 4 °C.
- Opti-MEM (Life Technologies).
- Transfection reagent; here Lipofectamine 2000 (Invitrogen).
- Plasmids coding for Halo- and SNAP-tagged stress granule proteins; here pSems HaloTag-G3BP1 and pSems SNAP-tag-IMP1 (*see* **Note 3**) [5].
- Nerve growth factor (NGF): 10 μg/ml 7S mouse NGF (Alomone Laboratories) in phenol red-free 1% serum/DMEM, store at −80 °C.
- HaloTag Ligand-TMR (HTL-TMR) (Promega): 250 μM Stock solution in DMSO and 2.5 μM working solution in DMSO. Store sealed with parafilm and protected from light at −20 °C.
- SNAP-SiR (here SNAP-Cell 647-SiR; New England Biolabs): 200 μM Stock solution in DMSO and 50 μM working solution in DMSO. Store sealed with parafilm and protected from light at −20 °C.
- 50 mM Sodium arsenite.
- Coverslip-mounting microscopy chamber [15] (Fig. 2b).

2.3 Equipment for Image Acquisition

TIRF microscope (e.g., Olympus IX81 with a four-line motorized TIR condenser (cellTIRF)) equipped with a 150× oil-immersion objective with NA 1.45 (e.g., Olympus oil, UAPON 150×/1.45) enclosed in an incubation chamber that can maintain 37 °C and 10% CO_2 and 561 and 640 nm lasers as well as corresponding filters (e.g., Semrock BrightLine HC 600/37 for TMR and BrightLine HC 697/58 for SiR). High-speed recordings require an appropriate digital camera (e.g., Hamamatsu ORCA-Flash4.0 V2 C11440-22CU) and image acquisition software (e.g., Olympus CellSens 1.14).

2.4 Software and Hardware for Image Processing and Analysis

- Operating system: Any system that is able to run a Java environment, Matlab, and Python like Microsoft Windows 7, Microsoft Windows 10, any Mac OSX or any Linux distribution.
- Hardware requirements: The PC should be equipped at least with 16 GB of RAM due to the processing of large image files. In order to keep the processing time in a reasonable range, we recommend a CPU equivalent to an Intel Core i7-3770 with 3.4 GHz.
- Software requirements: Fiji [16, 17] for cropping of images and saving them as tiff files; Matlab (ideally version 2013a) with the image processing toolbox, the optimization toolbox, the parallel computing toolbox, and the statistics toolbox in order to work

with the custom-written graphical user interface SLIMfast (*see* **Note 4**), which is implementing well-established localization and tracking algorithms [18, 19]; Python (ideally version 2.7) with NumPy 1.11.0 and SciPy 0.17.0 packages [20] (*see* **Note 5**); Lifetime.py file; Scientific data analysis software of your choice (e.g., Origin for Windows, SciDAVis for Linux) (*see* **Note 6**).

3 Methods

3.1 Glass Coverslip Coating

The following steps should take place under a sterile workbench at room temperature. The incubator in use operates at 37 °C and 10% CO_2.

1. Pass the glass coverslip three times through the flame of a Bunsen burner and place it into a 35 mm culture dish (*see* **Note 7**).
2. Add 1.5 ml ethanol (99.8%) to the culture dish and incubate for 5 min at room temperature.
3. Wash the coverslip two times with 1.5–2 ml sterile ddH_2O.
4. Aspirate the ddH_2O and add 0.5 ml PLL onto the coverslip. Incubate for 30 min at room temperature.
5. Replace the PLL with 1.5–2 ml sterile ddH_2O and incubate for 60 min at room temperature. Repeat this procedure a second time (*see* **Note 8**).
6. Aspirate the ddH_2O and add 0.5 ml collagen onto the coverslip. Incubate for 45 min in the incubator.
7. Wash the coverslip two times with 1.5–2 ml PBS. The PBS should remain in the culture dish until cell plating.

3.2 PC12 Cell Culture

The following steps should take place under a sterile workbench at room temperature. The incubator in use operates at 37 °C and 10% CO_2. The timeline for a typical experiment is shown in Fig. 2a.

3.2.1 Transfection

1. Pipette 120 μl Opti-MEM into a 15 ml polystyrene conical tube and add 6 μl of Lipofectamine 2000. Incubate for 5 min at room temperature.
2. Pipette 120 μl Opti-MEM into a 1.5 ml reaction tube and add 2.5 μg of the pSems HaloTag-G3BP1 plasmid and 2.5 μg of the pSems SNAP-tag-IMP1 plasmid.
3. Transfer the Opti-MEM/DNA mixture from the reaction tube into the polystyrene tube and mix well by pipetting up and down for ten times. Incubate for 45 min at room temperature. The resulting transfection mix is sufficient for one coverslip (*see* **Note 9**).

3.2.2 Cell Plating

PC12 cells are cultivated in a 10 cm tissue culture dish with 10 ml 15% serum/DMEM. Splitting is performed after the cells reach 80–90% confluence (usually every 3–4 days). After 25 passages a new batch of cells is grown from a frozen stock.

1. When the cells reach 80–90% confluence, aspirate the medium and detach the cells with 5 ml fresh and pre-warmed 15% serum/DMEM.
2. Transfer 10 μl of the suspension into a 1.5 ml reaction tube (for cell counting) and the rest into a 15 ml conical tube. Store the conical tube in the incubator with the cap closed loosely.
3. Pipette 10 μl of trypan blue into the reaction tube and mix by pipetting up and down. Load 10 μl of the suspension in a hemocytometer and determine the cell concentration.
4. Prepare 1.5 ml suspension with 1.2×10^5 cells/ml in a reaction tube using pre-warmed 15% serum/DMEM and the appropriate amount of cell suspension from the conical tube (*see* **Note 10**).
5. Transfer the cell solution in the appropriate concentration into the polystyrene tube containing the transfection mix prepared in Subheading 3.2.1 and incubate it for 5 min in the incubator.
6. Aspirate the PBS from the coverslips prepared in Subheading 3.1 and transfer the cell/transfection mix into the coverslip-containing culture dish. Gently swivel the culture dish in order to evenly distribute the cells. Incubate for at least 5 h in the incubator.

3.2.3 Neuronal Differentiation

1. On the next day after plating, aspirate the medium and replace it with 1.5 ml pre-warmed 1% serum/DMEM.
2. Add 15 μl mouse NGF and swivel the culture dish gently. Incubate for 2 days in the incubator.
3. Replace the NGF-containing medium with fresh 1.5 ml NGF-containing 1% serum/DMEM and incubate for another day in the incubator.

3.2.4 Labeling

Image acquisition should be performed on the day after the second NGF treatment. Sample preparation includes substoichiometric labeling of the transfected cells with HTL-TMR and SNAP-SiR as well as stress granule induction via sodium arsenite.

1. Prepare 1 ml of pre-warmed 1% serum/DMEM containing 0.5 nM HTL-TMR and 25 nM SNAP-SiR.
2. Aspirate the medium from the cell culture dish containing the coverslip and replace it with the TMR/SiR/DMEM mixture. Incubate for 20 min in the incubator.

3. Aspirate the medium and replace it with 1 ml pre-warmed 1% serum/DMEM. Incubate for 5 min in the incubator. Repeat this procedure two more times.
4. Mount the coverslip into the microscopy chamber (Fig. 2b), add 1 ml of pre-warmed 1% serum/DMEM, and check if everything is sealed correctly.
5. Prepare 1 ml of pre-warmed 1% serum/DMEM containing 0.5 mM sodium arsenite.
6. Aspirate the medium in the microscopy chamber and replace it with the arsenite/DMEM mixture. Incubate for 20 min in the incubator (*see* **Note 11**).
7. Aspirate the medium containing arsenite and replace it with pre-warmed 1% serum/DMEM.

3.3 Image Acquisition

The heating system of the microscope should be turned on at least 2–3 h prior to imaging to ensure uniform temperature distribution throughout the setup, which is important for image acquisition as well as cell viability (*see* **Note 12**). For transmission bright-field images a standard halogen lamp is used. Single-molecule imaging is performed by operating the TIRF microscope in the highly-inclined laminar optical sheet (HILO) mode [14] using a 561 nm (TMR channel) and 640 nm (SiR channel) laser. The pixel size should be close to 100 nm and depending on the camera and objective combination it might be necessary to lower the overall image resolution via binning. Scanning of the coverslip is performed at 33 fps (i.e., frame time of 30 ms) with a laser output of ~30 W/cm^2 and ~100 W/cm^2 for the 561 nm and 640 nm laser, respectively.

1. Use bright-field illumination to focus on the cells.
2. Switch to the SiR channel and scan the coverslip for a positively transfected cell that shows granular accumulations (*see* **Note 13**).
3. Switch to the TMR channel to confirm double transfection of the cell.
4. Increase the 561 nm laser output to 2.74 mW and bleach the dye until single molecules can be distinguished (*see* **Note 14**).
5. Take a snapshot in the SiR channel with an increased laser intensity of 2.74 mW. Afterwards, record a time series in the TMR channel at 100 fps (i.e., frame time of 10 ms) with a length of up to 8000 frames (*see* **Note 15**).
6. Save the snapshot and the time series separately and switch back to the SiR channel to scan the coverslip for the next cell.

3.4 Image Processing

The acquired time series can consist of up to 8000 frames with a size of over 6 GB. In order to decrease the necessary processing power and time for the single-molecule localization and tracking, it is advised to duplicate/crop the desired region of interest, e.g., the stress granule.

1. Open the saved SiR snapshot as well as the TMR time series with Fiji.
2. Select a region of interest (ROI) containing a stress granule in the SiR snapshot and add it to the "ROI Manager" by pressing the hotkey "T" or open the ROI manager via "Analyze → Tools → ROI Manager" and press the "Add" button.
3. Select the TMR time series and apply the ROI by clicking on the ROI ID in the "ROI Manager."
4. Right-click into the ROI and select "duplicate" from the context menu. Confirm in the upcoming window that you want to duplicate the whole stack. Repeat this step for the SiR snapshot.
5. Save the snapshot and the time series as TIFF files.

3.5 Single-Molecule Localization and Tracking

1. Open Matlab and navigate to the SLIMfast folder.
2. Open SLIMfast by running the SLIMfast Matlab file.
3. Select "Project → Create" to create a new project. In the upcoming window select the "+" button and select the SiR snapshot and TMR time series to open. Confirm by clicking on the "OK" button.
4. SLIMfast will create two projects with the name "unknown." Drag and drop the snapshot into the project with the times series or vice versa. Rename the project and the files for easier orientation (Fig. 3a) (*see* **Note 16**).
5. Select the time series and click on the "SHOW" button to open it. Open the "UNIT MANAGER" from the toolbar and adjust the pixel size, frame count, and photon count (*see* **Note 17a**). Now open the "LOCALIZATION MANAGER" and adjust "localization range" (*see* **Note 17b**) as well as the "PSF Radius" (*see* **Note 17c**). Check the "Use Multiple Cores" box and press the "Localize" button.
6. After the localization process is finished a new window appears showing the accumulated localizations. Rename the file in the main SLIMfast window and click on the "track single emitter" button in the toolbar of the new window. In the upcoming "TRACKING MANAGER" window several options need to be adjusted according to Fig. 3b. Finally, press the "Track" button to initialize the tracking.

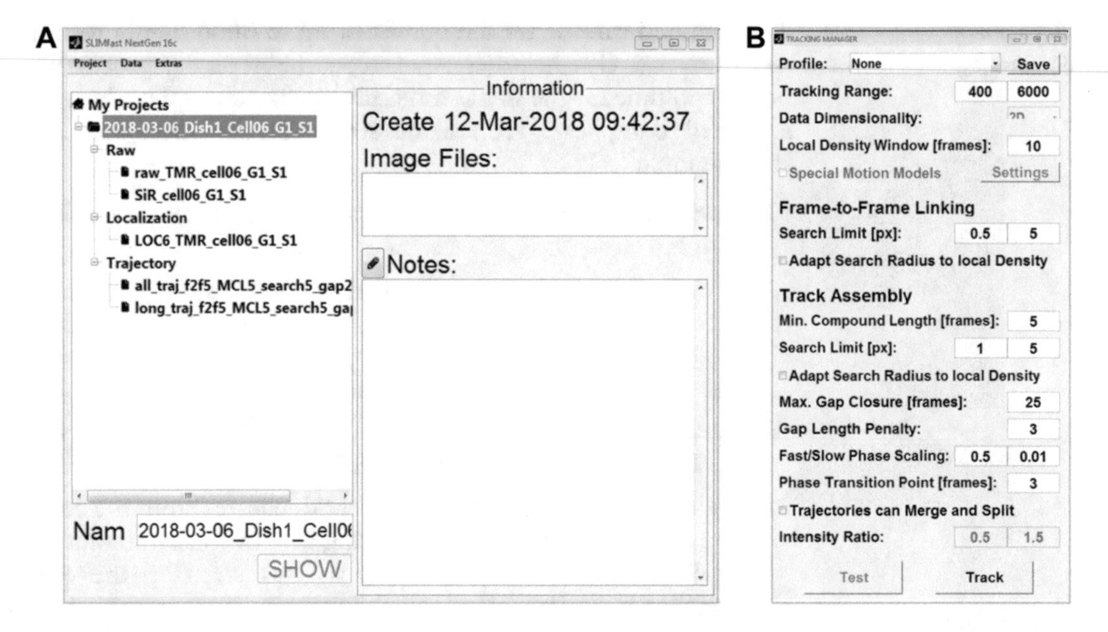

Fig. 3 Screenshot of the SLIMfast and TRACKING MANAGER windows. (**a**) SLIMfast is a custom-written graphical user interface used for single-molecule localization and tracking. The main window of SLIMfast with a loaded project depicting the general organization of a project is shown. (**b**) The tracking manager is used to fine-tune the area in which the algorithm connects single localizations into tracks. For more information about the different options, hover with the mouse over the respective names to see a tooltip

7. After the tracking is finished a new window appears showing all trajectories in a list. Press the "LIST↑MAP" button to show the 2D representation of the trajectories inside of the granule (Fig. 4a). In order to exclude all trajectories that are shorter than 100 ms, press the "LIFE" button in the toolbar to open the "DISTRIBUTION MANAGER." Press the button in between the "SET" and "DATA" icons of the toolbar and exclude all trajectories that are shorter than 0.1 s. Press the "Data" button in the toolbar to filter the trajectories and a new window will appear showing only trajectories that are longer than 100 ms (Fig. 4b). Rename the new window in the main SLIMfast window (*see* **Note 18**).

3.6 Determination of Bound and Free Fraction

The trajectory set that has been obtained contains in the case of G3BP1 trajectories that are short-lived, while exploring wide areas, and trajectories that only explore a compact area over an extended period of time. Analysis of the distribution of instantaneous diffusion constants with a two-component Gaussian filter can be used to visualize and filter these fractions.

1. Open the window that contains all trajectories that are longer than 100 ms and press the "DIFF." button in the toolbar. In the upcoming "MSD CURVE ESTIMATOR" change the "Fit Range" to 1–15 and the "Model" to Anomalous Diffusion,

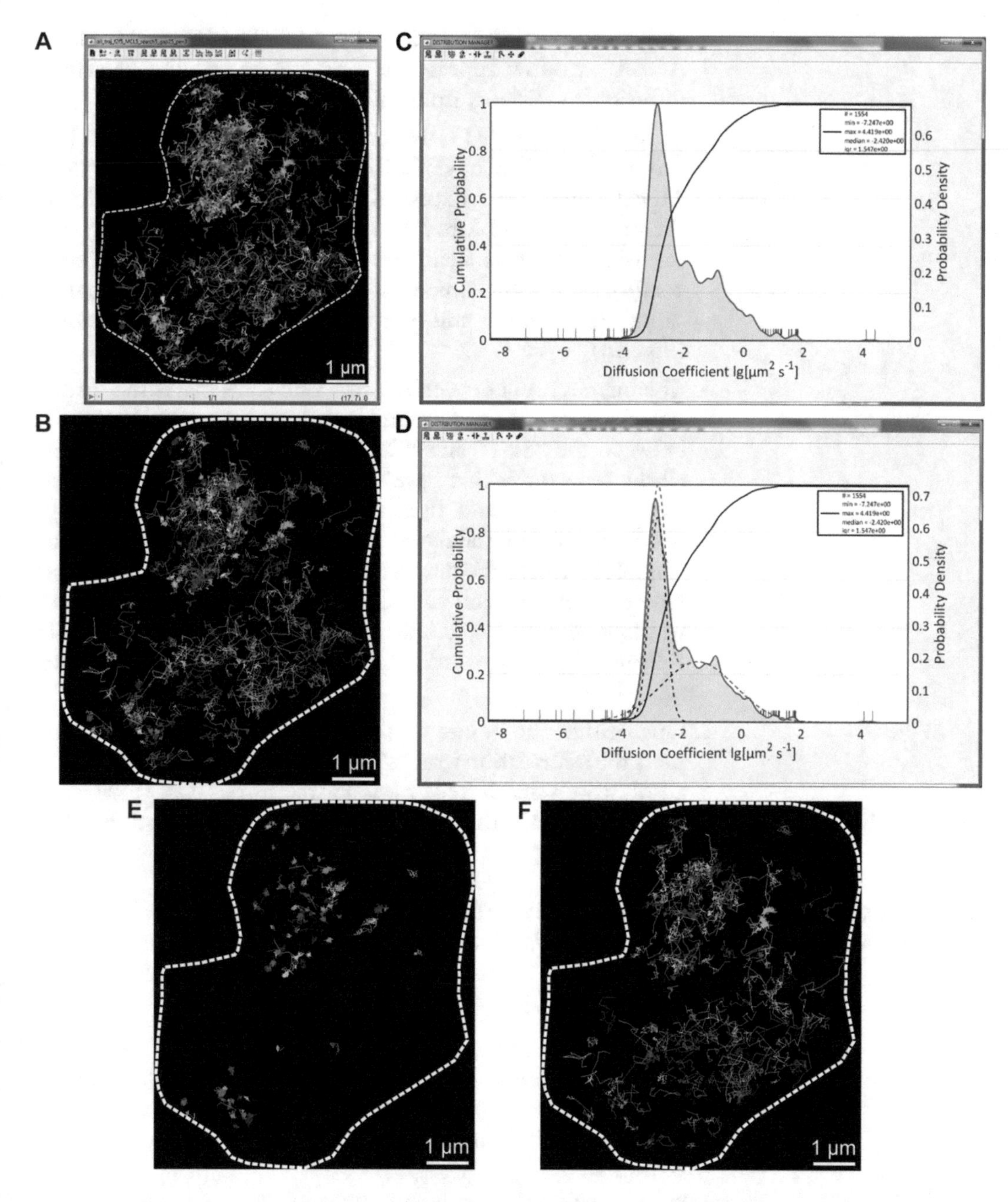

Fig. 4 Processing of single-molecule trajectories for quantitative determination of the fraction of bound and mobile molecules. After performing tracking of the single molecules shown in Fig. 1a, a new window appears showing all detected tracks (**a**). In order to eliminate randomly connected trajectories, tracks with a lifetime of less than 100 ms are excluded afterwards via the "DISTRIBUTION MANAGER" that can be opened with the "LIFE" button. The resulting tracks (**b**) can be used to determine the bound and mobile fraction of molecules via analysis of the distribution of instantaneous diffusion constants (**c**). Fitting of these diffusion constants with a two-component Gaussian filter indicates the presence of two fractions (**d**). The fractions can be filtered and visualized as the bound fraction (**e**) and the mobile fraction (**f**). The outline of the stress granule is indicated with a dashed line

and then press "Accept". The "DISTRIBUTION MANAGER" window appears showing the probability density of diffusion coefficients from the trajectories (Fig. 4c).

2. In the "DISTRIBUTION MANAGER" press the "FIT" button in the toolbar to open the "DISTRIBUTION ESTIMATOR." Check the "Expected # of Populations" box, change the value to 2 (*see* **Note 19**), and press "Estimate." The resulting window shows again the probability density of diffusion coefficients but this time with two Gaussian distributions indicating the bound and the mobile fraction of trajectories (Fig. 4d).
3. The intersection between the two Gaussians marks the threshold for the free and bound fraction. Press the button in between the "SET" and "DATA" icons of the toolbar, select all the trajectories that have smaller diffusion coefficients than the intersection, and then press the "DATA" button. The upcoming window shows all trajectories of the bound fraction (Fig. 4e). Repeat the process but select all the trajectories that have a larger diffusion coefficient than the intersection to get a window with all trajectories of the mobile fraction (Fig. 4f). Rename the new windows in the main SLIMfast window accordingly.
4. Open either the bound or the mobile trajectory window and press the information icon "i" in the toolbar to get access to the absolute number of free or bound trajectories. Use these numbers to calculate the relative amount of the bound and free fraction.

3.7 Lifetime Determination

The lifetimes of all trajectories of the bound fraction can be plotted as histograms and subsequently fitted with a single exponential function in order to determine the average lifetime of a binding event (*see* **Note 20**).

1. Open the window that contains the trajectories of the bound fraction and press the "DATA" button in the toolbar. In the upcoming window select "traj ID" and export the data as a text file ending with "10ms_Traj_ID.txt" into an empty folder (*see* **Note 21a**).
2. Execute the python script "lifetime_granules.py" with the respective input parameters (*see* **Note 21b**). The resulting "lifetimes.dat" file contains the lifetimes of the individual trajectories.
3. Import the "lifetimes.dat" file into the data analysis software of your choice and create a histogram by performing a frequency count with relative probability output. Plot the lifetime

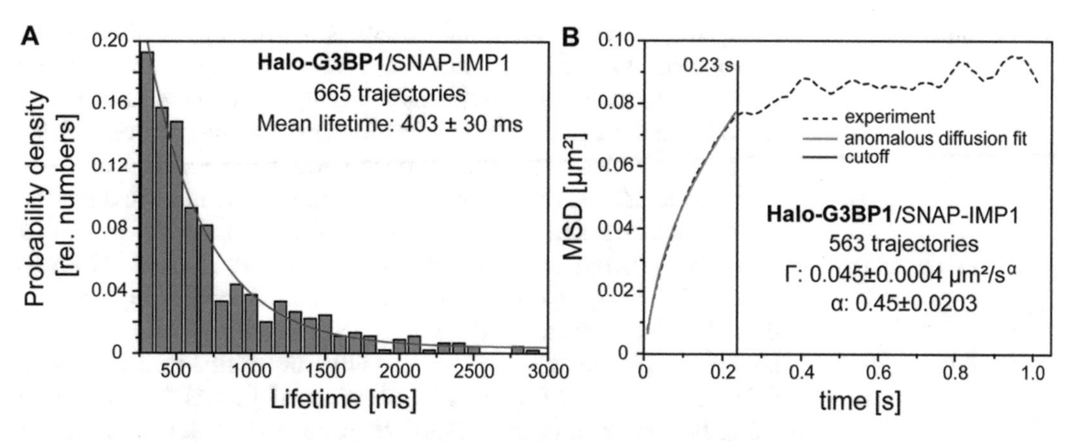

Fig. 5 Determination of the lifetime of RBP binding in nanocores and the diffusion behavior in the liquid phase of SGs. (**a**) The trajectories of the bound fraction are plotted as histograms and fitted with a single exponential function (red line) to determine the average lifetime of a binding event. (**b**) Determination of the mean diffusion constant (Γ) and the anomaly factor (α) is performed via fitting of the mean squared displacement over time (green line) that is derived from the trajectories of the mobile fraction. Prior to fitting it is necessary to perform a cutoff (black line) before the exponential curve reaches a plateau. Proper fitting of the lifetime as well as the mean diffusion constant requires a high number of trajectories. In order to provide such a high number it is advised to pool data of several stress granules. The depicted figures are derived from three independent experiments with 14 cells and 25 stress granules

histogram with a single exponential function to determine the mean lifetime (Fig. 5a) and subtract the estimated bleaching rate ($0.040 \pm 0.002\ s^{-1}$ for TMR) [5].

3.8 Diffusion Constant and Anomaly Factor Determination

The mobile fraction of trajectories can be used to determine the mean squared displacement (MSD) with time. Fitting of such a plot allows the calculation of the mean diffusion constant as well as the anomaly exponent α, which can indicate the diffusion type (superdiffusion/active transport $\alpha > 1$; normal diffusion $\alpha = 1$; subdiffusion/molecular crowding $\alpha < 1$).

1. Open the window that contains the trajectories of the mobile fraction and press the "STEP" button in the toolbar to open the "JUMPSIZE SERIES MANAGER." Press the "FIT" button in the toolbar, then check the "Expected # of Populations" box, and change the value to 1 in the upcoming "DISTRIBUTION ESTIMATOR" window. Open the export dialog by pressing the "DATA" button in the toolbar and save the "MSD Curve" data as a text file.
2. Import the "MSD Curve" text file into the data analysis software of your choice and visualize the data. Perform a cutoff where the curve reaches a first plateau and determine the diffusion constant Γ and the anomaly exponent α by fitting the remaining data with a nonlinear, anomalous diffusion fit (Fig. 5b) (*see* **Note 22**).

3.9 Summary

Analyzing the spatiotemporal dynamics of proteins within SGs is a demanding task that cannot be tackled by diffraction-limited microscopy techniques. Utilization of single-molecule tracking on the other hand enables the localization and visualization of individual protein trajectories, thus allowing the determination of intragranular fractions, the lifetime of binding events, and the diffusion constants of tracked proteins. In case of G3BP1 we can observe a biphasic partition inside of SGs with a bound fraction of ~50% (Fig. 4d). Further analysis of the bound fraction (Fig. 4e) as described in Subheading 3.7 reveals that an average binding event lasts for ~400 ms (Fig. 5a). Assessing the diffusion constant of G3BP1 inside of SGs requires plotting of the MSD with time using the trajectories of the mobile fraction (Fig. 4f), as shown in Subheading 3.8. Fitting of the curve displays an anomalous diffusion of G3BP1 ($\alpha = 0.45$) with a diffusion constant of 0.045 μm/s$^{\alpha}$ (Fig. 5b). The presented data was acquired by pooling the trajectories of 14 cells containing 25 SGs gathered in three independent experiments.

Although we describe the methodology for tracking only a single protein it is essentially possible to observe two different proteins simultaneously at a single-molecule level, e.g., HTL-TMR-tagged G3BP1 and SNAP-SiR-tagged IMP1. However, utilization of two distinct fluorophores reduces the recording time in the single-molecule regime due to varying dye properties, such as bleaching rate. Furthermore, it requires additional equipment for true simultaneous recording or can only be performed in a quasi-simultaneous fashion (*see* **Note 15**).

4 Notes

1. In general, any kind of glass coverslip for microscopy can be used, as long as it can be mounted into a microscopy chamber that allows the addition of 1% serum/DMEM.
2. Instead of using PLL and collagen it is also possible to coat the coverslips with poly-L-lysine-graft (polyethylene glycol)-copolymer functionalized with RGD [21]. However, PC12 cells seem to have an increased tolerance against detaching during stress conditions when plated on PLL and collagen-coated surfaces.
3. Here the pSems vector is used for transfection; however essentially every vector plasmid should be applicable as long as the used cell line tolerates plasmid delivery and expression rate of the exogenous protein.
4. SLIMfast was written using Matlab 2013a and therefore works most stable with this version. Later iterations of Matlab can cause minor problems but are nonetheless operable. SLIMfast

was developed and is being maintained by Christian P. Richter from the Department of Biophysics of the University of Osnabrück. For more information or access to the software contact him via christian.richter@biologie.uni-osnabrueck.de.

5. WinPython is a package for windows that contains Python, NumPy, and Scipy as a bundle.
6. People proficient with Matlab can perform the analysis as well as statistical evaluation in Matlab.
7. Instead of passing the coverslips through a flame, it is also possible to use a plasma cleaner (e.g., Diener electronic FEMTO) for 15 min and subsequently wash the coverslips 2–3 times with ddH_2O.
8. Instead of directly using the PLL-coated coverslip for collagen coating, it is possible to store it up to 1 month at 4 °C. Aspirate the ddH_2O and let the dish dry in the bench at room temperature until all of the remaining water is evaporated. Seal the culture dish with parafilm and store at 4 °C.
9. In case more than one coverslip needs to be transfected, it is possible to scale up the amount of reagents and prepare the transfection mix in a single polystyrene tube. The transfection protocol described is specific for utilization of Lipofectamine 2000. If another transfection reagent is used the protocol might vary significantly and the description of the manufacturer should be consulted.
10. If more than one coverslip needs to be prepared, it is possible to scale up the amount of cell suspension (*see* also **Note 9**).
11. PC12 cells tend to detach easily after the sodium arsenite treatment. The vibrations from carrying the culture dish to the microscope can lead to a substantial cell loss. In order to avoid this problem, the arsenite treatment, incubation, and washing should be performed directly at the microscope.
12. In addition to preheating the microscope it is advised to manually focus the lasers with fluorescent beads (e.g., Invitrogen TetraSpeck Microspheres) before image acquisition. In order to obtain optimal HILO conditions, all used laser lines need to be focused perfectly into the backfocal plane of the objective. The fluorescent beads need to be attached to a PLL-coated glass coverslip. For TetraSpeck Microspheres: Vortex the beads for 1 min and ultrasonicate for 10 min afterwards. Dilute 1:100 in ddH_2O and spread 5 μl of the dilution on a PLL-coated coverslip. Let the coverslip dry and wash it afterwards with ddH_2O. Aspirate the water and let the coverslip dry in the sterile bench. Mount the prepared coverslip into the microscopy chamber, add 1 ml pre-warmed 1% serum/DMEM, focus beads and perform manual laser focusing and alignment.

13. Scanning for an appropriate cell is performed using the total internal reflection mode. Upon finding a positively transfected cell the HILO mode is utilized via adjustment of the laser angle for proper illumination of the granules. Fine readjustments of the laser angle are necessary depending on the size and form of the cell as well as the distribution of granules. In general, it can be difficult to properly adjust focal plane and laser angle combination due to the three-dimensional structure of the granules, resulting in a heavy fluctuation in the *Z*-axes and thus a low signal-to-noise ratio.
14. The fluorescence signal prior to bleaching is very bright due to excess of TMR-tagged G3BP1. The resulting signal-to-noise ratio and spatial distribution of tagged proteins do not allow for the identification of single molecules. In order to properly localize single emitters using SLIMfast, the recorded images need to contain a distinguishable amount of tagged proteins, which can be achieved via initial bleaching of TMR.
15. In our hands TMR coupled with a HaloTag ligand is the most suitable dye/tag combination for single-molecule tracking in granules allowing recordings over an extended period of time with a relatively high signal-to-noise ratio (alternatively use Janelia Fluor 549-HTL). It is, however, possible to acquire quasi-simultaneous recordings via filter switching every 100–200 frames or actual simultaneous recordings using a QuadView (e.g., Photometrics QuadView QV2) with the combination of HTL-TMR and SNAP-SiR. Nonetheless, the overall recording time is reduced when compared to HTL-TMR-only recordings due to the inferior photostability of SNAP-SiR. Furthermore, it is rather challenging to achieve the single-molecule regime in both channels simultaneously, thus additionally reducing the timeframe that can be used for single-molecule localization.
16. It is possible to select an additional ROI with SLIMfast in case that the imported image created with Fiji still contained parts of the cell plasma (e.g., rectangular ROI). In order to create the ROI, open the SiR image and press the "ROI" button in the toolbar of the upcoming window. Select "Freehand" in the "Shape" field and check all three boxes at the bottom of the window. Press the "Create" button and draw the ROI while pressing the left mouse button. Release the mouse button to finalize the ROI and right-click into the created ROI to save it or to deactivate the position info. Then open the TMR image, press the "ROI" button in the toolbar, and load the created ROI in the upcoming window.

17. (a) The value for the photon count is specific for the camera that is being used in the imaging setup. Consult the manual, contact the manufacturer, or ask the designated supervisor of the microscope/s for more information. For example, the camera of the microscopy setup outlined in Subheading 2 has a photon count of 0.44. (b) Depending on the imaging process, the first frames of the time series can be too bright for the localization of single molecules, while the last frames might be already bleached extensively. The "localization range" allows for adjusting which part of the time series is suitable for the localization, thus decreasing the processing time. (c) The "PSF (point spread function) radius" varies depending on the dye that is used for labeling and can be adjusted by clicking on the button to the right, which is displaying a number. The "PSF ESTIMATOR" window that will appear allows to either choose a dye from a list in the upper half or estimate the "PSF radius" based on the loaded time series in the lower half of the window. If imaging is performed in the TIRF mode, it is sufficient to choose the used dye from the list. In case of utilization of the HILO mode or labeling with a dye not included in the list, it is advised to perform an estimation of 3–5 time series/granules/cells and use the mean PSF for the following experiments.
18. It is advised to perform the single-molecule localization and tracking for all granules prior to further analysis. Afterwards, it is possible to pool all trajectories that are longer than 100 ms and evaluate them as described in Subheadings 3.6–3.8. Open all projects via "Project → Load → Select Project File" and select all trajectory files that should be pooled. Then select "Data → Trajectory → Pool" to create a new project containing a file with all trajectories.
19. Analysis of other stress granule components might reveal a different distribution of diffusion coefficients. SLIMfast can additionally perform an estimation of the number of populations. Choose the "Test # of Populations" box instead of the "Expected # of Populations" box in the "DISTRIBUTION ESTIMATOR" window and choose a reasonable number (the preset of 4 is fine). After the estimation process select the most probable number in the "Choose final Complexity" field and press on the "OK" button to apply the Gaussian filter.
20. The measured lifetimes can be distorted due to bleaching of the dye, resulting in very similar lifetimes across different conditions/samples. In order to verify if the measured lifetimes are bleaching limited, it is possible to perform imaging with different cycle times or varying laser intensities and analyze if the lifetimes change accordingly (e.g., decreased lifetimes with increased laser intensity or cycle time).

21. (a) Depending on the chosen exposure time of the time series during image acquisition the filename ending needs to be modified, e.g., "20ms_Traj_ID.txt" in case of a 20 ms exposure. The python script also includes exposure time options for 25 ms, 33 ms, 35 ms, as well as 36 ms and would need to be modified for deviating values. (b) The usage of python scripts is quite simple on Linux operating systems or Linux virtual machines and is thus preferable. Open the terminal emulator and enter the command "python lifetime_granules.py -i /path/to/folder/with/trajectory/ID/text/files." The input folder can contain several text files that are all processed. The resulting "lifetimes.dat" file is created in the directory containing the python script.
22. The resulting diffusion constant needs to be divided by 4 in case of two dimensional diffusion as $\langle r^2 \rangle = 4\Gamma\tau^{\alpha}$, where $\langle r^2 \rangle$ is the MSD, Γ is the diffusion constant, τ is the time, and α is the anomaly exponent.

Acknowledgments

We thank Lidia Bakota for critical reading and preparation of the schematic model. The work was supported by Deutsche Forschungsgemeinschaft Grant SFB 944, Project P1 (to R.B.), and the Z-project of the SFB. The authors declare no competing financial interests.

References

1. Banani SF, Lee HO, Hyman AA et al (2017) Biomolecular condensates: organizers of cellular biochemistry. Nat Rev Mol Cell Biol 18:285–298
2. Alberti S, Hyman AA (2016) Are aberrant phase transitions a driver of cellular aging? Bioessays 38:959–968
3. Kroschwald S, Maharana S, Mateju D et al (2015) Promiscuous interactions and protein disaggregases determine the material state of stress-inducible RNP granules. elife 4:e06807
4. Jain S, Wheeler JR, Walters RW et al (2016) ATPase-modulated stress granules contain a diverse proteome and substructure. Cell 164:487–498
5. Niewidok B, Igaev M, Pereira Da Graca A et al (2018) Single-molecule imaging reveals dynamic biphasic partition of RNA-binding proteins in stress granules. J Cell Biol 217:1303–1318
6. Kedersha N, Panas MD, Achorn CA et al (2016) G3BP-Caprin1-USP10 complexes mediate stress granule condensation and associate with 40S subunits. J Cell Biol 212:845–860
7. Kedersha N, Anderson P (2007) Mammalian stress granules and processing bodies. Methods Enzymol 431:61–81
8. Moschner K, Sundermann F, Meyer H et al (2014) RNA protein granules modulate tau isoform expression and induce neuronal sprouting. J Biol Chem 289:16814–16825
9. Tourriere H, Chebli K, Zekri L et al (2003) The RasGAP-associated endoribonuclease G3BP assembles stress granules. J Cell Biol 160:823–831
10. Coletta A, Pinney JW, Solis DY et al (2010) Low-complexity regions within protein sequences have position-dependent roles. BMC Syst Biol 4:43
11. Uversky VN (2015) Intrinsically disordered proteins and their (disordered) proteomes in neurodegenerative disorders. Front Aging Neurosci 7:18

12. Garcia-Mayoral MF, Hollingworth D, Masino L et al (2007) The structure of the C-terminal KH domains of KSRP reveals a noncanonical motif important for mRNA degradation. Structure 15:485–498
13. Greene LA, Tischler AS (1976) Establishment of a noradrenergic clonal line of rat adrenal pheochromocytoma cells which respond to nerve growth factor. Proc Natl Acad Sci U S A 73:2424–2428
14. Tokunaga M, Imamoto N, Sakata-Sogawa K (2008) Highly inclined thin illumination enables clear single-molecule imaging in cells. Nat Methods 5:159–161
15. Appelhans T, Busch K (2017) Single molecule tracking and localization of mitochondrial protein complexes in live cells. Methods Mol Biol 1567:273–291
16. Linkert M, Rueden CT, Allan C et al (2010) Metadata matters: access to image data in the real world. J Cell Biol 189:777–782
17. Schindelin J, Arganda-Carreras I, Frise E et al (2012) Fiji: an open-source platform for biological-image analysis. Nat Methods 9:676–682
18. Jaqaman K, Loerke D, Mettlen M et al (2008) Robust single-particle tracking in live-cell time-lapse sequences. Nat Methods 5:695–702
19. Serge A, Bertaux N, Rigneault H et al (2008) Dynamic multiple-target tracing to probe spatiotemporal cartography of cell membranes. Nat Methods 5:687–694
20. Oliphant TE (2007) Python for scientific computing. Comput Sci Eng 9:10–20
21. Wedeking T, Lochte S, Birkholz O et al (2015) Spatiotemporally controlled reorganization of signaling complexes in the plasma membrane of living cells. Small 11:5912–5918

Chapter 4

Live-Cell Single-Molecule Imaging with Optogenetics Reveals Dynamics of a Neuronal Activity-Dependent Transcription Factor

Hironobu Kitagawa, Noriyuki Sugo, and Nobuhiko Yamamoto

Abstract

Powerful imaging techniques have been developed to investigate the spatiotemporal dynamics of molecular players that are involved in various biological functions. A fine-tuned single-molecule imaging technique allows us to study the movement of transcription factors in the cell nucleus. Our technique combined with optogenetics enables us to reveal neuronal activity-dependent dynamics of transcription factors in living cortical neurons. Here, we describe the detailed experimental procedures to study the transcriptional activity with physiological stimulation in living CNS neurons.

Key words Single-molecule imaging, Highly inclined and laminated optical (HILO) sheet microscopy, Optogenetics, Activity-dependent gene expression, Transcription factor, Cortical neuron

1 Introduction

Transcription is a highly dynamic process with protein-DNA and protein-protein interactions [1]. The transcription regulatory components (transcription factors/cofactors, epigenetic modulators, and basal transcriptional machinery) are recruited to their target genes, and their interactions play a key role in regulating transcription [2]. In this process, the dynamics of these molecules is thought to determine the timing and level of transcription, which regulate various physiological and developmental events [1, 3].

In general, binding of transcription factors to the promoter regions of their target genes is an essential step in initiating transcription, together with recruitment of basal transcriptional machinery components such as RNA polymerase II and cofactors [4, 5]. In neuronal cells, firing and synaptic activities induce the activation of a number of transcription factors, and binding of the activated transcription factors to the target sites contributes to expression of a subset of downstream genes, which are involved in

Nobuhiko Yamamoto and Yasushi Okada (eds.), *Single Molecule Microscopy in Neurobiology*, Neuromethods, vol. 154, https://doi.org/10.1007/978-1-0716-0532-5_4,

many developmental events including neuronal survival, axon/dendrite growth, and synaptic plasticity [6–9]. An important issue is how the kinetics and dynamics of transcription factor recruitment and binding to their target sites regulate the activity-dependent transcription.

To date, biochemical studies have demonstrated the kinetics of transcription factors by measuring the equilibrium-binding constants of purified transcription factors to the target DNA at distinct time points [10, 11]. Direct tracking of transcription factors in living cells would provide a much better understanding of their binding and dissociation properties. Indeed, fluorescent correlation spectroscopy (FCS) and fluorescence recovery after photobleaching (FRAP) experiments have shown the intracellular dynamics of transcription factors by measuring the average movement of fluorescence-tagged transcription factors [12, 13].

Recent advances in single-molecule imaging and labeling technologies allow us to further investigate the spatiotemporal dynamics of individual molecules in various biological contexts [14–16]. Total internal reflection fluorescence (TIRF) microscopy has been developed for single-molecule imaging [17, 18], and is utilized to analyze the dynamics of molecules that are distributed close to the plasma membrane of cultured cells. For example, TIRF microscopy has revealed dynamic aspects of membrane receptors [19–21] and cell adhesion molecules [22]. However, this illumination technique cannot detect the behavior of transcription factors, because these molecules are present in the nucleus and thus located out of the illumination range. Highly inclined and laminated optical sheet (HILO) illumination enables us to display such molecular dynamics by overcoming this limitation [23]. Indeed, HILO illumination has revealed the assembly of glucocorticoid receptor, a transcription factor, and its cofactors in the nucleus [24].

We have succeeded in single-molecule analysis of cAMP-response element-binding protein (CREB) in living neuroblastoma cells and cortical neurons [25, 26]. One of the most interesting issues is the neuronal activity-dependent function of CREB. To address this, a stimulus technique is an important component. In biochemical studies, pharmacological treatments have been widely used, but they cannot stimulate particular cells. The optogenetic method is groundbreaking in terms of specifying the source of stimulation, and can apply various kinds of patterned stimulation. This chapter describes the methods for single-molecule imaging combined with an optogenetic tool to investigate the activity-dependent dynamics of transcription factors in the nucleus of cortical neurons, focusing on CREB dynamics in living neurons [26]. Representative results of spatiotemporal analysis of single CREB molecules also demonstrate the relevance and significance of this method.

2 Materials and Equipment

2.1 Chemicals and Reagents

2.1.1 Cortical Cell Culture

- Poly-L-ornithine (Sigma, P3655).
- Hanks' balanced salt solution (HBSS).
- Ca^{2+}-, Mg^{2+}-free PBS.
- Ca^{2+}-, Mg^{2+}-free HBSS.
- 0.125% Trypsin (Thermo Fisher Scientific, 27250-018) and 0.02% EDTA in Ca^{2+}-, Mg^{2+}-free PBS.
- DMEM/F12 (Thermo Fisher Scientific, 11320033).
- Fetal bovine serum (FBS) (HyClone).
- B27 supplement (Thermo Fisher Scientific).
- Ca^{2+}-, Mg^{2+}-, and phenol red-free HBSS (Thermo Fisher Scientific, 14025092).

2.1.2 Plasmids

- pTet-On Advanced/TRE-Tight HaloTag-CREB.
- pTet-On Advanced/TRE-Tight HaloTag-CREB (L318/325V).
- pTet-On Advanced/TRE-Tight HaloTag-CREB (R301L).
- pTα1-EGFP.
- pCAGGS-hChR2(H134R)-EYFP (kindly gifted by Dr. Naoyuki Uesaka).

2.1.3 Pharmacological Study

- KCl depolarization solution (170 mM KCl, 1.3 mM $MgCl_2$, 0.9 mM $CaCl_2$, 10 mM HEPES, pH 7.4).
- Tetrodotoxin solution (100 μM TTX diluted with sterile distilled water; Sankyo).

2.1.4 Drug Treatment

- Doxycycline (1 mg/ml diluted with sterile distilled water; Clontech, 631311).

2.1.5 Fluorescent Ligand

- HaloTag tetramethyl rhodamine (TMR) Direct Ligand (diluted with Ca^{2+}-, Mg^{2+}-free PBS; Promega, G2991).

2.2 Culture Dish

- 35 mm Glass-bottom dish (Greiner Bio-One).

2.3 Microscopy and Equipment

- An inverted microscope (Ti-E, Nikon) on an anti-vibration stage with N_2 gas pressure.
- Fluorescence illumination (488 nm, Nikon).
- 561 nm laser (20 mW; Coherent).

- EMCCD camera (iXon 897, Andor Technology).
- A stage-top incubator (Tokai Hit).
- Solis software (Andor Technology).
- NIS Element software linked to a deconvolution module (Nikon).

2.4 Electroporation

- Plate electrode (LF513-5; BEX).
- Square-pulse generator (CUY21EX; BEX).

2.5 Photostimulation

- Pulse generator (Master-8, AMPI).
- Solid-state illuminator (475 nm wavelength; Lumencor SPECTRA).

3 Methods

3.1 Neuronal Culture

Monolayer cultures are suitable to study CREB dynamics at the single-molecule level, as HILO illumination (*see* below) should be applied to cell nuclei containing the transcription factor through the bottom of the culture dish (Figs. 1 and 2). Moreover, the culture bottom should consist of thin glass such as a cover glass. To study dynamics of CREB, we utilize cortical cell cultures, since activity-dependent aspects of CREB are well characterized in these neurons.

3.1.1 Preparation of Poly-L-Ornithine-Coated Glass-Bottom Dishes

1. Dissolve 10 mg of poly-L-ornithine in 1 ml of 0.15 M borate buffer, pH 8.5. Dilute this solution 100 times with sterile distilled water (0.1 mg/ml).
2. Coat a 35 mm glass-bottom dish (a glass region at the center) with 0.2 ml of 0.1 mg/ml poly-L-ornithine solution for at least 1 h at room temperature.
3. Discard the solution and wash three times with 1 ml of sterile distilled water (*see* **Note 1**).
4. Dry the dish completely in a clean bench (*see* **Note 2**).

3.1.2 Preparation of Primary Cortical Neurons

1. Anesthetize an embryonic day 16 (E16) pregnant mouse with pentobarbital (50 mg/kg). Noon of the day on which the vaginal plug is detected in the morning is designated embryonic day 0 (E0).
2. Remove embryos under caesarian dissection and place them in a 100 mm dish containing ice-cold HBSS. Do not use premature embryos, which have a smaller and whitish body.

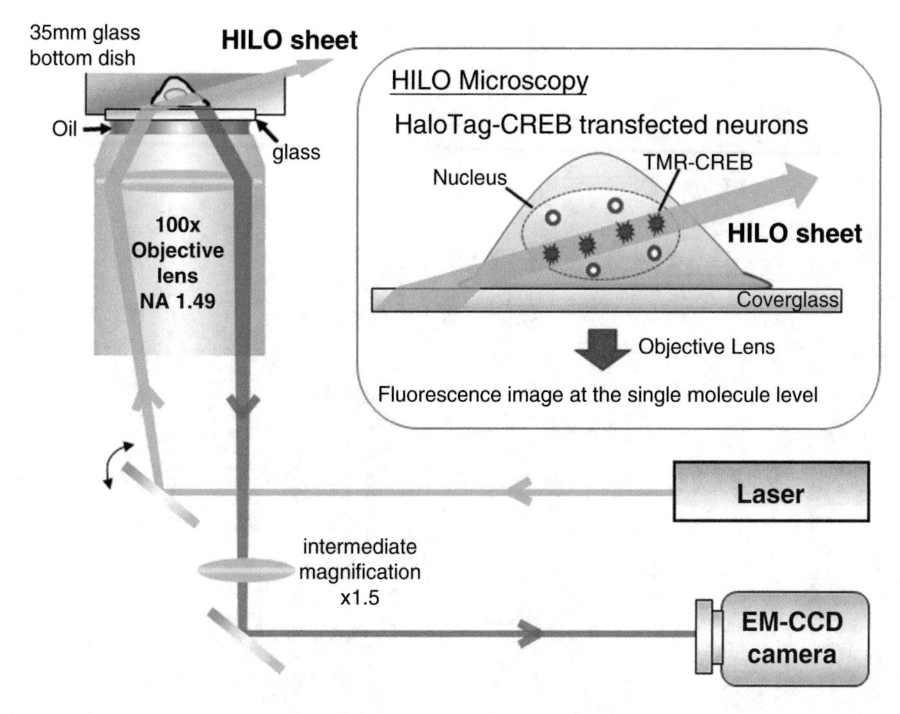

Fig. 1 Schematic drawing of the HILO microscope setup. The laser beam enters the edge of the 100× oil-immersion objective lens (NA 1.49) through a moving mirror, which can modify the angle of the laser beam, and illuminates the specimen from the side through an oblique and thin optical sheet. Fluorescence light from the specimen is collected by the same objective lens, and is then captured by an EMCCD camera

3. Remove the heads by decapitation with scissors and transfer them to fresh ice-cold HBSS.
4. Remove skin and skull, and take the whole brain with a spatula.
5. Dissect cortical hemispheres and place them in 35 mm dishes containing about 0.5 ml of ice-cold Ca^{2+}-, Mg^{2+}-free PBS.
6. Mince cortical tissues thoroughly with fine scissors. In this step, cortical tissues should be cut into evenly sized fine pieces as quickly as possible.
7. Transfer the minced tissues to a 15 ml test tube.
8. Incubate the tissues with 5 ml of 0.125% trypsin and 0.02% EDTA in Ca^{2+}-, Mg^{2+}-free PBS for 5 min at 37 °C in a water bath. Invert the test tube a few times during incubation (*see* **Note 3**).
9. Add 5 ml of DMEM/F12 medium containing 10% FBS to block the trypsin reaction.
10. Gently pass the finely minced tissues through a fire-polished glass Pasteur pipette 7–10 times to dissociate cells.
11. Remove cell clusters with a Pasteur pipette.

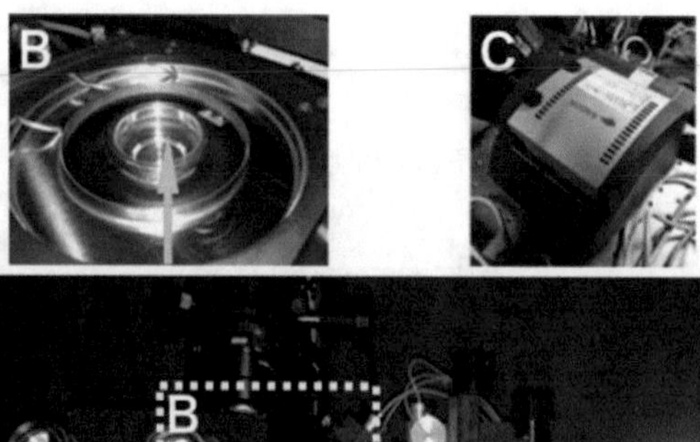

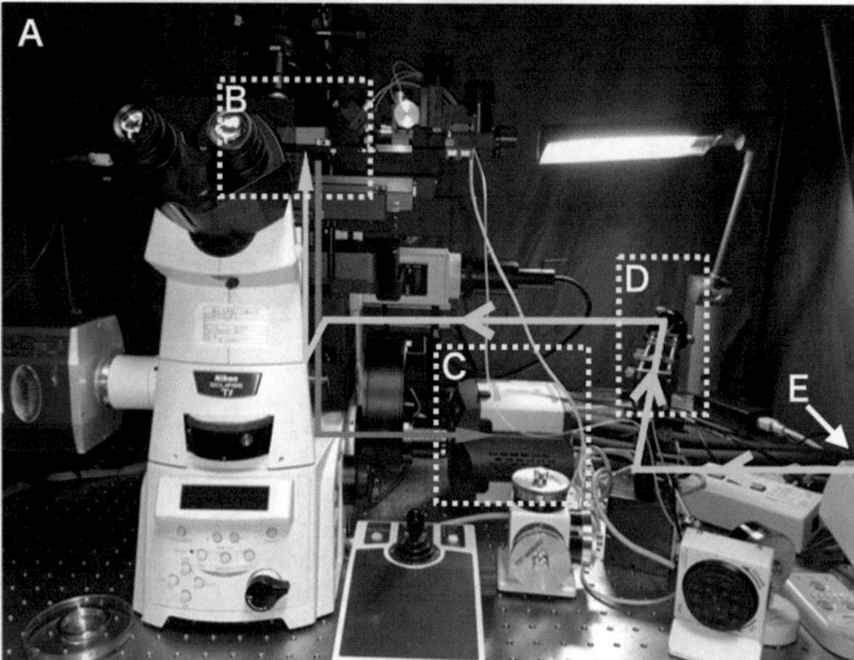

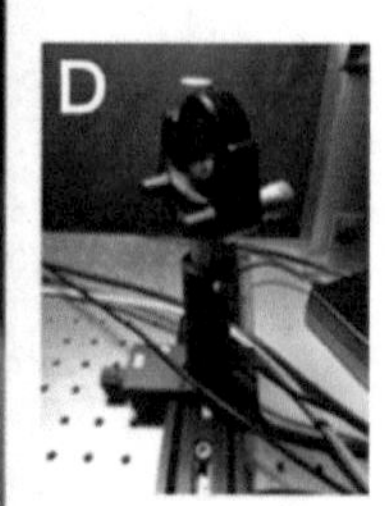

Fig. 2 Setup for the single-molecule imaging system using HILO microscopy. Overview of the microscopy and equipment for single-molecule imaging (**a**). The optical path from laser (**e**) to specimen (**b**) through a moving mirror (**d**) is shown by the green line. The optical path of fluorescence light from specimen to EM-CCD camera (**c**) is shown by the red line. The cell culture dish is mounted on a stage-top incubator (**b**)

12. Remove the supernatant after centrifugation at 180 × *g* for 5 min.
13. Resuspend the cells in DMEM/F12 medium supplemented with 2% B27 and 5% FBS. Count the cell number (*see* **Note 4**).
14. Seed the cell suspension to adjust the cell density to 3.3×10^4 cells/cm^2 on the poly-L-ornithine-coated glass-bottom dish (*see* **Note 5**).
15. Replace the culture medium after incubation for 1 h at 37 °C in an environment of 5% CO_2 and humidified 95% air (*see* **Note 6**).
16. Maintain the culture in the CO_2 incubator. Replace a half volume of the medium with fresh culture medium before becoming acidic (*see* **Note 7**).

3.2 Plasmids

3.2.1 pTet-On Advanced/TRE-Tight HaloTag-CREB

Many labeling techniques have been developed for single-molecule imaging. In particular, the HaloTag is a self-labeling protein tag that can bind to specific ligands irreversibly and covalently in a one-to-one manner. This feature is very suitable for single-molecule imaging. Moreover, the binding of HaloTag ligands with HaloTag is more stable and rapid than for other self-labeling protein tags, such as SNAP-tag.

We have demonstrated the dynamics of CREB molecules at the single-molecule level in neuroblastoma Neuro2a cells transfected with CMV promoter-driven HaloTag-CREB expression vector [25]. However, overexpression causes excessive competition between HaloTag-CREB and endogenous CREB for binding to CRE sites. Therefore, a low level of HaloTag-CREB expression is required to investigate the genuine behavior of CREB molecules in living cells. For this, a tetracycline (Tet)-inducible expression system can be used (Fig. 3). By using either a Tet-On or a Tet-Off system, expression levels of target genes can be controlled by the concentration of doxycycline. However, the Tet-Off system requires fresh doxycycline to be added at least every 2 days to keep the expression level low, as the half-life of doxycycline is about 24 h. In contrast, the Tet-On system can rapidly (<1 day) induce low-level expression of HaloTag-CREB with a small amount of doxycycline (*see* below). Indeed, the Tet-On system can generate a low level of HaloTag-CREB expression, which is close to the endogenous level (Fig. 3 and *see* also Fig. 9) [26].

1. To generate pTet-On Advanced/TRE-Tight vector, ligate a XhoI-digested fragment containing the P_{Tight}-inducible promoter from pTRE-Tight (Clontech) to a XhoI-digested fragment containing the CMV promoter with a rtTA2^{s}-M2 transactivator from pTet-On Advanced (Clontech).
2. Amplify HaloTag-human CREB cDNA from pFN21AB5414 (Promega) by PCR using the following primers:
   Primer forward: 5′-ATGGCAGAAATCGGTACTGGC-3′.
   Primer reverse: 5′-CGTTTAAACATCTGATTTGTGGCAG-3′.

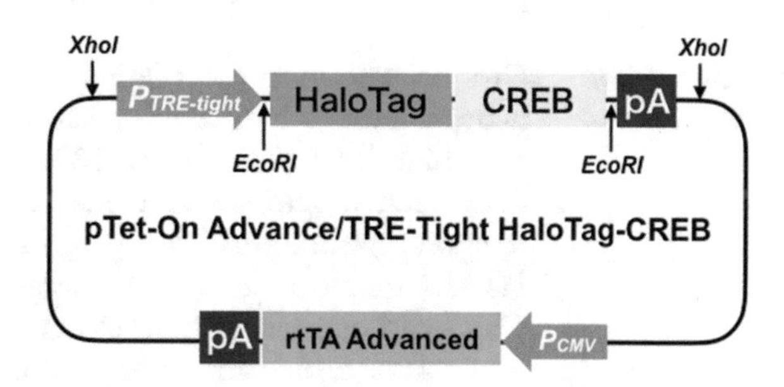

Fig. 3 A plasmid map of pTet-On Advanced/TRE-Tight HaloTag-CREB

3. Ligate the PCR product into pGEM T-easy vector (Promega) using TA cloning, and confirm that the amplicon has the correct DNA sequence.
4. Ligate an EcoRI fragment of pTet-On Advanced/TRE-Tight with an EcoRI-digested fragment containing the HaloTag-CREB of pGEM T-easy HaloTag-CREB (Fig. 3).
5. Transform *E. coli* cells with the obtained plasmid, and collect a large amount of the plasmid with a maxi-prep method.

3.2.2 pTet-On Advanced/TRE-Tight HaloTag-Mutant CREBs (L318/325V or R301L)

To compare the single-molecule behavior of wild-type CREB with that of mutant CREBs, pTet-On Advanced/TRE-Tight HaloTag-mutant CREBs (L318/325V or R301L) can be used; these mutants cannot bind to CRE sites due to an inability to dimerize (L318/325V) [27] or due to disruption of the DNA-binding domain despite dimerization ability (R301L) [28].

1. Introduce point mutations into pFN21AB5414 by PCR-mediated site-directed mutagenesis using the following mutagenic primers:
   Primer pair for L318/325V: 5′-TTTAGAAAACAGAGTGG CAGTGGTTGAAAATCAAAACAAGACAGTGATTGAGGA GCTA-3′ and its complementary oligonucleotide.
   Primer pair for R301L: 5′-GAAGCAGCTCGAGAGTG TCCTAGAAAGAAGAAAGAATATG-3′ and its complementary oligonucleotide.
2. Ligate an AsiSI/PmeI fragment of pTet-On Advanced/TRE-Tight HaloTag with an AsiSI/PmeI fragment containing the mutant CREBs (L318/325V or R301L) of pFN21AB5414 (L318/325V or R301L).
3. Transform *E. coli* with the resultant plasmid, and collect a large amount of the plasmid with a maxi-prep method.

3.2.3 pTα1-EGFP

The Tα1 (βIII tubulin) promoter is active in postmitotic cells of neuronal lineage [29, 30]. pTα1 vector contains part of 5′ flanking Tα1 promoter region and is a neuron-specific driver [31]. Tα1 promoter-driven EGFP expression is useful to identify neurons in primary dissociated cortical neurons.

3.2.4 pCAGGS-hChR2 (H134R)-EYFP

Channel rhodopsin 2 (ChR2) is widely used in contemporary neuroscience to induce firing activity of ChR2-expressing neurons with blue light illumination [32]. To date, many ChR2 mutants have been developed to enhance photocurrent and to improve temporal photosensitivity [33]. In particular, ChR2 with the H134R mutation generates a larger photocurrent than wild-type ChR2, although its channel closing is slower [34]. ChR2 (H134R) is the first choice for optogenetic experiments to induce firing activity in cultured cortical cells. Indeed, blue light stimulation can efficiently induce action potentials in pCAGGS-hChR2(H134R)-EYFP-transfected cortical neurons [35].

3.3 Electroporation

Cultured cortical neurons are transfected with pTet-On-Advanced/TRE-Tight HaloTag-CREB and pTα1-EGFP to image simple CREB dynamics. To study specific binding of CREB to its target DNA site (CRE site), pTet-On-Advanced/TRE-Tight HaloTag-mutant CREB is used. Moreover, cortical cells are transfected with pTet-On-Advanced/TRE-Tight HaloTag-CREB and pCAGGS-hChR2(H134R)-EYFP to image activity-dependent aspects of CREB. The following electroporation method is carried out at 3 days in vitro (DIV), by which time neurons have completely established neuronal polarity.

1. Soak the electrodes of the plate electrode (Fig. 4) in 70% ethanol for a few minutes and dry the plate electrode completely in a clean bench.
2. Remove the culture medium from the cortical culture dish and retain it during electroporation.
3. Add a sufficient amount of the plasmid solution (0.5–1.0 μg/μl DNA in Ca^{2+}-, Mg^{2+}-, and phenol red-free HBSS) to the glass-bottom dish to cover the cells (*see* **Note 8**).
4. Place the plate electrode, which connects to a square-pulse generator, on the glass region (Fig. 4).

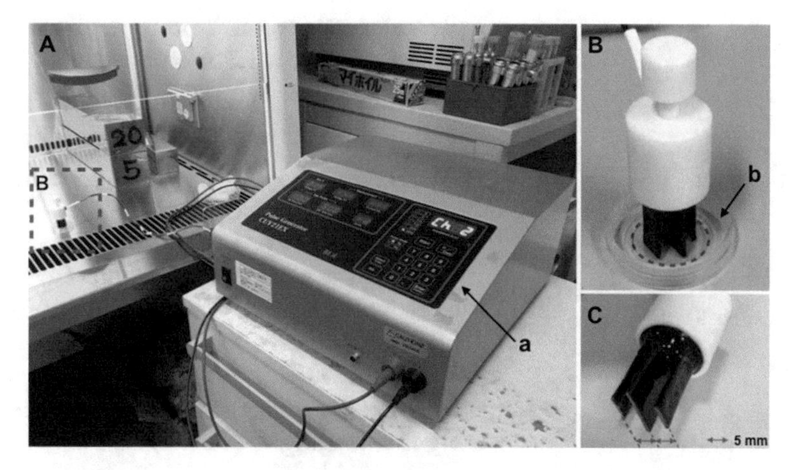

Fig. 4 Setup for in vitro electroporation (**a**). Square electric pulses are delivered from an electroporator (a). A plate-type electrode is connected into the electroporator. Electroporation experiments are performed in a clean bench (red dashed rectangle, B). The electrode is placed on a glass region (dashed circle, b) of the 35 mm glass-bottom culture dish (**b**). The electrode is composed of three plates that are arranged at equal intervals of 5 mm (**c**)

5. Deliver electrical pulses: one 275 V pulse of 10-ms duration and five 30 V driving pulses of 50-ms duration, with 50-ms intervals. Then, deliver the same driving pulses after polarity change. The following values will be obtained: voltage, 265–270 V; current, 0.7–0.9 A; and resistance, 0.3–0.4 kΩ.
6. Remove the plasmid solution and wash the cells with 1 ml of HBSS (Ca^{2+}-, Mg^{2+}-, and phenol red-free) several times to completely remove the plasmid solution. Keep the plasmid solution, as it can be reused several times.
7. Return the retained culture medium to the electroporated culture, and maintain it in a CO_2 incubator.

   Wash the plate electrode with sterile distilled water after the electroporation experiment, and dry it completely in a clean bench. Clean the electrodes; otherwise the transfection efficiency will be drastically reduced (*see* **Note 9**).

3.4 Induction of HaloTag-CREB Expression

The Tet-On promoter is activated in the presence of the antibiotic tetracycline or one of its derivative doxycycline. Doxycycline is recommended for all experiments using Tet-inducible systems because a 100-fold lower concentration of doxycycline than of tetracycline is sufficient for completely inducing the expression. Moreover, the half-life of doxycycline (~24 h) is longer than that of tetracycline (~12 h) in culture medium. As the single-molecule imaging will be finished within 24 h, medium exchange is not necessary during the experiment when doxycycline is used. Doxycycline can be added to the culture medium 1 day before observation of HaloTag-CREB.

1. Transfer half of the culture medium from the electroporated culture dish to a 1.5 ml tube.
2. Add doxycycline to the tube, and return the mixture to the culture dish (final concentration, 0.05 ng/ml; *see* below).
3. Maintain the doxycycline-treated cultures in a CO_2 incubator for 15–17 h (*see* **Note 10**).

3.5 Staining for HaloTag-CREB

Vital staining is an important component of live imaging. TMR-HaloTag ligand is a one-to-one cell-permeable ligand for HaloTag-CREB. Other colored fluorescent ligands can be used, but TMR is most brilliant, tolerant to bleaching, and less harmful in terms of photodynamic damage. Indeed, its fluorescence will persist stably during the imaging session (*see* below).

1. Transfer the doxycycline-containing culture medium to a tube and retain it until after the live staining.
2. Add 1.5 ml of fresh culture medium containing 10 nM HaloTag TMR Direct Ligand.

3. Incubate the culture in the CO_2 incubator for 15 min, which is needed to stabilize the covalent bind of HaloTag TMR Direct Ligand to HaloTag-CREB.
4. Replace the staining medium with the retained culture medium.
5. Maintain the stained cells in the CO_2 incubator for 30 min before observation, to allow them recover after the medium changes.
6. Mount the culture on a stage-top incubator (Fig. 2) maintained at 37 °C in an environment of 5% CO_2 and humidified 95% air (*see* **Note 11**).

3.6 The Image Capture for CREB Dynamics Using HILO Microscopy

HILO microscopy is suitable for single-molecule imaging of the molecular dynamics within cells. This technology can illuminate fluorophores with a high signal-to-background ratio even in the nucleus, owing to thin and oblique optical sheet generated by the large refraction at the glass specimen surface (Figs. 1 and 2). To capture clear and fine images in the nucleus at the single-molecule level using HILO microscopy, at least a 100× objective lens with high numerical aperture (>1.4) is required. Although HILO illumination is limited to a small imaging area and depth, the binding event of transcription factors to their target DNA sites can be adequately observed (*see* below).

1. Place one drop of immersion oil on the objective lens and mount the sample on the microscope stage.
2. Search for EGFP-labeled cells by fluorescence microscopy (excitation, 488 nm).
3. Roughly adjust the focus, using autofocus systems in the Ti-E inverted microscope.
4. Perform laser irradiation (561 nm). Examine whether TMR-CREB spots appear in the nuclei of EGFP-labeled cells.
5. Adjust the angle, depth, and aperture of the HILO laser manually to find the optimal conditions for single-molecule imaging of TMR-CREB spots (Fig. 5) (*see* **Note 12**).
6. Capture real-time images of the single-molecule dynamics at 10 frames per second with 256 × 256 region of interest (ROI) for 1–2 min using the EMCCD camera with Solis software (Figs. 5 and 6) (*see* **Note 13**).

3.7 Pharmacological Study

Pharmacological treatments are widely used to alter neuronal activity in cultured cells. Addition of KCl depolarization solution (final concentration, 50 mM, [37, 38]) depolarizes the cultured cells, while application of a sodium channel blocker, TTX (100 nM) [25, 39], prevents action potential occurrence. These pharmacological treatments are carried out after adding the HaloTag ligand.

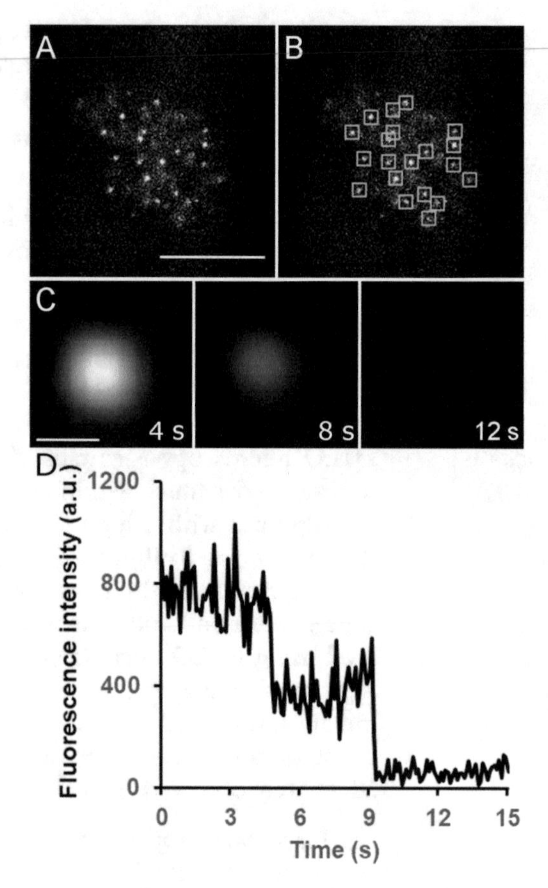

Fig. 5 Visualization of TMR-CREB spots in living cortical neurons at the single-molecule level. (**a**) HILO microscopy shows individual TMR-CREB spots in the nucleus of a living cortical neuron at 7 DIV. (**b**) Tracking of TMR-CREB spots with the ImageJ software plug-in PTA. Scale bar, 10 μm. (**c**) Stepwise disappearance of TMR-CREB spots by continuous excitation. Scale bar, 0.5 μm. (**d**) Time course of fluorescence intensity of the spot in (**c**). The spot exhibits two-step reduction of its fluorescence intensity

3.7.1 KCl Treatment

1. Add 0.41 volumes of KCl depolarization solution to the cell culture medium at 7 DIV, when spontaneous firing activity is low in this culture condition.
2. Maintain the treated cell cultures in the CO_2 incubator for 1 h.

3.7.2 TTX Treatment

1. Cultures around 14 DIV are suitable for TTX treatment, as spontaneous firing activity is prominent at this stage.
2. Transfer half of the culture medium (0.75 ml) at 14 DIV to a 1.5 ml tube.
3. Add 1.5 μl of 100 μM TTX to the culture medium in the tube.
4. Return the TTX-containing medium to the culture dish.
5. Maintain the TTX-treated cell culture in the CO_2 incubator for 24 h.

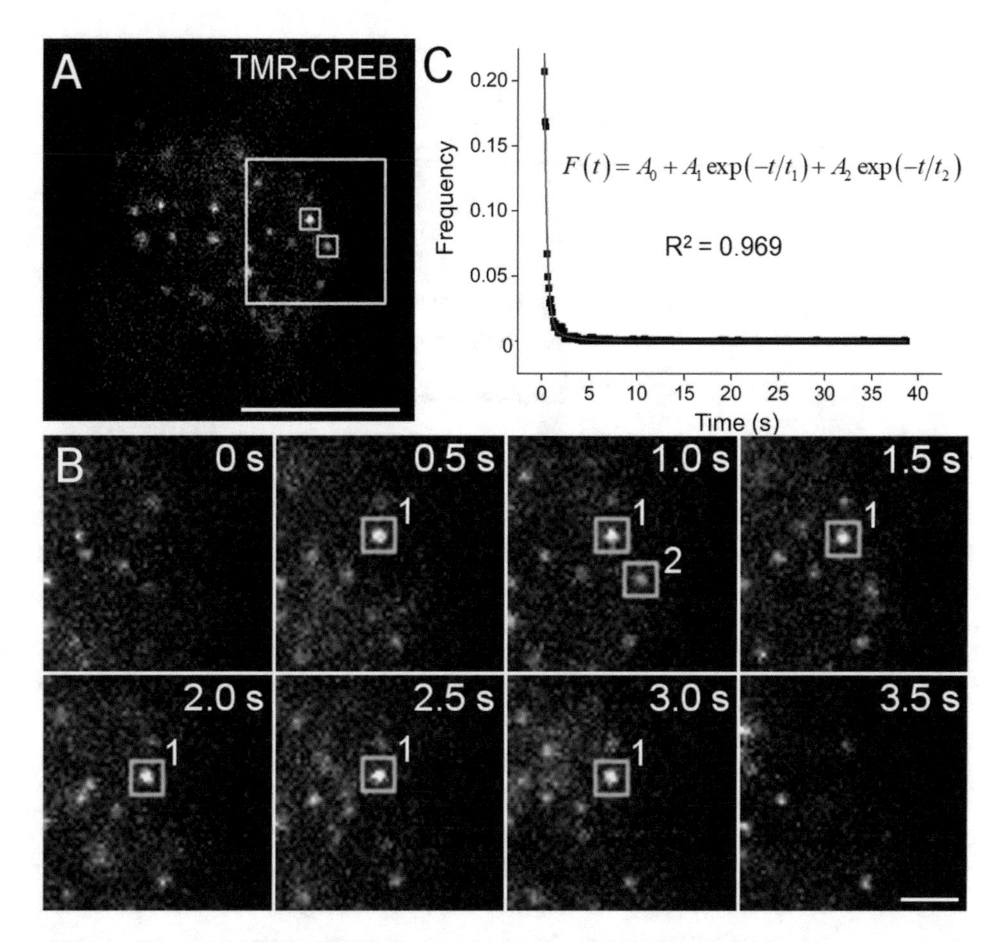

Fig. 6 Binding dynamics of TMR-CREB in the nuclei of living neurons. (**a**) HILO microscopy shows individual TMR-CREB spots in a nucleus at 5 DIV. Scale bar, 10 μm. (**b**) Time-lapse images of the yellow box in (**a**) are shown. The blue boxes show TMR-CREB spots with long (spot 1) and short (spot 2) residence times at fixed positions. Scale bar, 2 μm. (**c**) An example of curve fitting to the residence time distribution of TMR-CREB. The red line shows a biexponential function curve. Black dots represent the frequency of spots with certain residence times

3.8 Photostimulation

1. Cotransfect cultured cortical cells with pCAGGS-hChR2 (H134R)-EYFP (*see* above) and pTet-On-Advanced/TRE-Tight HaloTag-CREB. Add doxycycline and HaloTag ligand for single-molecule imaging of CREB (*see* above).
2. Search for HaloTag-CREB and ChR2-EYFP-cotransfected cells using the 561 nm laser. Do not use blue light illumination in this step, because this illumination induces neuronal firing before application of patterned stimulation by optogenetics.
3. Capture real-time images of the single-molecule dynamics for 1 min as described above.

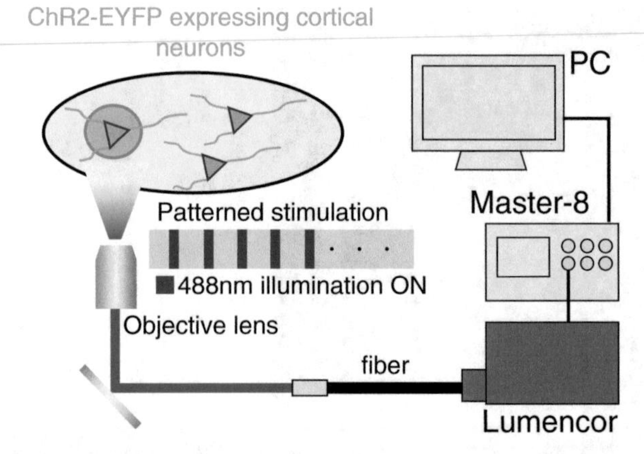

Fig. 7 Schematic drawing of the optogenetic experiment. Patterned blue light (475 nm) is emitted from the illuminator, which receives patterned signals from a pulse generator. The microscope aperture is adjusted so that the blue light illuminates only one cell body

4. Connect the solid-state illuminator to the microscope so that blue light illumination (475 nm wavelength) can be applied to the sample through the objective lens (Fig. 7).
5. Adjust the microscope aperture manually to illuminate only the cell body.
6. Apply patterned blue light illumination to the transfected neuron using the pulse generator, which controls the duration and frequency of light stimulation (Fig. 7).
7. Reconnect the laser system to the microscope and capture the real-time images of the single-molecule dynamics for 1 min again.

3.9 Analysis for Single-Molecule Tracking

A huge number of single-fluorescent spots will be obtained in the experiment. To analyze the appearance and disappearance of each spot, we use a self-made ImageJ plug-in, Particle Tracking Analysis (PTA) (Fig. 6). We also use Origin 9.1 software (OriginLab) and Excel 2010 (Microsoft) for data fitting and statistical analyses.

3.9.1 Analysis of the Residence Time of Fluorescence Spots

1. Import the data for single-molecule images (.sif file obtained from Solis software) into ImageJ and open the PTA plug-in.
2. Determinate the searching pixel size, pixel size of maximal and minimal spots, and detection methods (centroid or two-dimensional Gaussian distribution fitting for a single spot) and obtain all data (1200 frames; 2 min) for the residence time of each spot.

   The detection of a single spot is performed under the following parameters:

   Detection methods: two-dimensional Gaussian distribution fitting.

Roi size (x, y) of searching area: 9 × 9 pixels.

Pixel size of the minimal spot: 3 pixels.

Nearest particle range: 0.3 pixel.

Maximum miss frame: 2 frames.

Threshold: MaxEntropy.

Other parameters are the same as the default.

3. Open Origin 9.1 software.
4. Plot the residence time distribution from the obtained data, and fit the distribution with the following biexponential function curve (Fig. 6):

$$F(t) = A_0 + A_1 \exp(-t/t_1) + A_2 \exp(-t/t_2) \qquad (1)$$

where t is time and A_0 is constant. A_1 and A_2 are the fractions with dissociation rate constants $1/t_1$ and $1/t_2$ for the short- and long-residence components, respectively.

3.9.2 Analysis of Spot Distribution in Cell Nuclei

1. Obtain the data for the two-dimensional distribution of spots from the PTA plug-in.
2. Open Origin 9.1 software.
3. Count the number of spots appearing in two-dimensional distribution inside a single cell.
4. Calculate the rate of spot appearance in each section.
5. Make a heat map of the distribution of spot appearance in a cell (Fig. 8).

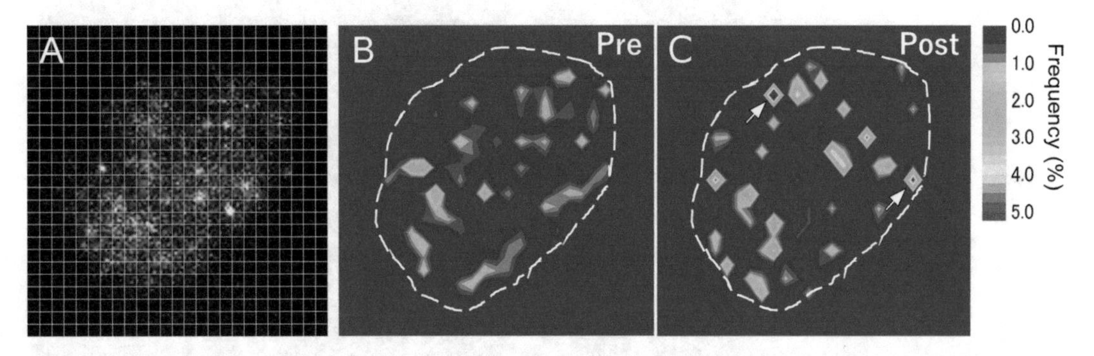

Fig. 8 Spatial dynamics of TMR-CREB spots having long residence times (>1 s) in a living cortical neuron. (**a**) The whole nucleus of a cortical neuron is divided into 625 (25 × 25) equal sub-areas (0.8 × 0.8 μm) to analyze the spatial distribution of TMR-CREB spots. (**b**, **c**) Color heat maps show the spatial distribution of the appearance frequency of TMR-CREB spots having long residence times (>1 s) in the living neuron, shown before (**b**) and after (**c**) patterned photostimulation (0.1 Hz, 1-s duration, 30 pulses). In (**c**), arrow indicates subnuclear areas (0.8 × 0.8 μm) in which >5% of the total TMR-CREB spots with longer residence time (>1 s) are accumulated (hot spots). The dashed line represents the perimeter of the nucleus

4 Typical Results

It is necessary to determine the optimal concentration of doxycycline for single-molecule imaging. Figure 9 shows fluorescence images of TMR-CREB in cultured cortical neurons after addition of different concentrations of doxycycline (0, 0.05, 5 ng/ml). HILO microscopy demonstrates spotlike fluorescence signals of TMR-CREB in the presence of the lower concentration of doxycycline (0.05 ng/ml) (Fig. 9). The expression level of HaloTag-CREB in cortical neurons treated with this concentration of doxycycline is very low compared to that of endogenous CREB [26]. Furthermore, leak expression of HaloTag-CREB in pTet-On Advanced/TRE-Tight HaloTag-CREB-transfected neurons is never observed in the absence of doxycycline (Fig. 9). Thus, the Tet-inducible expression system works well for single-molecule imaging.

TMR-CREB spots observed by HILO microscopy can be analyzed by the PTA plug-in software (Fig. 6). From the tracking data, the fluorescence intensities in individual TMR-CREB spots exhibit one- or two-step reduction (Fig. 5). The two-step disappearance does not take place in the L318/325V mutant CREB, which lacks the ability to dimerize [27]. Therefore, this method can visualize

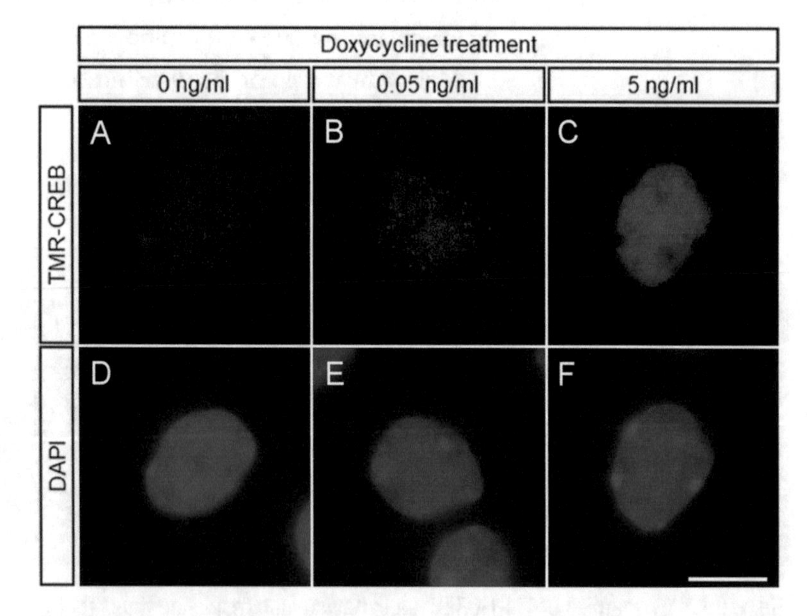

Fig. 9 The optimal concentration of doxycycline in Tet-inducible HaloTag-CREB-expressing neurons for single-molecule imaging. (**a–c**) Fluorescence images of TMR-CREB in pTet-On Advanced/TRE-Tight HaloTag-CREB-transfected cortical neurons at 5 DIV, after addition of different concentrations of doxycycline (0, 0.05, 5 ng/ml). HaloTag-CREB is visualized by 10 nM TMR-HaloTag ligand. Nuclei are stained by DAPI after fixation (**d–f**). Scale bar: 10 μm

not only molecular behavior but also dimerization of proteins in cortical neurons at the single-molecule level. Real-time movement of TMR-CREB spots is detected clearly by capturing images at 10 frames per second. Figure 6 shows that TMR-CREB spots reside at fixed positions for a certain time period. Many TMR-CREB spots disappear within 1 s (Fig. 6, spot 2), while a small fraction of them stay at the same positions for several seconds (Fig. 6, spot 1). The residence time distribution of TMR-CREB spots can be fitted well to biexponential function (Eq. 1 in Fig. 6). The fact that the number of spots with longer residence times significantly decreases in mutant CREB, which cannot bind CRE sites [25, 26], suggests that the spots with a longer residence time (>1 s) represent specific interactions of TMR-CREB with CRE sites [25, 26]. Overall, this single-molecule imaging method enables us to uncover the temporal interactions between transcription factors and their target DNA sites in living cortical neurons.

One might expect that neuronal activity would alter the residence time distribution of TMR-CREB spots. However, the distribution is not obviously changed by either KCl or TTX treatment. Application of repetitive stimulation with the optogenetic technique also does not affect the residence time distribution. An interesting aspect is that the dynamics of the spatial distribution is altered by optogenetic stimulation. Figure 8 shows the neuronal activity-dependent dynamics in the spatial distribution of TMR-CREB spots. Micro regions where >5% out of the total spots with long residence time (>1 s) accumulate (defined as "hot spots") are very rare in the nucleus at 5 DIV, when spontaneous activity is very low (Fig. 8). However, after repetitive photostimulation (0.1 Hz, 1-s duration, 30 pulses), some hot spots appear in ChR2-EYFP-expressing cells, while the residence time distribution is not altered (Fig. 8) [26]. Hot spots can also appear upon photostimulation with different frequencies in ChR2-EYFP-expressing cortical neurons [26]. It is likely that neuronal activity promotes the formation of hot spots, where TMR-CREB binds more frequently to specific genome locations. Thus, the combination of single-molecule imaging with optogenetics allows us to reveal spatiotemporal dynamics of transcription factors under many physiological conditions.

5 Conclusion

The live-cell single-molecule imaging technique using HILO microscopy is efficient to uncover the spatiotemporal dynamics of transcription factor behavior in primary neurons. Recent findings have also demonstrated dynamic aspects of intranuclear events, such as nucleosome dynamics [40, 41], chromatin remodeling [42], DNA repair [43], DNA replication, RNA splicing [44], and transcription [36, 45–48]. This imaging method combined with optogenetic techniques will contribute to a better understanding of neuronal activity-dependent molecular features.

6 Notes

1. Sterile distilled water should be at room temperature, because cool water is inadequate for complete washing.
2. Poly-L-ornithine-coated dishes can be kept for at least one month at room temperature. Laminin coating enhances neuronal differentiation including process outgrowth, but is not absolutely required for cellular development with high-density cell culture.
3. Do not allow the trypsin reaction to exceed 5 min. Prolonged reaction may cause cell damage.
4. Five percent FBS is a critical concentration for neuronal culture to minimize proliferation of glial cells. Under this condition, neuronal cells can survive and grow well for at least 14 days.
5. Considering the effect of electroporation (see below), the high cell density (3.3×10^4 cells/cm^2) is better than the lower cell density used for normal cell cultures (~1.5×10^4 cells/cm^2).
6. Confirm that dissociated cells are attached to the culture substratum before the medium is exchanged. For pH equilibration, the fresh culture medium should be maintained in the CO_2 incubator before the medium change.
7. More than ten million cortical cells can be obtained from about 10 mouse embryos. This number may be too large to use for a single experiment. These dissociated cortical cells can be stored frozen. To do so, resuspend cells in the culture medium containing 10% FBS and 10% dimethyl sulfoxide or in cell banker (Nippon Zenyaku Kogyo; CB011), make aliquots in freezer vials, and put them in a deep freezer. Transfer these vials to the liquid nitrogen tank for long-term storage.
8. The total plasmid concentration should not exceed 1.0 μg/μl even when more than one plasmid is being electroporated. The optimal volume of plasmid solution is 200 μl for the glass region (Fig. 4b) to which cells are attached.
9. Chemical transfection by lipofection can be used for cell lines, but is not efficient for primary neuronal cells. The electroporation technique with plate electrodes is much more efficient. The cells near the electrodes will be well transfected, although cells immediately under the electrodes will be damaged. To improve transfection efficiency, a second electroporation can be applied by rotating the electrode by 90°.

10. It is not necessary to wash out doxycycline because the HaloTag-CREB expression induced by doxycycline changes little until the next day. Moreover, the survival and development of neuronal cells are not influenced by the concentration used in the present study.
11. Observation should be completed within half a day after staining, as the level of free TMR-HaloTag ligand will be increased by turnover of HaloTag-CREB protein in the cells.
12. These procedures should be completed very quickly. Otherwise, fluorescent spots excited by the laser will be diminished by photobleaching because the laser power must be high in order to obtain clear images of single molecules. Indeed, TMR signals will be almost abolished by laser illumination of about 10 min.
13. This level of time resolution is necessary to characterize the movement of transcription factors interacting with DNA [25, 26, 36]. Time resolution can be modified for other proteins of interest and physiological events.

Acknowledgments

This research was supported by MEXT KAKENHI on Innovative Areas "Mesoscopic Neurocircuitry" (No. 23115102) to N.Y., "Dynamic Regulation of Brain Function by Scrap and Build System" (No. 16H06460) to N.Y., "Cross-talk between moving cells and microenvironment as a basis of emerging order in multicellular system" (No. 23111516) to N.S., and Grant Nos. 20200009 to N.S. and 20300110 and 2330018 to N.Y. We thank Dr. Ian smith for critical reading of the manuscript and Dr. Yoshiyuki Arai for providing the PTA plug-in. We also thank Dr. Masatoshi Morimatsu and Dr. Toshio Yanagida for setting up the single-molecule microscopy system.

References

1. Hager GL, Mcnally JG, Misteli T (2009) Transcription dynamics. Mol Cell 35:741–753
2. Ptashne M, Gann A (1997) Transcriptional activation by recruitment. Nature 386:569–577
3. Bentovim L, Harden TT, DePace AH (2017) Transcriptional precision and accuracy in development: from measurements to models and mechanisms. Development 144:3855–3866
4. Hampsey M (1998) Molecular genetics of the RNA polymerase II general transcriptional machinery. Microbiol Mol Biol Rev 62:465–503
5. Spitz F, Furlong EEM (2012) Transcription factors: from enhancer binding to developmental control. Nat Rev Genet 13:613–626
6. Impey S, McCorkle SR, Cha-Molstad H et al (2004) Defining the CREB regulation: a genome-wide analysis of transcription factor regulatory regions. Cell 119:1041–1054
7. Flavell SW, Cowan CW, Kim T et al (2006) Activity-dependent regulation of MEF2 transcription factors suppresses excitatory synapse number. Science 311:1008–1012
8. Zhou Z, Hong EJ, Cohen S et al (2006) Brain-specific phosphorylation of MeCP2 regulates

activity-dependent Bdnf transcription, dendritic growth, and spine maturation. Neuron 52:255–269

9. Lin Y, Bloodgood BL, Hauser JL et al (2008) Activity-dependent regulation of inhibitory synapse development by Npas4. Nature 455:1198–1204
10. Riggs AD, Bourgeois S, Cohn M (1970) The lac repressor-operator interaction. III. Kinetic studies. J Mol Biol 53:401–417
11. Kim JG, Matthews BW, Anderson WF (1987) Kinetic studies on Cro repressor-operator DNA interaction. J Mol Biol 196:149–158
12. Mayr BM, Guzman E, Montminy M (2005) Glutamine rich and basic region/leucine zipper (bZIP) domains stabilize cAMP-response element-binding protein (CREB) binding to chromatin. J Biol Chem 280:15103–15110
13. Michelman-ribeiro A, Mazza D, Rosales T et al (2009) Direct measurement of association and dissociation rates of DNA binding in live cells by fluorescence correlation spectroscopy. Biophys J 97:337–346
14. Xia T, Li N, Fang X (2013) Single-molecule fluorescence imaging in living cells. Annu Rev Phys Chem 64:459–480
15. Liu Z, Lavis LD, Betzig E (2015) Imaging live-cell dynamics and structure at the single-molecule level. Mol Cell 58:644–659
16. Shao S, Xue B, Sun Y (2018) Intranucleus single-molecule imaging in living cells. Biophys J 115:181–189
17. Ishijima A, Yanagida T (2001) Single molecule nanobioscience. Trends Biochem Sci 26:438–444
18. Funatsu T, Harada Y, Tokunaga M et al (1995) Imaging of single fluorescent molecules and individual ATP turnovers by single myosin molecules in aqueous solution. Nature 374:555–559
19. Ulbrich MH, Isacoff EY (2007) Subunit counting in membrane-bound proteins. Nat Methods 4:319–321
20. Sako Y, Minoghchi S, Yanagida T (2000) Single-molecule imaging of EGFR signalling on the surface of living cells. Nat Cell Biol 2:168–172
21. Zhang W, Jiang Y, Wang Q et al (2009) Single-molecule imaging reveals transforming growth factor-beta-induced type II receptor dimerization. Proc Natl Acad Sci U S A 106:15679–15683
22. Iino R, Koyama I, Kusumi A (2001) Single molecule imaging of green fluorescent proteins in living cells: E-cadherin forms oligomers on the free cell surface. Biophys J 80:2667–2677
23. Tokunaga M, Imamoto N, Sakata-sogawa K (2008) Highly inclined thin illumination enables clear single-molecule imaging in cells. Nat Methods 5:159–161
24. Paakinaho V, Presman DM, Ball DA et al (2017) Single-molecule analysis of steroid receptor and cofactor action in living cells. Nat Commun 8:1–14
25. Sugo N, Morimatsu M, Arai Y et al (2015) Single-molecule imaging reveals dynamics of CREB transcription factor bound to its target sequence. Sci Rep 5:1–9
26. Kitagawa H, Sugo N, Morimatsu M et al (2017) Activity-dependent dynamics of the transcription factor of cAMP-response element binding protein in cortical neurons revealed by single-molecule imaging. J Neurosci 37:1–10
27. Dwarki VJ, Montminy M, Verma IM (1990) Both the basic region and the "leucine zipper" domain of the cyclic AMP response element binding (CREB) protein are essential for transcriptional activation. EMBO J 9:225–232
28. Walton KM, Lochner JE, Rehfuss RP et al (1992) A dominant repressor of cyclic adenosine 3′,5′-monophosphate (cAMP)-regulated enhancer-binding protein activity inhibits the cAMP-mediated induction of the somatostatin promoter in vivo. Mol Endocrinol 6:647–655
29. Gloster A, Wu W, Speelman A et al (1994) The Td function at-tubulin promoter specifies gene expression as a of neuronal growth and regeneration in transgenic mice. J Neurosci 14:7319–7330
30. Wang S, Wu H, Jiang J et al (1998) Isolation of neuronal precursors by sorting embryonic forebrain transfected with GFP regulated by the Tα1 tubulin promoter. Nat Biotechnol 16:196–201
31. Hatanaka Y, Murakami F (2002) In vitro analysis of the origin, migratory behavior, and maturation of cortical pyramidal cells. J Comp Neurol 454:1–14
32. Boyden ES, Zhang F, Bamberg E et al (2005) Millisecond-timescale, genetically targeted optical control of neural activity. Nat Neurosci 8:1263–1268
33. Lin JY (2010) A user's guide to channelrhodopsin variants: features, limitations and future developments. Exp Physiol 96:19–25
34. Nagel G, Brauner M, Liewald JF et al (2005) Light activation of channelrhodopsin-2 in excitable cells of Caenorhabditis elegans triggers rapid behavioral responses. Curr Biol 15:2279–2284
35. Malyshevskaya O, Shiraishi Y, Kimura F, Yamamoto N (2013) Role of electrical activity in

horizontal axon growth in the developing cortex: a time-lapse study using optogenetic stimulation. PLoS One 8:e82954

36. Morisaki T, Müller WG, Golob N et al (2014) Single-molecule analysis of transcription factor binding at transcription sites in live cells. Nat Commun 5:4456
37. Bading H, Ginty DD, Greenberg ME (1993) Regulation of gene expression in hippocampal neurons by distinct calcium signaling pathways. Science 260:181–186
38. Sugo N, Oshiro H, Takemura M et al (2010) Nucleocytoplasmic translocation of HDAC9 regulates gene expression and dendritic growth in developing cortical neurons. Eur J Neurosci 31:1521–1532
39. Uesaka N, Hirai S, Maruyama T et al (2005) Activity dependence of cortical axon branch formation: a morphological and electrophysiological study using organotypic slice cultures. J Neurosci 25:1–9
40. Hihara S, Pack CG, Kaizu K et al (2012) Local nucleosome dynamics facilitate chromatin accessibility in living mammalian cells. Cell Rep 2:1645–1656
41. Nozaki T, Imai R, Tanbo M et al (2017) Dynamic organization of chromatin domains revealed by super-resolution live-cell imaging. Mol Cell 67:282–293
42. Zhen CY, Tatavosian R, Huynh TN et al (2016) Live-cell single-molecule tracking reveals co-recognition of H3K27me3 and DNA targets polycomb Cbx7-PRC1 to chromatin. Elife 5:e17667
43. Yang G, Liu C, Chen S-H et al (2018) Super-resolution imaging identifies PARP1 and the Ku complex acting as DNA double-strand break sensors. Nucleic Acids Res 46:3446–3457
44. Martin RM, Rino J, Carvalho C et al (2013) Live-cell visualization of pre-mRNA splicing with single-molecule sensitivity. Cell Rep 4:1144–1155
45. Gebhardt JCM, Suter DM, Roy R et al (2013) Single-molecule imaging of transcription factor binding to DNA in live mammalian cells. Nat Methods 10:421–426
46. Chen J, Zhang Z, Li L et al (2014) Single-molecule dynamics of enhanceosome assembly in embryonic stem cells. Cell 156:1274–1285
47. Groeneweg FL, Van Royen ME, Fenz S et al (2014) Quantitation of glucocorticoid receptor DNA-binding dynamics by single-molecule microscopy and FRAP. PLoS One 9:1–12
48. Hipp L, Beer J, Kuchler O et al (2019) Single-molecule imaging of the transcription factor SRF reveals prolonged chromatin-binding kinetics upon cell stimulation. Proc Natl Acad Sci U S A 116:880–889

Chapter 5

Single-Molecule Imaging of Recycling Synaptic Vesicles in Live Neurons

Merja Joensuu, Ramon Martínez-Mármol, Mahdie Mollazade, Pranesh Padmanabhan, and Frédéric A. Meunier

Abstract

The capacity of neurons to communicate and store information in the brain critically depends on neurotransmission, a process which relies on the release of chemicals called neurotransmitters stored in synaptic vesicles at the presynaptic nerve terminals. Following their fusion with the presynaptic plasma membrane, synaptic vesicles are rapidly reformed via compensatory endocytosis. The investigation of the endocytic pathway dynamics is severely restricted by the diffraction limit of light and, therefore, the recycling of synaptic vesicles, which are roughly 45 nm in diameter, has been primarily studied with electrophysiology, low-resolution fluorescence-based techniques, and electron microscopy. Here, we describe a recently developed technique we named subdiffractional tracking of internalized molecules (sdTIM) that can be used to track and study the mobility of recycling synaptic vesicles in live hippocampal presynapses. The chapter provides detailed guidelines on the application of the sdTIM protocol and highlights controls, adaptations, and limitations of the technique.

Key words Endocytosis, Hippocampal neurons, Synaptic vesicles, Super-resolution microscopy, Single-particle tracking, Vesicle-associated membrane protein 2 (VAMP2-pHluorin), Nanobodies

1 Introduction: Activity-Dependent Internalization of Nanobodies

The recycling of synaptic vesicles is fundamental to maintain synaptic activity [1], and to promote neuronal survival [2] and presynaptic homeostasis [3]. Synaptic vesicles are highly enriched in presynaptic nerve terminals. Following Ca^{2+} influx, these vesicles fuse with the presynaptic plasma membrane releasing neurotransmitters into the synaptic cleft, thereby transmitting the information to the postsynaptic neuron. Following fusion, the vesicle membrane is retrieved from the plasma membrane by compensatory endocytosis to replenish the recycling synaptic vesicle pool. Despite appearing morphologically similar in electron micrographs, synaptic vesicles are not functionally identical, and vesicles that share similar properties (such as release probabilities) can be categorized

Nobuhiko Yamamoto and Yasushi Okada (eds.), *Single Molecule Microscopy in Neurobiology*, Neuromethods, vol. 154, https://doi.org/10.1007/978-1-0716-0532-5_5, © Springer Science+Business Media, LLC, part of Springer Nature 2020

into distinct vesicle pools. While varied descriptions of the pools and accompanying nomenclature exist [4, 5], three major functional synaptic vesicle pools are commonly described based on their ability to undergo regulated fusion: the readily releasable pool, the recycling pool, and the reserve pool [4–9]. Synaptic vesicles can be further subcategorized into populations such as the superpool that transits rapidly between neighboring synaptic boutons and the surface pool of vesicular proteins that resides on the plasma membrane following synaptic vesicle fusion and neurosecretion [5, 10–13].

Owing to their small size [14], the details of the heterogeneous mobility patterns of individual synaptic vesicles can only be resolved by super-resolution imaging techniques [10–13, 15–20]. A detailed investigation into the cellular processes associated with these heterogeneous populations requires direct and continual tracking of multiple synaptic vesicles simultaneously, in order to collect sufficient data to robustly infer the various mobility states and parameters characterizing vesicle recycling [21]. We recently developed a novel pulse-chase-based method called *s*ub*d*iffractional *t*racking of *i*nternalized *m*olecules (sdTIM) [21, 22] that allows investigators to acquire thousands of relatively long-lasting single-molecule trajectories from recycling synaptic vesicles in the crowded live hippocampal nerve endings and adjacent axons.

sdTIM relies on the activity-dependent internalization of fluorescent ligands into endocytic compartments, such as synaptic vesicles [3, 21, 22] and endosomes [2, 22], and can also be used to track retrogradely transported autophagosomes [3]. This book chapter describes one of the applications of the sdTIM protocol to study the mobility of individual recycling synaptic vesicles in nerve terminals of live hippocampal neurons by overexpressing pHluorin-tagged vesicle-associated membrane protein 2 (VAMP2-pHluorin [23]). The activity-dependent internalization of externally applied anti-green fluorescent protein (GFP) Atto647N-labeled nanobodies (Atto647N-NBs), which bind specifically to pHluorin, is achieved by inducing depolarization with high-potassium (K^+) buffer (referred to as the "pulse"). pHluorin moiety is a pH-sensitive GFP variant [23–25] that, when fused to the C-termini of VAMP2, is quenched in the acidic intravesicular environment (pH 5.5). Following the exocytic fusion of the synaptic vesicle with the plasma membrane, the intravesicular pHluorin of VAMP2-pHluorin is exposed to the neutral (pH 7.0) extracellular space and unquenched, thereby emitting fluorescence. VAMP2-pHluorin can, therefore, be used as an optical indicator to monitor secretion and synaptic transmission and to discriminate active nerve terminals from the adjacent axonal segments. Following stimulation, neurons are then washed with low-K^+ buffer to remove any free Atto647N-NBs and incubated in low-K^+ buffer for 10 min (referred to as the "chase"). VAMP2-pHluorin-bound Atto647N-

NBs internalize into recycling synaptic vesicles, and the pHluorin is re-quenched following the subsequent reacidification of the synaptic vesicles [23–25]. Neurons are then imaged under oblique illumination to detect and track single molecules of internalized VAMP2-pHluorin-bound Atto647N-NBs. In oblique illumination, the laser beam passes through the sample at an angle slightly smaller than the critical angle which is used in total internal reflection fluorescence (TIRF) microscopy. Thus, unlike TIRF illumination which only excites a thin optical section close to the coverslip, oblique illumination allows image acquisition from a thicker optical section that extends beyond the basal surface of the neuron while not illuminating the molecules outside the evanescent wave [26, 27]. Individual synaptic vesicles containing Atto647N-NBs can then be tracked using single-particle tracking algorithms until the Atto647N-NBs either bleach or exit the field of view.

The sdTIM protocol involves the following main steps: (1) neuronal dissection and culturing from embryonic rats (or mice) (Fig. 1), (2) induction of activity-dependent internalization of Atto647N-NBs in recycling synaptic vesicles (Figs. 2 and 3),

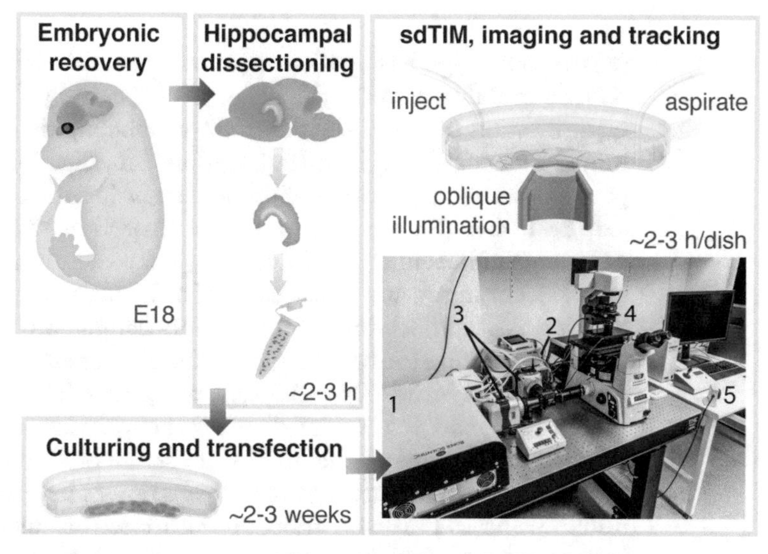

Fig. 1 The main steps of sdTIM protocol and expected duration. The protocol starts with embryonic (E18) recovery from rats, followed by dissection of the hippocampi. Hippocampal neurons are plated and cultured on super-resolution-compatible glass-bottom dishes and transfected with VAMP2-pHluorin. Induction of activity-dependent internalization of Atto647N-NBs is then performed under oblique illumination, using custom-made experimental perfusion and aspiration system [22], to image single molecules of VAMP2-pHluorin-bound Atto647N-NBs internalized in recycling synaptic vesicles. Single molecules are then tracked with single-molecule tracking software PALM Tracer in the MetaMorph software. *Bottom right*, our Roper iLas2 microscope single-particle imaging setup: lasers (1), iLas^2 (2), EMCCD cameras (3), microscope (4), and computer for image acquisition software (5) are indicated

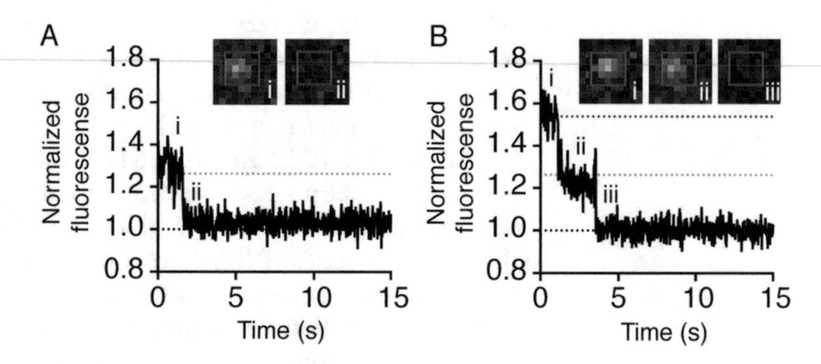

Fig. 2 Obtaining Atto647N emission curves of the anti-GFP nanobodies. (**a**) A representative intensity-time (s) trace of the time recording of Atto647N-NB deposited on a glass surface at pH 7.4. The graph shows a representative fluorescence trace from 5 $\times$ 5 pixels (red square) around the center of the Atto647N-NB during excitation (i) and after bleaching (ii), indicating a single-emission step. (**b**) The graph shows a trace of Atto647N-NB during excitation (i) and after first bleaching step (ii) and second bleaching step (iii), indicating two emission steps. The intensity-time traces are normalized to background. Dotted orange and purple lines indicate the detected emission steps. The number of internalized Atto647N-NBs inside synaptic vesicles in fixed (Subheading 3.8) and live (Subheading 3.9) neurons can be assessed similarly

(3) live-cell imaging (Fig. 3), and (4) tracking and analysis of single-particle mobility patterns (Figs. 4–7). We begin by describing the quantification of Atto647N-fluorophore labeling density of nanobodies in a cell-free system (Subheading 3.1, Fig. 2). The sdTIM protocol starts with coating of the super-resolution-compatible cell culture dishes (Subheading 3.2) and dissection of hippocampal neurons (Subheading 3.3), followed by neuronal plating (Subheading 3.4), culturing, and transfection of differentiated neurons (Subheading 3.5). Following imaging (Subheading 3.6) and single-molecule tracking (Subheading 3.7), the assessment of the number of internalized Atto647N-NBs inside synaptic vesicles in fixed (Subheading 3.8) and live (Subheading 3.9) neurons can be performed. For the validation of the sdTIM experimental setup, we also describe how to evaluate the correct targeting of internalized nanobodies using sdTIM in correlation with electron microscopy and how to assess neuronal maturity (Subheading 3.10, Fig. 8). Finally, we discuss potential sdTIM adaptations (Subheading 3.11) to study endocytosis of various biological processes using different fluorescent labels and ligands, and in combination with other super-resolution imaging techniques (e.g., single-particle tracking photo-activated localization microscopy, sptPALM).

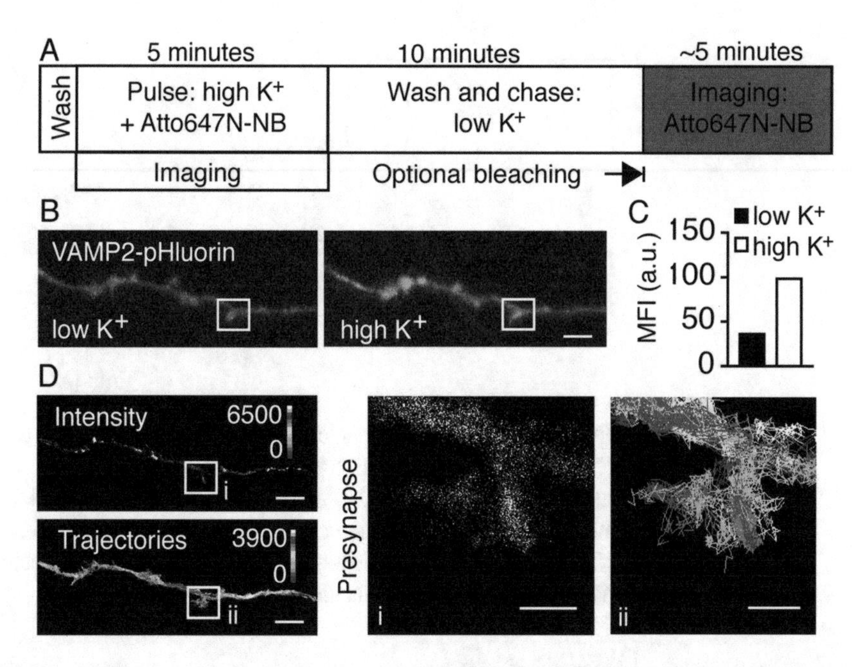

Fig. 3 Induction of activity-dependent internalization of Atto647N-NBs in recycling synaptic vesicles. (**a**) sdTIM experimental timeline. Hippocampal neurons expressing VAMP2-pHluorin are washed with low-K^+ buffer, stimulated in anti-GFP Atto647N-NB (3.2 pg/μl) containing high-K^+ buffer for 5 min (pulse), during which the unquenching of the VAMP2-pHluorin is monitored (imaging). Neurons are then washed and chased for 10 min in low-K^+ buffer. Prior to imaging, the internalized Atto647N-NBs can be partially bleached with short laser illumination (optional) to help discriminate single recycling synaptic vesicles in the highly crowded presynaptic environment. After the chase, single-molecule imaging of internalized Atto647N-NB in recycling SVs is performed under oblique illumination microscopy. (**b**) Wide-field image of hippocampal neuron expressing VAMP2-pHluorin subjected to sdTIM showing fluorescence unquenching at hippocampal presynapse (boxed ROI) following high-K^+ stimulation. (**c**) A representative measurement of the VAMP2-pHluorin fluorescence intensity (*MFI* mean fluorescence intensity, *a.u.* arbitrary unit) before stimulation (low-K^+ buffer) and following high-K^+ stimulation from the indicated presynapse in (**b**). (**d**) Corresponding super-resolved average intensity image shows the density map of internalized VAMP2-pHluorin-bound Atto647N-NB localizations. The colored bar represents localization densities, and the colder colors indicate higher density. Trajectory color coding refers to acquisition frame number. Bars, 3 μm for (**b**) and (**d**), and 750 nm for the magnified presynaptic ROIs (i, ii) on bottom right

2 Materials and Equipment

2.1 Neuronal Dissection, Culturing, and Transfection

- *Coating of the glass-bottom cell culture dishes*:

  Use poly-L-lysine hydrobromide (PLL) and boric acid (both from Sigma-Aldrich) to prepare filter-sterilized 1 mg/ml PLL in 0.1 M borate buffer pH 8.5. The 1 mg/ml PLL solution can be stored at −20 °C for up to 2 months. Use UltraPure DNase/RNase-free dH2O (Invitrogen-Thermo Fisher Scientific) for washes and storage of coated dishes.

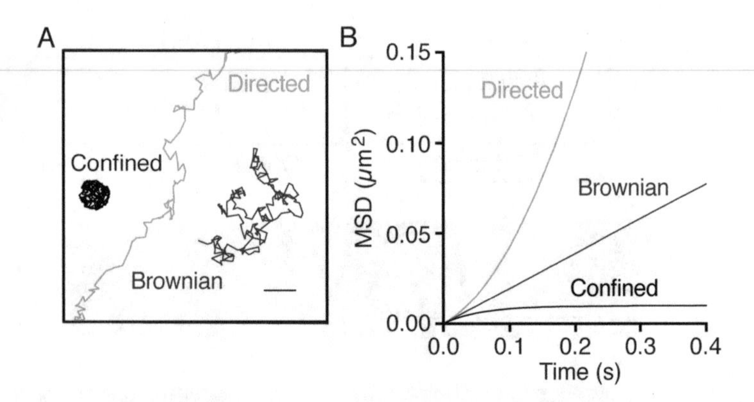

Fig. 4 In silico simulations of anticipated synaptic vesicle motion patterns. (**a**) Examples of simulation of two-dimensional particle trajectories undergoing confined (black), brownian (purple), and directed (orange) motion. Bar 200 nm. (**b**) Average MSDs of simulated particle trajectories undergoing different motion patterns. Each MSD curve is obtained from 100 simulated trajectories with 200 time steps sampled at 10-ms interval

- *Dissection medium*:

  1× Hanks' buffered salt solution without calcium, magnesium, and phenol red (HBSS) and 10 mM HEPES pH 7.3 (both from Gibco-Thermo Fisher Scientific) and 100 U/ml penicillin-100 μg/ml streptomycin (Invitrogen-Thermo Fisher Scientific). Filter-sterilize the buffer with a sterile disposable filter unit and store the buffer at 4 °C for up to 1 month.

- *Neuronal plating medium*:

  Neurobasal medium, 100 U/ml penicillin-100 μg/ml streptomycin, 1× GlutaMAX supplement, and 5% fetal bovine serum (FBS) all from Gibco-Thermo Fisher Scientific. Mix the solution well, filter-sterilize if necessary, and store at 4 °C up to the manufacturer's expiration date.

- *Neuronal culture medium*:

  Neurobasal medium, 100 U/ml penicillin-100 μg/ml streptomycin, 1× GlutaMAX supplement, and 1× B27 (Gibco-Thermo Fisher Scientific). Mix the solution well, filter-sterilize if necessary, and store at 4 °C up to the expiration date recommended by the manufacturer. Optional: The medium can be supplemented with 20% neuroglial conditioned medium (for details, *see* Joensuu et al. [22]). Furthermore, 5 μM cytosine β-D-arabinofuranoside (Ara-C; Sigma-Aldrich) can be added to the culture dish on day 3 in vitro (DIV3) to decrease the proportion of glial cells in the culture.

- *Neuronal maintenance medium*:

  Neurobasal medium, 100 U/ml penicillin-100 μg/ml streptomycin, 1× GlutaMAX supplement, and 1× B27. Mix

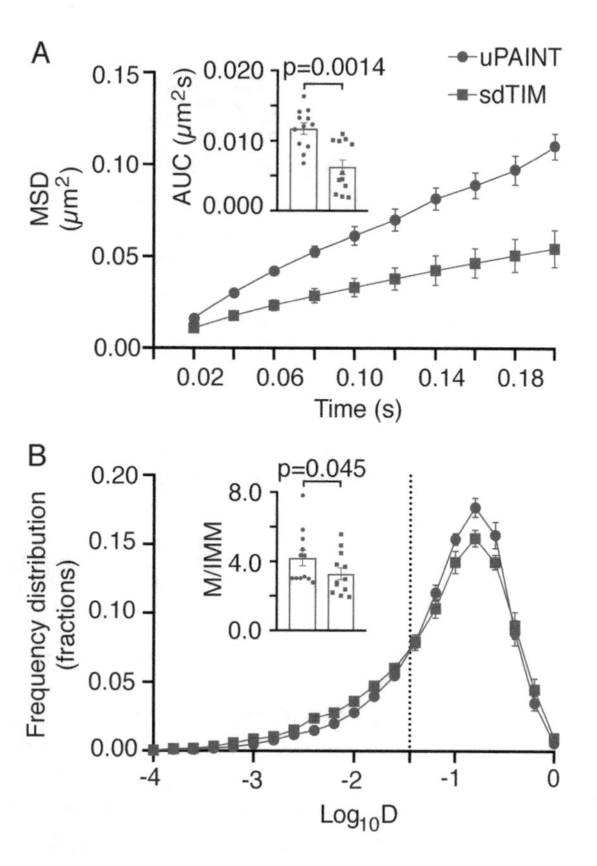

Fig. 5 Comparison of single-molecule mobility of VAMP2-pHluorin-bound Atto647N-NBs on the plasma membrane and following internalization. The comparison of VAMP2-pHluorin-bound Atto647N-NB mobility following uPAINT (i.e., on the plasma membrane; blue) and sdTIM (i.e., internalized into recycling synaptic vesicles; red) is shown as (**a**) MSD (μm^2) over time (s) and AUC (μm^2 s; inset), as well as (**b**) frequency distribution of the mean $\mathrm{Log}_{10}D$ and mobile-to-immobile (M/IMM) ratio. The threshold for the immobile ($\mathrm{Log}_{10}D \leq -1.45$) and mobile ($\mathrm{Log}_{10}D > -1.45$) fraction is indicated with a dotted line in (**b**). $n = 12$ hippocampal neurons in individual cultures following sdTIM (97,400 trajectories) and uPAINT (52,800 trajectories). Statistical analyses of independent experiments were performed using Mann-Whitney *U* test (inset in **a**) and the Student's *t* test (inset in **b**). Single-molecule tracking was performed with PALM-Tracer software operating in the MetaMorph. The graphs are modified from Joensuu et al., Journal of Cell Biology 2016 with a permission from Rockefeller University Press

the solution well, filter-sterilize if necessary, and store at 4 °C up to the recommended expiry date.

- *Neuronal transfection*:

  Transfection of VAMP2-pHluorin (kindly provided by J. Rothman, Yale University [23]) is performed according to

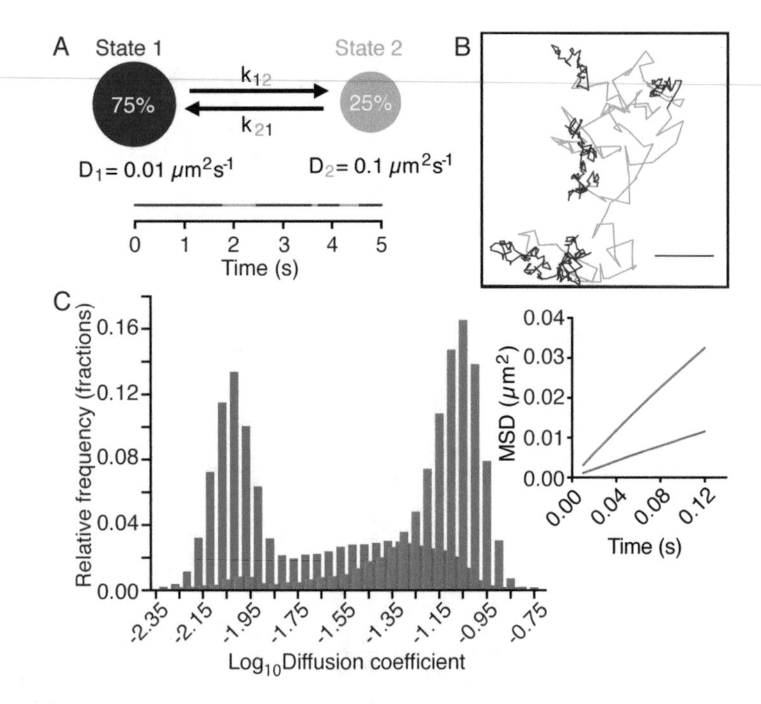

Fig. 6 Predictions of a model with two diffusive states. (**a**) A schematic of a two-state model where particle trajectories switch between two distinct diffusive states in a circular region with 500 nm radius. D_1 and D_2 are the diffusion coefficients of state 1 (blue) and state 2 (green), respectively; k_{12} is the rate of transition from state 1 to state 2; and k_{21} is the rate of transition from state 2 to state 1. Circles represent different states, and their area is proportional to the indicated percentage state occupancy over 5-s-long simulation shown in (**a**). *Bottom*, timeline showing the temporal sequence of in silico simulated particle switching between the two states shown in (**a**). (**b**) An example of a simulation of one particle switching between two diffusive states (blue and green). Bar, 200 nm. (**c**) *Left*, distribution of diffusion coefficients and, *right*, average MSDs of particles switching between the two states with $D_1 = 0.01\ \mu m^2\ s^{-1}$, $D_2 = 0.1\ \mu m^2\ s^{-1}$, $k_{12} = 1\ s^{-1}$, and $k_{12} = 4\ s^{-1}$ (red), and $D_1 = 0.01\ \mu m^2\ s^{-1}$, $D_2 = 0.1\ \mu m^2\ s^{-1}$, $k_{12} = 4\ s^{-1}$, and $k_{12} = 1\ s^{-1}$ (gray). In (**c**), each MSD curve and distribution of diffusion coefficients were obtained from 20,000 simulated trajectories with 50 time steps sampled at 10-ms intervals

the manufacturer's instructions (Lipofectamine 2000 Transfection Reagent, Invitrogen) on DIV13–16.

2.2 Atto467N Nanobodies, Beads, and Buffers for sdTIM

- *Low-K$^+$ buffer*:

  Prepare 0.5 mM $MgCl_2$, 2.2 mM $CaCl_2$, 5.6 mM KCl, 145 mM NaCl, 5.6 mM D-glucose, 0.5 mM ascorbic acid, 0.1% (wt/vol) bovine serum albumin (BSA), and 15 mM HEPES, pH 7.4. Measure the osmolarity using an osmometer (the osmolarity should be 290–310 mOsm). Filter-sterilize and store at 4 °C for up to 2 weeks.

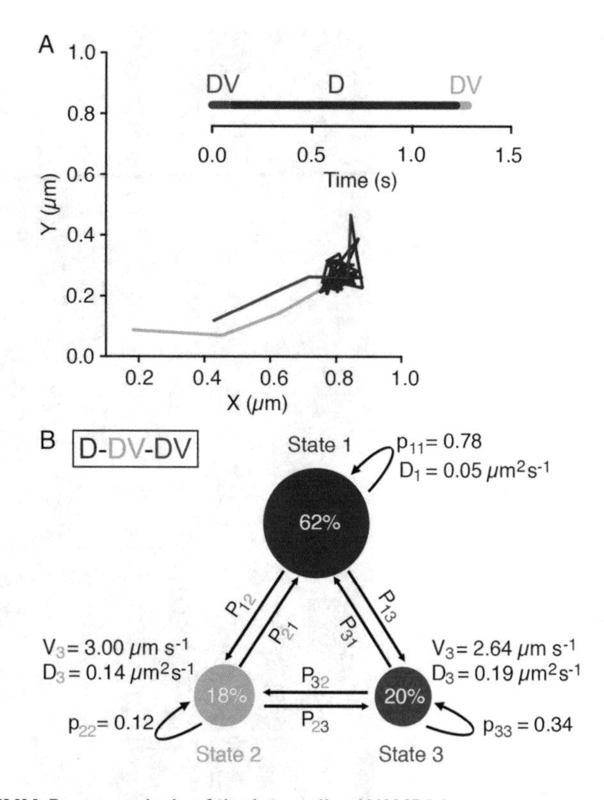

Fig. 7 HMM-Bayes analysis of the internalized VAMP2 in presynapses and axonal segments. (**a**) An example of recycling synaptic vesicle trajectory displaying D–DV–DV (D, diffusive state; DV, transport state) motion inferred by HMM-Bayes analysis. Timeline shows the order and duration (s) of the motion states in the trajectory. (**b**) Example of a three-state D–DV–DV model in a presynapse, inferred from a set of internalized VAMP2-pHluorin-bound Atto647N nanobody trajectories. Circles represent the different states, and their area is proportional to the percentage occupation of VAMP2-pHluorin-bound Atto647N nanobodies in the respective state. P_{ij}, where $i = 1, 2$, or 3 and $j = 1, 2$, or 3, are the transition probabilities, with $i = j$ indicating the probability of staying in the same state and $i \neq j$ indicating the probability of switching to another state. Apparent diffusion coefficients (D) and velocity magnitudes (V) for the states 2 and 3 are indicated. $P_{12} = 0.22$, $P_{21} = 0.30$, $P_{13} = <0.001$, $P_{31} = 0.16$, $P_{23} = 0.58$, and $P_{32} = 0.50$. Images are modified from Joensuu et al., Journal of Cell Biology 2016 with a permission from Rockefeller University Press

- *High-K⁺ buffer*:

  Prepare 0.5 mM $MgCl_2$, 2.2 mM $CaCl_2$, 56 mM KCl, 95 mM NaCl, 5.6 mM D-glucose, 0.5 mM ascorbic acid, 0.1% (wt/vol) BSA, and 15 mM HEPES, pH 7.4. Measure the osmolarity using an osmometer (the osmolarity should be 290–310 mOsm). Filter-sterilize and store at 4 °C for up to 2 weeks.

- *Fluorescent beads for drift correction*:

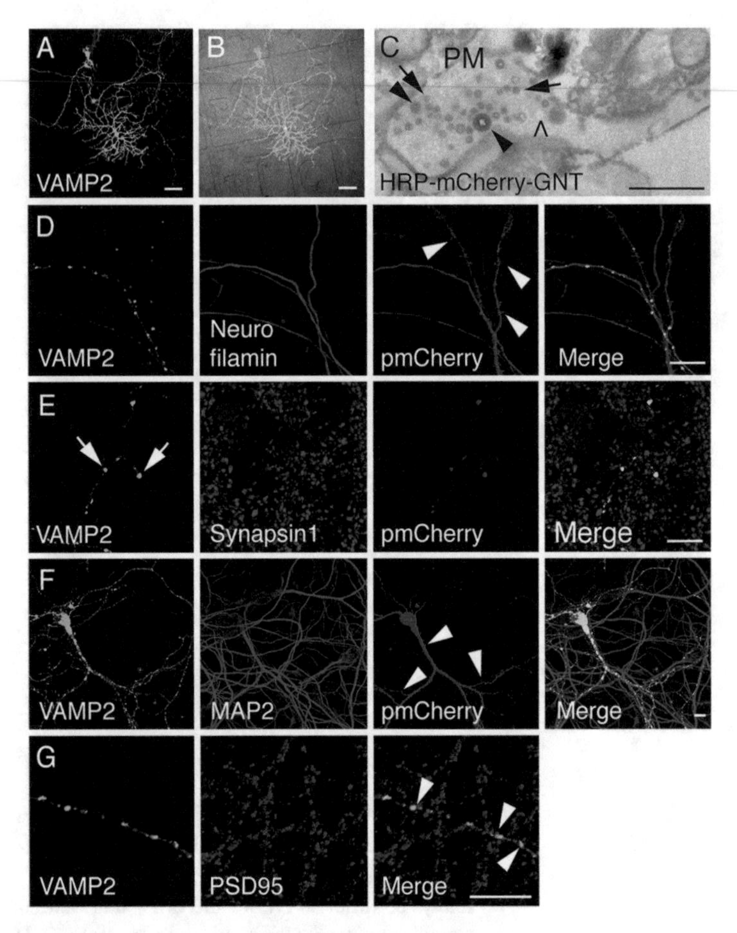

Fig. 8 Evaluation of the correct targeting of anti-GFP nanobodies with CLEM and the assessment of neuronal maturity. (**a**) Alpha numeric localization of live hippocampal neuron expressing VAMP2-pHluorin on glass-bottom grids is recorded (**b**) and neurons are then subjected to sdTIM using HRP-mCherry-GNT. (**c**) After fixation and cytochemical staining, neurons are processed for EM. The same neuron in (**a**) is correlated to the electron micrograph by following the recorded location of the neuron on the grid. HRP precipitate in recycling synaptic vesicles (arrows) and in endosomes (arrowheads) is indicated along with unstained synaptic vesicles (open arrowhead). Residual staining can be observed on the plasma membrane (PM). Bars 10 μm in (**a**, **b**, **d**–**g**) and 500 μm in (**c**). Hippocampal neurons transfected with VAMP2-pHluorin (**d**–**g**) and mCherry-C1 (**d**–**f**) and immunostained against endogenous (**d**) neurofilamin, (**e**) synapsin1, (**f**) MAP2, or (**g**) PSD95

Prepare a dilution of TetraSpeck™ Microspheres (Thermo Fisher Scientific) in low-K^+ buffer.

- *Atto647N-labeled anti-GFP nanobodies*:

  Heavy-chain antibodies from *Camelidae* sp. recognize antigens via their variable domain (referred to as V_HH or nanobody) and lack the antibody light chains [28].

A variety of labeled (e.g., Atto647N and Atto565) camelid single-domain nanobodies (also known as sdABs) with different target specificity are commercially available from Synaptic Systems. Alternatively, unlabeled anti-GFP V_HH-purified nanobodies can be purchased, for example from Chromotek, which can then be labeled with Atto647N (ATTO-TEC Gmbh) or other fluorophores as we have described previously [21, 29]. These nanobodies can recognize and bind to any protein of interest with an extracellular GFP. It will also recognize GFP-related fluorescent protein tags such as mCerulean, pHluorin, and YFP but not mCherry or red fluorescent protein (RFP). Fluorescently labeled nanobodies are light sensitive and should therefore be protected by wrapping the nanobody-containing solution tubes in aluminum foil. We recommend avoiding repetitive cycles of freezing and thawing by storing the nanobodies in small aliquots at −80 °C for up to 12 months. For best results, always prepare fresh nanobody dilution for each experiment and protect it from light.

- *Quantifying Atto647N labeling of the nanobodies*:

  Reagents required to prepare the flow chamber: 1 M potassium hydroxide, Milli-Q water, PBS, and hydrochloric acid for titration. The required volume of nanobody dilution for a flow chamber is 200 μl. Prepare 3.20 pg/μl anti-GFP Atto647N-NB dilution in phosphate-buffered saline (PBS) at pH 7.4 (or 7.0). For acidic environment tests, titrate PBS with HCl to pH 5.0 (or 5.5) and use immediately. Prepare the dilution fresh, protect it from light, and keep the dilution on ice until proceeding with the cell-free flow-chamber imaging and analysis of the Atto647N labeling density of the nanobodies (Subheading 3.1).

- *Determining the appropriate Atto467N-nanobody dilution for sdTIM*:

  Super-resolution techniques such as Universal Point Accumulation Imaging in Nanoscale Topography (uPAINT) [30, 31], which is ideal for the investigation of plasma membrane protein mobility, require a continuous and stochastic protein labeling at the plasma membrane, using externally applied nanobodies. In contrast, after initial binding to VAMP2-pHluorin at the plasma membrane and following their internalization into synaptic vesicles using sdTIM, the appropriate concentration of the Atto647N-NBs is determined by the balance between the bleaching rate of Atto647N-fluorophores and the renewal of the Atto647N-NBs in the imaging area through the mobility of labeled endocytic structures [21, 22]. In our experience, ideal sdTIM labeling density allows simultaneous tracking of 4 ± 1 recycling synaptic vesicles per presynapse at any given time which can be achieved by preparing a fresh 3.20 pg/μl dilution of anti-GFP Atto647N-NBs in prewarmed (37 °C) high-K^+ buffer (protect the dilution from light prior to sdTIM experiments) (Subheading 3.6). Note that other fluorescent nanobodies, such as Atto565-labeled nanobodies (Atto565-NBs), can be used at the same concentration for sdTIM. Quantitation of the fluorophore labeling density of the nanobodies (Subheading 3.1), and the number of internalized nanobodies per vesicle (Subheadings 3.8 and 3.9), as well as titrating the nanobody concentration, is highly recommended to ensure optimal imaging conditions, high localization precision (in our studies 36 ± 1 nm for Atto647N-NBs [21] and 32 ± 4 nm for Atto565-NBs [22]), and minimal background when using sdTIM adaptations. Note that ideally the nanobody concentration should be chosen based on presynaptic vesicle labeling density, as the nerve terminals are a far more crowded environment than the axons.

- *Assessment of the number of internalized Atto647N-NBs in fixed neurons*:

  Prepare 4% paraformaldehyde (PFA; e.g., Electron Microscopy Sciences, EMS) in PBS for fixation. Note that PFA is a tissue fixative and should be handled in a fume hood using appropriate protective clothing, gloves, and goggles, and subsequently discarded in accordance with institutional regulations.

2.3 sdTIM Validation and Controls

- *Correlative Light Electron Microscopy (CLEM)*:

  To validate the correct targeting of internalized nanobodies, transfect neurons with VAMP2-pHluorin (using Lipofectamine 2000), and use horseradish peroxidase (HRP)-tagged GFP Nanotrap (HRP-mCherry-GNT) instead of fluorescent nanobodies [22]. For fixation, prepare a fresh solution of 2% glutaraldehyde (e.g., EMS) and 1.5% PFA in PBS pH 7.4. Note that glutaraldehyde is a tissue fixative and must be handled as described above for PFA. For cytochemical staining of the HRP, prepare 1 mg/ml 3,3′-diaminobenzidine tetrahydrochloride (DAB; e.g., Sigma-Aldrich) in PBS, vortex for 5 min, and filter through a 0.2 μm syringe filter. For contrasting, prepare 1% osmium tetroxide (OsO_4) in 0.1 M sodium cacodylate buffer (e.g., Sigma-Aldrich) prior to dehydration and embedding in LX-112 resin (Ladd Research). Note that osmium tetroxide and sodium cacodylate are toxic and should be handled according to the manufacturer's safety instructions and discarded according to institutional regulations.

- *Assessing neuronal maturity and synaptic transmission*:

  For co-transfection (using Lipofectamine 2000), use VAMP2-pHluorin and pmCherry-C1 constructs. Fix with 4% PFA and 4% sucrose (Sigma-Aldrich) in PBS. Permeabilize with 0.1% Triton X-100 (Sigma-Aldrich) in PBS. Blocking solution:

3% goat serum (Gibco) with 0.025% Triton X-100 (Sigma-Aldrich) in PBS. Primary antibody solution: 5% BSA, 1% FBS, and 0.025% Triton X-100 (all from Sigma-Aldrich) in PBS. Secondary antibody solution: 1% goat serum (Gibco) and 0.025% Triton X-100 (Sigma-Aldrich) in PBS. All the washing steps are done with PBS. Mounting: ProLong mounting medium (Thermo Fisher Scientific). Primary antibodies: mouse anti-synapsin-1 and guinea pig anti-MAP2 (both from Synaptic Systems), mouse anti-PSD95 (Abcam), and mouse anti-neurofilament (BioLegend). Secondary antibodies: goat anti-mouse Alexa Fluor 647 and goat anti-guinea pig Alexa Fluor 647 (both from Thermo Fisher Scientific).

2.4 Equipment

- *Neuronal dissection*:

  Laminar flow hood equipped with a dissecting microscope, cell culture incubator at 37 °C with a humidified 5% CO_2, a 37 °C water bath, microcentrifuge, hemocytometer for counting the cells, timer, and a vacuum system for aspiration. Sterile dissecting tools including scissors, forceps, fine-tipped tweezers, and scalpel handle blade. Sterile plasticware including 35 mm glass-bottom cell culture dishes (Cellvis, cat. no. D35-20-1.5N or equivalent), serological pipettes, 60 mm tissue culture dishes, centrifuge tubes, filter tips, disposable filter units (Thermo Fisher Scientific), and glass Pasteur pipettes.

- *Assessment of Atto647N labeling of nanobodies*:

  Glass slides and coverslips (Thermo Fisher Scientific), coverslip tweezers (e.g., Dumont), double-sided tape, compressed air, and a sonicator.

- *Super-resolution microscopy*:

  The method can be implemented on a standard single-molecule microscope operating with oblique illumination. Our super-resolution imaging setup comprises a Roper iLas^2 microscope (Roper Scientific) CFI Apo TIRF, 100×/1.49 NA oil-immersion objective (Nikon) with ×1.5 additional magnification, Piezo Z-drive and Nikon perfect focus system, two electron-multiplying charge-coupled device (EMCCD) cameras (Photometrics), QUAD beam splitter and QUAD band emitter (Semrock), and MetaMorph software for image acquisition (Molecular Devices). Note that other microscopes with a suitable single-molecule detection acquisition rate (here 50 Hz) can be used to implement the sdTIM technique (the requirement for the imaging rate depends on the time scale of the biological phenomenon). For sdTIM setup, we recommend using a custom-made experimental perfusion and aspiration system designed for 35 mm glass-bottom cell culture dishes (for detailed instructions for 3D printing of the custom-made perfusion lid and construction of the perfusion chamber, *see* [22]) or, alternatively, an equivalent commercially available perfusion system can be used (e.g., PeCon, POC Cell Cultivation System).

- *CLEM*:

  Gridded glass-bottom dishes (P35G-1.5-14-CGRD; MatTek Corporation) and an inverted point-scanning laser confocal microscope (e.g., LMS 510 META, Zeiss) with a Plan-APO 10× 0.45 NA objective (e.g., Zeiss) at 37 °C, argon laser, and Zen 2009 software are used to record the localizations of the VAMP2-pHluorin-positive neurons on gridded dishes. Electron microscopy: BioWave tissue processing system (Pelco), an ultramicrotome (e.g., Leica Biosystems) for thin sectioning, and a transmission electron microscope (e.g., JEOL) equipped with an appropriate camera (e.g., Olympus).

- *Analysis programs*:

  PALMTracer program in MetaMorph [30, 32] or TrackMate plug-in [33] for Fiji/ImageJ (National Institutes of Health) with an additional analysis routine to obtain single-molecule mobility parameters (e.g., MATLAB routine provided in [22]).

3 Methods

3.1 Quantification of the Fluorophore Labeling of the Nanobodies

Nanobodies are emerging as a powerful tool for biological research. The anti-GFP-nanobody, which consists of a variable domain of the heavy chain of HCAbs (V_HH), developed by immunization of a llama [28, 34], has previously been used to track single-glutamate receptors in live neurons [30]. Different fluorophores have different chemical characteristics, and cellular compartments provide varying chemical and physiological environments which can affect the fluorophore brightness, lifetime, and signal-to-noise ratio. We recommend quantifying the fluorophore labeling of the nanobodies (Fig. 2) before performing sdTIM experiments to ensure optimal outcomes.

Due to their small size, high affinity for GFP binding, and a photostable organic fluorophore, anti-GFP Atto647N-NBs are ideal for labeling small endocytic structures such as recycling synaptic vesicles using the sdTIM technique. When studying synaptic vesicle mobility with sdTIM, the externally applied Atto647N-NBs are first exposed to the extracellular space where the pH is typically close to neutral, after which they are internalized into synaptic vesicles where the pH becomes acidic. We therefore recommend the assessment of the Atto647N-NB fluorescence emission to be performed in both neutral (pH 7.0–7.4) and acidic buffers (pH 5.0–5.5). After the quantification of the labeling density of the nanobodies and following the sdTIM experiments (Subheading 3.6) and single-particle tracking (Subheading 3.7), the number of internalized nanobodies within synaptic vesicles can be quantified (Subheadings 3.8 and 3.9).

1. Quantification of Atto647N-NB labeling density is performed in a purpose-built cell-free flow chamber (note that this can be adapted for the analysis of other fluorophores or probes). Construct the flow chamber by first cleaning the glass slides (22 × 10 × 1 mm) and the rectangular, super-resolution-compatible coverslips by sonication in a glass beaker containing 1 M KOH for 20 min at RT. Wash the glass slides and coverslips with Milli-Q water five times, ensuring that the coverslips do not stick together. Then, dry the glass slides and coverslips individually with a clean airflow. Assemble the flow chamber in a clean environment free of dust, such as a fume hood, by attaching double-sided tape along the length of the glass slide on each side, and align the cleaned rectangular coverslip with the strips of tape by gently pressing the coverslip to create a proper seal and prevent leakage. Handle the cleaned coverslips with tweezers.
2. Pipette the prepared 200 μl of a 3.20 pg/μl dilution of Atto647N-NBs in PBS at pH 7.4 (or pH 5.0, for acidic environment tests) into the flow chamber (*see* **Note 1**) and incubate for 15 min at room temperature (i.e., 22–25 °C) to allow nanobody attachment to the glass surface (*see* **Note 2**). Cover the flow chamber from light to minimize bleaching. Wash the flow chamber twice with PBS pH 7.4 (or pH 5.0 for acidic environment tests) (*see* **Notes 1** and **2**). We recommend avoiding working under direct light as the Atto647N-NBs are light sensitive and performing the recordings without delay to minimize bleaching.
3. Acquire time-lapse images from several regions with oblique illumination (50 Hz, 20-ms exposure, 16,000 frames by image streaming). We recommend to image regions of interest (ROI) that were not exposed to laser illumination prior to image acquisition to prevent bleaching the fluorophores and to allow a reliable estimation of the fluorophore labeling density. This can be done by focusing on an area adjacent to the ROI, then by switching off the laser, and moving to the imaging area without turning the laser on. The laser should be turned on during the first frame or while acquiring the movie. To obtain comparable emission curves, it is important to use the same imaging setup for both cell-free systems and live-cell experiments. Therefore, perform this step at 37 °C. To allow reliable estimation of the fluorophore labeling density, it is important

that the Atto647N-NBs are deposited on the glass surface as separate molecules (*see* **Notes 1** and **2**).

4. To quantify the number of Atto647N fluorophores per nanobody, open obtained images with Fiji/ImageJ software (or any equivalent image analysis program) to analyze the fluorescence emission steps of the Atto647N-NBs. Choose 5 × 5 pixels around the center of each fluorescent spot (to obtain meaningful data, >100 fluorescent spots originating from several regions should be analyzed) to obtain intensity-time traces using the "plot z-axis profile" function in Fiji/ImageJ or, alternatively, a custom-written routine as described previously [35].
5. Subtract the background locally by quantifying the average intensity of a 2 × 2 pixel region surrounding each fluorescent spot. Overlapping nanobodies and fluorophores that bleach within the first two frames of the movies should be excluded from the analysis. Plot fluorescence intensity versus time for all the remaining fluorescent spots. Count the number of emission steps in the intensity-time traces manually (representative intensity-time traces of one and two emission steps are shown in Fig. 2). The mean intensity of the smallest step should be at least twice the standard deviation (SD) of the background fluorescence. The steplike emission traces reflect the bleaching of the Atto674N-fluorophores, and if the Atto647N-NBs are deposited on the glass surface sparsely enough to distinguish separate molecules (for best results, aim for 50–100 Atto647N-NBs per 27 × 27 μm^2), the number of steps will correspond to the number of fluorescent molecules per nanobody (*see* **Notes 1** and **2**).
6. Correct for missed events. In our experience, two molecules of Atto647N-dyes will bleach within the same imaging frame in less than 2% of the traces (these steps are twice the size of single-bleaching steps and should be counted as two). Note that the signal-to-noise ratio should be at least 1.5 to allow accurate assessment of the bleaching steps.

In our preparations of Atto647N-NBs, the average number of Atto647N fluorophores per nanobody is 1.5 [21] indicating that, on average, nanobodies contain either one or two Atto647N fluorophores (Fig. 2). The time course of the fluorescence emission of other fluorescently labeled ligands is determined similarly.

3.2 PLL Coating of Glass-Bottom Cell Culture Dishes

We recommend coating the super-resolution-compatible glass-bottom cell culture dishes (D29-20-1.5N; Cellvis; glass coverslip area 314 mm^2) with PLL at least for 3–4 h. Perform the coating in a sterile fume hood using sterile consumables and reagents.

1. Pipette 300–500 μl of 1 mg/ml PLL in 0.1 M borate buffer pH 8.5 onto the glass surface of the dishes and incubate the dishes at 37 °C at least for 3–4 h.
2. Remove the PLL solution and rinse the coated glass-bottom dish three times with UltraPure DNase/RNase-free dH_2O or sterile PBS pH 7.4. Maintain the coated glass-bottom dish in water until plating the neurons. At this point, the dishes can be stored in a clean, dust-free, area (e.g., in a 4 °C fridge) for several days before use. It is important to avoid drying.

3.3 Hippocampal Neuron Dissections

Rats must be handled according to institutional and governmental ethical guidelines. All reagents and surgical instruments should be sterile. In order to improve the viability of neurons, the procedure must be carried out in ice-cold dissection medium. All experiments should be carried out in accordance with relevant institutional and governmental ethical guidelines and regulations.

1. Following euthanasia by cervical dislocation of a pregnant dam at embryonic days 17–18 (E17–18) (*see* **Note 3**), dissect the uterus, remove the fetuses, and place them in a sterile 60 mm tissue culture dish containing dissection medium. The following steps must be done under a sterile laminar flow cabinet. Using a dissecting microscope, remove the brains and place them in a sterile dish containing dissection medium (Fig. 1). The tissue must remain submerged in the dissection medium at all times.
2. Dissect both forebrains and use fine-tipped tweezers to remove the meninges carefully. Dissect out the hippocampi, gently collect the dissected tissue in an Eppendorf tube, and bring the volume up to 270 μl with dissection medium (Fig. 1). Add 30 μl of 2.5% (wt/vol) trypsin, incubate for 10 min at 37 °C with regular agitation (every 2–3 min), and then stop the trypsinization process by adding 150 μl of FBS (or horse serum) and mixing by tapping the Eppendorf tube. Add 50 μl of 1% (wt/vol) DNase I and incubate for 10 min at 37 °C with regular agitation (every 2–3 min). Then, triturate the tissue by pipetting up and down (40–50 times) using a P200 pipette until the solution is homogenous.
3. Concentrate the cells by centrifugation (120 × *g*, 8 min, at room temperature). Discard the supernatant and resuspend the pellet in 0.5 ml of prewarmed (37 °C) neuronal plating medium. Determine the neuronal density using a hemocytometer and a cell counter and check the cell viability using trypan blue exclusion. We expect 80–90% hippocampal neuron viability, with the number of hippocampal neurons per brain depending on the species (rat or mouse), strain, embryonic

state, and experience of the person performing the dissection. Once isolated, the neurons must be plated as soon as possible as their viability decreases rapidly. Adjust the neuronal density to 50,000–100,000 hippocampal neurons in 0.5 ml of medium.

3.4 Hippocampal Neuron Plating on PLL-Coated Glass-Bottom Cell Culture Dishes

1. Before plating the dissected neurons, aspirate the UltraPure water from the PLL-coated glass-bottom dishes (Subheading 3.2) and rinse the glass-bottom dish once with neuronal plating medium.
2. Aspirate neuronal plating medium and plate the dissected neurons immediately on the glass-bottom dishes (Fig. 1). We recommend doing this step one dish at the time to prevent drying of the PLL-coated glass surface. We use a density of 50,000–100,000 hippocampal neurons in 0.5 ml of medium per 314 mm^2 of the glass surface.
3. Place the neurons in a tissue culture incubator at 37 °C with a humidified 5% CO_2 atmosphere for 2–4 h.

3.5 Hippocampal Neuron Culturing and Transfection

1. Examine culture dishes under a transmission light microscope 2–4 h after plating. Majority of the neurons should be attached to the glass surface, with small filopodia and lamellipodia protruding from the cells [36].
2. Replace the plating medium with 2 ml of neuronal culture medium that can be supplemented with 20% of neuroglial conditioning medium (*see* [22] for further details). Optional: 5 μM Ara-C can be added to the culture dish on DIV3 to decrease the glial cell content (*see* **Note 3**).
3. Every 7 days, refresh the cultures by replacing 1/3 to 1/2 of the medium with fresh neuronal maintenance medium. We recommend performing the sdTIM experiments at DIV14–21 (Fig. 1).
4. Prior to sdTIM, at DIV13–16, transfect the neurons with VAMP2-pHluorin (or another appropriate construct) (Fig. 1). We commonly use Lipofectamine 2000 transfection reagent (other methods can also be used; please *see* **Note 4**) to transfect primary neuron cultures [22, 37]. Pipette 100 μl of room-temperature neurobasal medium into two 1.5 ml Eppendorf tubes. Add 2 μl of Lipofectamine 2000 transfection reagent to one tube and 2 μg of VAMP2-pHluorin to the other tube and mix well by pipetting up and down (we recommend optimization of the total amount and the transfection reagent:plasmid ratio when using other sdTIM adaptations). Incubate the solutions for 5 min at room temperature, and then combine the contents of tubes, mix well, and incubate for 15–30 min at room temperature. Carefully collect the culture medium from the glass-bottom dish in a sterile

tube, leaving a residual volume of 380 µl to cover the neurons in the glass-bottom dish, and pipette the transfection mixture to the neurons in a dropwise manner (only onto the glass-bottom dish). Incubate the neurons for 2–5 h at 37 °C in 5% CO_2 atmosphere. After incubation, remove all excess culture medium from the plastic portion of the dish, wash the neurons 3–5 times gently with fresh prewarmed (37 °C) neurobasal medium, and then replace the medium with the medium collected from the glass-bottom dishes prior to the transfection (make sure that the collected culture medium is prewarmed at 37 °C). Continue transfection of the neurons for 24–48 h.

3.6 sdTIM: Induction of Synaptic Activity and Internalization of Atto647N-NBs into Recycling Synaptic Vesicles

Induction of synaptic activity and uptake of Atto647N-NBs (Fig. 3a) in DIV14–21 hippocampal neurons expressing VAMP2-pHluorin (Fig. 3b) can be done in glass-bottom dishes equipped with a custom-made perfusion lid (details of the 3D printing can be found in ref. 22), which allows a gentler exchange of liquid compared to pipetting straight onto the neurons and allows the liquid exchange at the microscope without having to remove the dish from the dish holder. Using the perfusion lid allows liquid exchange on the plastic-bottom surface of the cell culture dish, leaving the glass-bottom portion of the dish submerged (remnant volume ~380 µl) at all times, which protects the neurons from drying. We recommend taking into account the remaining volume in the glass bottom of the dish when calculating the nanobody dilutions. To induce synaptic activity and internalization of Atto647N-NBs into recycling synaptic vesicles, follow the steps below:

1. Prewarm the low-K^+ and high-K^+ buffers to 37 °C, prepare a 3.20 pg/µl dilution of anti-GFP Atto647N-NBs in prewarmed high-K^+ buffer, and cover it from light.
2. Remove the culture medium by aspiration and gently wash the neurons 4–5 times with 2 ml of prewarmed low-K^+ buffer (Fig. 3a). Immediately transfer the dish to the microscope and replace the culture dish lid with the perfusion lid. Attach two syringes to the perfusion lid tubing: an empty 10 ml syringe for aspiration and another 10 ml syringe filled with 2 ml of prewarmed high-K^+ buffer containing 3.20 pg/µl of Atto647N-NBs for the injection. Protect the Atto647N-NB solution from light using foil. Optional step: We recommend adding fluorescent beads diluted in low-K^+ buffer during the last washing step to augment drift correction (*see* **Note 5**).
3. Locate a representative hippocampal neuron expressing VAMP2-pHluorin (Fig. 3b) (*see* **Note 4**). Aspirate the low-K^+ buffer from the dish carefully (removed volume ~1.6 ml), leaving the glass-bottom portion of the dish submerged in the

low-K^+ buffer. Then, add the high-K^+ buffer containing 3.20 pg/μl of Atto647N-NBs to the dish to induce synaptic activity while simultaneously recording the unquenching of the VAMP2-pHluorin fluorescence (Fig. 3b, c; *see* **Note 5**). Acquire images at 50 Hz and 20-ms exposure time over 3000–5000 frames and stimulate the neurons for 5 min in total at 37 °C without removing the plate from the microscope (Fig. 3b). Alternatively, record a short movie or take a snapshot of VAMP2-pHluorin fluorescence prior to and immediately after stimulation, to compare the VAMP2-pHluorin mean fluorescence intensity (MFI) before and after the pulse. This will provide adequate information to identify active presynapses. As the unquenching of VAMP2-pHluorin fluorescence occurs quickly after the stimulation, the imaging should therefore be done immediately.

4. After 5-min stimulation, aspirate the high-K^+ buffer from the dish (remove ~1.6 ml) and carefully inject 2 ml of prewarmed low K^+ into the dish. Repeat this step five times to remove any traces of unbound Atto647N-NBs and incubate the plate for 10 min (chase) in total (Fig. 3a) (*see* **Notes 5** and **6**).
5. After the 10-min chase, acquire time-lapse images of the VAMP2-pHluorin-bound Atto647N-NBs internalized into recycling synaptic vesicles with oblique illumination microscopy (Fig. 3a) (50 Hz, 20-ms exposure, 16,000 frames by image streaming) using the same microscope settings as described in **step 5** of Subheading 3.1 (*see* **Notes 5** and **6**). The imaging should be done in the same ROI in which the VAMP2-pHluorin fluorescence was recorded (**step 3** of Subheading 3.6) (Fig. 3b) to correlate the single-molecule acquisition to VAMP2-pHluorin fluorescence information (Fig. 3d).
6. Proceed with single-molecule tracking (Subheading 3.7).

3.7 Single-Molecule Tracking with TrackMate

Different software packages are available for single-molecule detection and tracking [38]. The PALM-Tracer software [30, 32] operating in the MetaMorph software, which is used in our laboratory, employs a combination of wavelet segmentation [39] and optimization of multiframe object correspondence by simulated annealing [40], and allows single-molecule tracking. Alternatively, freely available software packages can be used for single-molecule tracking. Below we describe a single-molecule detection and tracking routine using TrackMate software [33] in conjunction with an additional analysis routine in MATLAB [22], which computes mean-square displacement (MSD) and diffusion coefficients (available for download at [22]).

1. Open the time-lapse acquisitions of internalized Atto647N-NBs (Subheading 3.6) using TrackMate software. If desired, a

presynaptic and axonal segment ROIs can be drawn at this stage.

2. Run TrackMate plug-in (select the "LoG detector," "median filter," and "subpixel localization" in the particle detection process). Adjust the linking distance, which depends on the time resolution and mobility of fluorescent spots, in the Simple LAP tracker while keeping the gap closing at zero to link single-particle mobility into tracks. We recommend choosing an exposure time which prevents the Atto647N-NBs from moving distances greater than the diameter of the point spread function of the microscope between adjacent frames of the acquisition. Then, run the tracker.
3. Select particle traces as appropriate using the spot-filtering options of the TrackMate. Generate the "Spots in Tracks Statistics" text file containing the *x* and *y* coordinates, as well as the spot intensity values, of the fluorescent spots linked into each track.
4. A custom-written MATLAB code and detailed instructions on how to use the scripts are available to download from [22]. Presynaptic and axonal ROIs can be manually selected before the tracking, which allows the discrimination of single-particle tracks of synaptic vesicles originating from these distinct regions [21] (active presynapses can be discriminated from axons by following the acquired VAMP2-pHluorin fluorescence videos obtained in Subheading 3.6; Fig. 3b, c). Run the downloaded MATLAB routine to compute the mean square displacement (MSD) curves and diffusion coefficients.

Here, using in silico simulations, we show the expected MSD curves of particles undergoing Brownian, directed, or confined motion (Fig. 4a, b). To validate the sdTIM technique [21], single-molecule mobility of VAMP2-pHluorin-bound Atto647N-NBs internalized in recycling synaptic vesicles in live hippocampal neurons was compared to that of those transiting on the plasma membrane using uPAINT technique [30, 31] (Fig. 5). Our results showed that the mobility of VAMP2-pHluorin-bound Atto647N-NBs internalized in recycling synaptic vesicle pool (imaging using sdTIM) differs significantly from the mobility of those located on the plasma membrane (imaging using uPAINT) [21].

Recycling synaptic vesicles switch stochastically between distinct motion states [21]. Here, the MSD and distribution of diffusion coefficients of particles switching between two distinct diffusive states are shown (Fig. 6) using MCell tool simulation [41]. By applying Bayesian model selection applied to hidden Markov modeling (HMM-Bayes), we investigated the anomalous and subdiffusive events of recycling synaptic vesicle mobility in live hippocampal neurons, and discovered that, in most nerve terminals, SVs stochastically switch between purely diffusive (D) and transport mobility (DV) states (Fig. 7) [21].

3.8 Procedure to Quantify the Number of Internalized Nanobodies Within Recycling Synaptic Vesicles in Fixed Neurons

Ideally, only one anti-GFP nanobody with a single Atto647N-tag would get internalized into an individual synaptic vesicle. Since VAMP2 is an abundant synaptic SNARE protein and, on average, up to 70 copies of VAMP2 can be found in a synaptic vesicle [42], with significant intervesicular variability [43, 44], it is possible that more than one Atto647N-NB may get trapped in a single synaptic vesicle, even when used at very low concentrations. To quantify the number of internalized VAMP2-pHluorin-bound Atto647N-NBs in fixed neurons, follow the steps below, and then continue to Subheading 3.9 to quantify the number of internalized VAMP2-pHluorin-bound Atto647N-NBs in live neurons.

1. Fix the neurons with freshly prepared 4% (wt/vol) PFA in PBS for 20 min in a fume hood following the 10-min chase in sdTIM (Subheading 3.6). Wash the neurons three times with PBS, leaving the sample covered in PBS. Continue with imaging without delay.
2. Image the internalized Atto647N-NBs using oblique illumination microscopy as described in Subheading 3.6.
3. Calculate the fluorescence emission steps of internalized Atto647N-NBs in fixed neurons by opening the time-lapse movies in ImageJ software and following the instructions in Subheading 3.1. Internalized VAMP2-pHluorin-bound Atto647N-NBs in synaptic vesicles will exhibit a similar steplike emission to that observed in the cell-free flow chamber experiments (Fig. 2a; Subheading 3.1). Calculate the number of emission steps originating from single synaptic vesicles manually as outlined in Subheading 3.1. Continue with quantification of the number of internalized Atto647N-NBs in recycling synaptic vesicles in live hippocampal neurons (Subheading 3.9).

Reflecting the number of Atto647N fluorophores internalized within synaptic vesicles, in our experiments the average number of emission steps originating from single synaptic vesicles is 1.5 (ranging from one to three), with the majority of synaptic vesicles exhibiting a single emission step [21]. Based on these quantifications, the majority of synaptic vesicles contain a single internalized Atto647N-NB with either one or two Atto647N fluorophores attached to it.

3.9 Procedure to Quantify the Number of Internalized Nanobodies Within Recycling Synaptic Vesicles in Live Neurons

To quantify the number of internalized VAMP2-pHluorin-bound Atto647N-NBs in live neurons, follow the steps below:

1. Following the 10-min chase in sdTIM, image internalized VAMP2-pHluorin-bound anti-GFP Atto647N-NBs in live neurons using oblique illumination microscopy as described in Subheading 3.6.
2. To determine the number of synaptic vesicles that reside within the nerve terminal at any given time, two methods can be used (*see* **steps 3** and **4**).
3. In the first approach, open the acquired time-lapse movies in ImageJ software and choose an ROI to cover the whole presynaptic terminal (presynapses can be discriminated from adjacent axonal segments based on the round morphology of the bouton and the VAMP2-pHluorin unquenching), and measure the Atto647N-NB fluorescence intensity over time as described in Subheading 3.1. The internalized Atto647N-NBs exhibit similar steplike emission to that in fixed neurons, with the number of steps reflecting the number of Atto647N fluorophores within that region. The average number of emission steps in presynapses is 6.0 ± 1.6 (SD) in our experiments [21].
4. In the second approach, record the fluorescence intensity of each internalized Atto647N-NB along its trajectories. This can be done after the single-molecule tracking of VAMP2-pHluorin-bound anti-GFP Atto647N-NBs in live neurons (Subheading 3.7). Open the "Spots in Tracks Statistics" text file containing the fluorescence intensity assigned to each spot in a track. Plot the fluorescence intensity assigned to each spot against time and count the number of observed steps. The Atto647N-NB fluorescence level along the trajectories remains unchanged until either the Atto647N-NB bleaches and the fluorescence intensity trace drops to the background level or a second recycling synaptic vesicle enters the same diffraction-limited presynaptic ROI causing a steplike increase in fluorescence. The total number of emission steps from the trajectories within the same presynapse (5.4 ± 1.3 on average in our experiments [21]) reflects the number of Atto647N fluorophores within that region and should be similar to the quantification results in **step 3**.
5. The number of simultaneously detected synaptic vesicles in presynapses at any given time can then be estimated from the number of Atto647N fluorophores per nanobody (Subheading 3.1), the number of internalized Atto647N-NBs per synaptic vesicle (Subheading 3.8), and the number of fluorescence emission steps in the presynapses (above).

In our experiments, the number of synaptic vesicles we can detect at any given time within a presynapse, by tracking VAMP2-pHluorin-bound anti-GFP Atto647N-NBs, is 4 ± 1 [21].

3.10 sdTIM Validation and Controls

3.10.1 CLEM

The activity-dependent internalization of the anti-GFP Atto647N-NBs depends on their binding to VAMP2-pHluorin. Although Atto647N-NBs exhibit high affinity to GFP, and are therefore likely to enter solely into VAMP2-pHluorin-expressing cells, the following points may need to be considered when performing sdTIM experiments: (1) overexpression of any construct may cause mislocalization artifacts, (2) bulk endocytosis contributes to compensatory endocytosis of recycling synaptic vesicles [45], and (3) nanobodies may get internalized in endocytic structures other than synaptic vesicles, such as endosomes, even when used at a very low concentration. To assess the correct localization of the internalized nanobodies in synaptic vesicles we performed CLEM as follows:

1. Grow hippocampal neurons from rats (E18) on PLL-coated (Subheading 3.1) gridded glass-bottom dishes (P35G-1.5-14-CGRD-D; MatTek Corporation; 154 mm^2 glass coverslip area) and transfect with VAMP2-pHluorin on DIV14–16 for 24–48 h (Fig. 8a, b). We use a density of 25,000–50,000 hippocampal neurons per 154 mm^2 of the glass surface. These specialized dishes have alphanumeric location markers etched into the glass coverslip in the bottom of the dish, allowing the correlation of light and electron microscopy images.
2. On DIV15–17, image the neurons live at 37 °C with an appropriate fluorescence microscope (such as on an inverted point-scanning laser confocal microscope, LSM 510 Meta, Zeiss) to record the location of VAMP2-pHluorin-positive neurons (Fig. 8a) on the cover grids (Fig. 8b).
3. Process the neurons for sdTIM as described in Subheading 3.6, except for using HRP-mCherry-GNT nanobodies instead of Atto647N-NBs. Omitting the live-cell imaging of the synaptic vesicles is optional.
4. Fix the neurons with 2% glutaraldehyde and 1.5% PFA in PBS for 1 h at room temperature.
5. Quench the free aldehyde groups with 20 mM glycine in PBS for 20 min at room temperature, and then wash the neurons with PBS.
6. For cytochemical staining, prepare 1 mg/ml DAB in PBS, add 1 ml of this equilibration buffer to each dish, and incubate for 5 min at room temperature. Add 3 μl of 30% H_2O_2 to 5 ml of 1 mg/ml DAB in PBS to prepare the reaction buffer and protect the tube from light as the solution is very light sensitive.

Replace the DAB equilibration buffer in each dish with 1 ml DAB reaction buffer and incubate in the dark for 20 min at room temperature. Wash 3 × 5 min with PBS.

7. Osmicate with 1% OsO_4 (aq) in 0.1 M sodium cacodylate buffer using the following settings: 80 W Vac ON for 2 min ON, 2 min OFF, and 2 min ON (performed in the Pelco BioWave). Repeat this step without a solution change, then remove the osmium, and rinse with Milli-Q water.
8. Replace the Milli-Q water and place the cell culture dish in the BioWave using 80 W Vac OFF for 40 s.
9. Dehydration series are performed in the BioWave as follows: Remove water and replace with 50% EtOH using 250 W Vac OFF for 40 s. Repeat using 70% and 90% EtOH ones and 100% EtOH twice.
10. Perform resin infiltration using LX112 resin. Prepare 33%, 50%, 66%, and 100% LX112 in 100% EtOH. Incubate cells in the BioWave in 33% LX112 resin using 250 W and Vac ON for 3 min. Repeat using 50% and 66% resin once and 100% resin twice.
11. Ensure that only a thin film of resin is left on the glass bottom of the dishes and polymerize the resin in a 60 °C oven for 24 h.
12. After polymerization, remove the dish from the oven and detach the polymerized block from the dish by plunging the disk into liquid nitrogen for 15–20 s. Use a razor blade to pry the plastic and glass away from the resin (additional liquid nitrogen immersion may be used if the resin does not detach easily).
13. Once the resin is separated, the imprinted grid squares from the dish will be visible as an impression on the resin. This is the monolayer where the cells reside. Using pliers, cut away the sides of the dish and all the corners until just a flat disk of resin and dish remains.
14. Identify the ROI using a dissecting microscope to view the grid squares. The correct location can be found using the fluorescence and phase-contrast images recorded earlier (**step 2**). Circle the grid squares of interest with a fine-tip sharpie. Use pliers and a fine-tooth saw to cut the individual grid squares (~3 grid squares may be isolated provided that they are not too close to one another).
15. Paying close attention to orientation, stick the resin grid squares to blank resin blocks that have been roughened up with wet and dry sandpaper. Use LX112 or Epon as the adhesive rather than super glue which may obscure the alphanumeric locators on the grid squares if applied too generously. Incubate the blocks in the 60 °C oven for 16 h.

16. When making a block face for sectioning, a nonsymmetrical shape should be trimmed to correspond to a particular area identified by fluorescence microscopy so that the electron micrographs can be aligned with the fluorescence data; once the sections are cut for electron microscopy the only points of registration will be the shape of the section as fluorescence data and the grid square identifiers will no longer be visible.
17. Flip the fluorescence image horizontally and print it. Once on the transmission electron microscope, the distribution of cells seen on this printout will provide the pattern to identify the same distribution of cells on the phosphor screen. When that pattern is identified, the VAMP2-pHluorin-positive cells can be located and imaged, and the localization of the internalized anti-GFP HRP-mCherry-GNT nanobodies can be assessed based on the electron-dense HRP precipitate (Fig. 8c).

Our results showed that HRP precipitate was only detected in neurons expressing VAMP2-pHluorin, indicating that HRP-mCherry-GNT binds specifically only to VAMP2-pHluorin [21]. In neurons expressing VAMP2-pHluorin, a great majority ($88.5 \pm 3.0\%$) of the HRP precipitate was found in recycling synaptic vesicles, while some residual staining was detected on the plasma membrane and in endocytic structures larger than 45 nm in diameter [21]. These results provide support to the activity-dependent internalization of the anti-GFP nanobodies into recycling synaptic vesicles.

3.10.2 Assessing Neuronal Maturity

In order to ensure the imaging of vesicles in functional synapses, it is essential to perform the sdTIM experiments in mature neurons. We recommend a combination of different morphological features and neuronal markers to assess the maturity of the hippocampal neurons. The simultaneous co-transfection with VAMP2-pHluorin and an empty vector (here, pmCherry-C1) will facilitate the analysis of the distribution of overexpressed VAMP2, allowing the simultaneous determination of the maturity state. Mature neurons show prominent spines along the dendritic arbor [46] (Fig. 8d, f) and axonal enlargements corresponding to synaptic boutons [47] (Fig. 8e). VAMP2-pHluorin is distributed along the neuronal soma and the axon, colocalizing with the axonal marker neurofilament (Fig. 8d) and to a lesser extent with microtubule-associated protein 2 (MAP2; Fig. 8f). VAMP2-pHluorin is expressed in presynaptic boutons together with another synaptic vesicle protein, synapsin-1 (Fig. 8e), forming connections with dendritic spines, here stained with postsynaptic density protein 95 (PSD95; Fig. 8g). To assess the maturity of hippocampal neurons, perform immunofluorescence assay as follows:

1. Grow hippocampal neurons from rats (E18) on PLL-coated (Subheading 3.2) glass-bottom dishes and co-transfect the

neurons with VAMP2-pHluorin and pmCherry-C1 on DIV14–16 for 24–72 h.

2. At DIV15–19, aspirate the neuronal medium and wash the neurons with PBS to remove dead cells and cell debris. In the following steps, as the neurons express VAMP2-pHluorin, it is important to protect the specimen from light by performing all the incubation in a dark box. It is also critical to ensure that the dishes never dry as this will damage the complex neuronal architecture.
3. Fix the neurons with PBS-PFA-sucrose for 15 min at room temperature.
4. Wash the samples three times with PBS.
5. Permeabilize the neurons with PBS-Triton X-100 for 10 min at room temperature.
6. Wash the samples once with PBS.
7. Perform the blocking step by incubating the neurons with blocking solution for 1 h at room temperature.
8. Wash the samples three times with PBS.
9. Incubate the neurons with appropriate primary antibodies for 3 h at room temperature or overnight at 4 °C. Due to the duration of this step, it is highly recommended that a humid chamber be created using soaked tissue to prevent evaporation of the antibody solution.
10. Wash the samples 3–5 times, 5 min each, with PBS.
11. Incubate the neurons with appropriate secondary antibodies for 1 h at room temperature.
12. Wash the samples 3–5 times, 5 min each, with PBS.
13. Mount the samples. Once the mounting medium is dry, the neurons are ready for imaging. We currently use an inverted point-scanning laser confocal microscope (LSM 510 Meta, Zeiss). The prepared samples can be stored for several weeks to months, protected from light, at 4 °C.

3.10.3 Monitoring Neuronal Secretion and Synaptic Transmission with a pH-Sensitive Fluorophore

Transfecting hippocampal neurons with VAMP2-pHluorin (other potential recycling synaptic vesicle protein targets are discussed in Subheading 3.11) is important to implement the sdTIM technique for the study of synaptic vesicle recycling. pH-sensitive indicators (such as pHluorin [23, 24]) can also be used to discriminate active presynapses from the adjacent axonal segments [21, 22]. Synaptic vesicles fuse with the plasma membrane in response to high K^+ stimulation, exposing VAMP2-pHluorin to the extracellular space and leading to fluorescence intensity increases in presynaptic sites of secretion (Fig. 3b), thereby confirming synaptic transmission [11, 21, 22].

3.11 Potential sdTIM Adaptations

The pulse-chase-based sdTIM technique can be relatively easily adapted to study any endocytic structure in various cell types. We first used sdTIM to investigate the role of presynaptic activity in the retrograde transport of autophagosomes (from the nerve terminal to the cell soma) using externally applied botulinum neurotoxin type-A (BoNT/A) [3]. Although autophagosomes can be imaged with conventional light microscopy, the detection of a very low concentration of BoNT/A in autophagosomes can only be achieved with single-molecule super-resolution microscopy, as only a few molecules are likely to enter each retrograde carrier. We then used sdTIM to localize retrogradely transported signaling endosomes in axons labeled with Alexa Fluor 647-conjugated cholera toxin subunit-B (Alexa647-CTB), using structured illumination microscopy (SIM) [2]. Most recently, we have used sdTIM to image recycling synaptic vesicles in their natural crowded presynaptic environment [21] and we have demonstrated the successful use of sdTIM technique for dual-color imaging of synaptic vesicles, using anti-GFP Atto565-NBs, in conjunction with signaling endosomes, using Alexa647-CTB [22], thereby opening a path for multicolor super-resolution imaging of different subdiffractional cellular structures. The current pulse-chase configuration of the sdTIM protocol (i.e. 5-min stimulation and 10-min chase) mainly labels the dynamic pool of synaptic vesicles, known as the recycling pool (the other two major pools being the readily releasable pool and the reserve pool; the classical and emerging roles of synaptic vesicle pools are reviewed in [48]). By adjusting the pulse (and hence the strength of the stimulus), the chase duration, and the timing of Atto647N-NB (or other) application or by reintroducing a second stimulus after the Atto746N-NB internalization, other synaptic vesicle pools can potentially be studied with sdTIM. Alternatively, evoked synaptic activity could be studied by combining sdTIM with electric field stimulation systems [49]. The mobility of inter-bouton synaptic vesicles, which form the so-called exchange or super-pool of synaptic vesicles, can also be studied based on the morphological information about the neuron by region-specific analysis of synaptic vesicles. The surface pool (i.e., synaptic vesicles following fusion with the plasma membrane) [21] can be studied with uPAINT [30, 31].

There are multiple potential adaptations of the sdTIM configuration to image cell endocytic pathways and cargo trafficking. Other potential synaptic vesicle markers such as synaptobrevins (Syb or VAMP) [50, 51], synaptotagmins (Syt) [52, 53], synaptophysins (SypHy) [52–54], vesicular glutamate transporter 1 (vGlut1) [52, 55], Vps10p-tail-interactor-1a (vti1a) [56] and synaptic vesicle 2 (SV2) [52, 53], and V-type ATPase [52], that can be tagged with a pHluorin moiety in the vesicle lumen, could potentially be used instead of VAMP2. Other fluorophores, such as

RFP, can also be used when employing anti-RFP nanobodies (commercially available from Synaptic Systems). sdTIM can also be combined with sptPALM to image fluorescently labeled endocytic cargoes in conjunction with intra- or extracellular overexpressed mEos-tagged (or analogue) proteins, thereby expanding the use of the sdTIM technique. In addition to adaptations to follow the internalization of other synaptic vesicle proteins or different plasma membrane proteins destined for endocytosis, sdTIM can potentially be adapted to study different endocytic cargoes (e.g., neurotropic factors, monoclonal antibodies, cell-penetrating peptides used for drug delivery), toxins (e.g., cholera and botulinum toxins [2, 3, 29, 57–59]), cellular endocytic and membrane-trafficking regulators (e.g., Rab- and Rho-GTPases), and pathogen internalization (e.g., poliovirus) [22].

We have also used sdTIM to study neurons cultured on custom-made microfluidic devices (for detailed instructions regarding the microfabrication, *see* [22]), which allows polarized culturing of neurons. Microfluidic devices expand the potential use of the sdTIM technique even further by enabling (i) region-specific application of pharmacological agents to the cell culture dish; (ii) transfection of the microfluidic soma and terminal wells with, e.g., pre- and postsynaptic markers, to study synaptic connections; and (iii) the study of other cell–cell connections, such as neuromuscular junctions, by culturing muscle cells and neurons in the microfluidic soma and terminal wells, respectively.

4 Notes

1. *Separate Atto647N-NBs cannot be discriminated in the flow chamber*:

   If the density of nanobodies on the glass surface is too high to discriminate single molecules (Subheading 3.1), this may be due to the concentration of the Atto647N-NBs being too high or insufficient washing. Wash the flow chamber again with PBS to remove excess Atto647N-NBs or prepare a higher dilution of Atto647N-NBs and repeat Subheading 3.1. Note that while a brief exposure to far-red laser can be used to bleach a subset of internalized nanobodies in live neurons (*see* **Note 6**), we do not recommend a similar step for the flow chamber system. Exposure to far-red laser prior to acquisition will preclude the quantitation of fluorophore labeling of the nanobodies.

2. *Atto647N-NBs move in the flow chamber during imaging*:

   This may indicate that the incubation time was too short to allow proper attachment to the glass surface. Allow a longer incubation time and perform the washing steps gently (Subheading 3.5).

3. *Glial content on the cell culture dish is too high*:

   Use E18 embryos instead of postnatal pups (Subheading 3.3). The neuronal culture medium can be supplemented with 5 μM Ara-C on DIV3 to decrease the glial content (Subheading 3.5).

4. *Low transfection efficiency*:

   Primary neurons are challenging to transfect (Subheading 3.4). We usually obtain a better transfection efficiency with younger cultured neurons (i.e., DIV14–16 or younger) than those that are more mature (i.e., >DIV16) using Lipofectamine 2000 transfection reagent. Note that some optimization may be required when using different expression constructs and/or transfection reagents. If the transfection efficiency is very low, consider transfecting younger neuronal cultures, changing the transfection time, and/or adjusting the DNA concentration of the transfection reaction mix. Note that if the duration of the transfection is prolonged, additional controls for possible overexpression artifacts may be required. When creating the transfection mixture, it is important to use neurobasal medium without any supplements, especially antibiotics, which will decrease the transfection efficiency. When performing the transfection, it is very important to never remove all the medium from the neurons, which may decrease the viability of the neurons. Prior to pipetting the transfection mix to the neurons, collect the culture medium from the plastic surface, leaving some residual medium (~380 μl) in the glass-bottom part to cover the neurons. It is very important to remove all transfection reagent from the plates after the incubation, to avoid cell death, by performing 3–5 washing steps. Alternatively, other transfection methods such as calcium phosphate transfection [60] or nucleofection [61] prior to plating the neurons can be used.

5. *The region of interest is lost during high-K^+ stimulation or low-K^+ washes*:

   We recommend using the perfusion system to carry out the sdTIM protocol (Subheading 3.6), as it does not require removing the plate from the microscope in between the protocol steps and allows a gentler liquid exchange on the plate to decrease possible drift issues. If the ROI is lost during the injection or aspiration of liquids, use less syringe pressure. If the neurons react to stimulation by moving, check that the high-K^+ buffer is made correctly, has the right pH and osmolarity, and is at the correct temperature. Image drift during the acquisition can be corrected using the fluorescent beads

(Subheading 3.6, **step 2**) and an appropriate software (e.g., PALMTracer program in MetaMorph [30, 32] or MIB [62].

6. *The concentration of Atto647N-NBs in the culture dish is too high, so that single synaptic vesicles cannot be discriminated*:

   The sdTIM technique is based on the uptake of externally applied Atto647N-NBs, followed by imaging with oblique illumination microscopy to detect only the bound Atto647N-NBs with minimal background. We stress that the concentration of the fluorescent probe must be titrated to allow optimal labeling of the endocytic structure of interest. For example, aiming for optimal labeling of synaptic vesicles in the highly crowded presynaptic environment, compared to in the adjacent axons, is a limiting factor. In our experience, ideal labeling density allows simultaneous tracking of 4 ± 1 recycling synaptic vesicles in a presynapse at any given time [21]. If single recycling synaptic vesicles cannot be discriminated on the plate and single molecules overlap, the washing steps may have been insufficient (Subheading 3.6) or the concentration of the Atto647N-NBs was too high. We recommend repeating Subheading 3.5 with a fresh neuronal culture, as well as increasing the number of washes with an adequate volume after the high-K^+ stimulation and decreasing the concentration of the Atto647N-NBs on the dish. Alternatively, the nerve terminals can be briefly exposed to far-red laser illumination to bleach some of the fluorophores (optional).

5 Conclusions

Understanding the molecular mechanisms underpinning synaptic vesicle release and recycling is an important aspect of modern neurobiology [63–66]. Traditionally, only electrophysiology, indirect fluorescent recovery after photobleaching (FRAP), and electron microscopy have allowed the study of synaptic vesicle recycling. Assessing the nanoscale synaptic environment and the mobility of individual synaptic vesicles is a powerful strategy to study presynaptic functions, synaptic connections, and neurotransmission. Super-resolution imaging techniques are essential to achieving such in-depth knowledge on the key molecular step of information transfer and storage. sdTIM is a recently developed localization microscopy-based super-resolution technique that allows the study of the recycling and exchange pools of synaptic vesicles, or other endocytic structures, with high spatiotemporal resolution (here, 20-ms time resolution and 30–40 nm localization precision for synaptic vesicles using Atto647N-NBs) [21, 22]. The main advantage of this technique is the production of a high density of trajectories for each presynapse, reaching the thousands for each neuron, thereby enabling the identification of discrete diffusive states. Importantly, this in turn allows hidden mobility parameters to be investigated, such as the number of distinct diffusional and transport states and switching between them, by applying hidden Markov models (HMMs) and Bayesian model selection (HMM-Bayes) to the sdTIM data [67, 68]. By employing HMM-Bayes to study synaptic vesicle mobility, we were able to annotate mobility along synaptic vesicle trajectories and discovered that, in most nerve terminals, recycling synaptic vesicles stochastically switch between purely diffusive and transport mobility states, and are relatively less likely to switch from diffusive states to transport states in either resting or stimulated neurons [21, 22]. The quantitative account of the heterogeneous nature of synaptic vesicle mobility in live presynapses demonstrated that the commonly used bulk measurements for the description of synaptic vesicle mobility provide only a limited view of the dynamic signature of synaptic vesicles.

Acknowledgments

The super-resolution imaging was carried out at the Queensland Brain Institute's (QBI) Advanced Microimaging and Analysis Facility. We thank all the authors of the original studies for their contribution [21, 22], and our collaborators for helpful discussions, and we would like to further extend our gratitude to N. Valmas for the schematic illustrations presented here, R. Amor for technical support on imaging, I. Morrow for support on EM, and R. Tweedale (QBI) for critical appraisal of the chapter. This work was supported by an Australian Research Council Discovery Project grant (DP150100539), an Australian Research Council Linkage Infrastructure, Equipment, and Facilities grant (LE130100078), and a National Health and Medical Research Council (NHMRC) grant (1120381) to F.A.M. M.J. is supported by an Academy of Finland Postdoctoral Research Fellowship (298124). F.A.M. is a NHMRC Senior Research Fellow (1060075).

References

1. Denker A et al (2011) A small pool of vesicles maintains synaptic activity in vivo. Proc Natl Acad Sci U S A 108(41):17177–17182
2. Wang T et al (2016) Flux of signalling endosomes undergoing axonal retrograde transport is encoded by presynaptic activity and TrkB. Nat Commun 7:12976
3. Wang T et al (2015) Control of autophagosome axonal retrograde flux by presynaptic activity unveiled using botulinum neurotoxin type a. J Neurosci 35(15):6179–6194
4. Fowler MW, Staras K (2015) Synaptic vesicle pools: principles, properties and limitations. Exp Cell Res 335(2):150–156
5. Denker A, Rizzoli SO (2010) Synaptic vesicle pools: an update. Front Synaptic Neurosci 2:135
6. Rizzoli SO, Betz WJ (2005) Synaptic vesicle pools. Nat Rev Neurosci 6(1):57–69

7. Chamberland S, Toth K (2016) Functionally heterogeneous synaptic vesicle pools support diverse synaptic signalling. J Physiol 594 (4):825–835
8. Alabi AA, Tsien RW (2012) Synaptic vesicle pools and dynamics. Cold Spring Harb Perspect Biol 4(8):a013680
9. Crawford DC, Kavalali ET (2015) Molecular underpinnings of synaptic vesicle pool heterogeneity. Traffic 16(4):338–364
10. Kamin D et al (2010) High- and low-mobility stages in the synaptic vesicle cycle. Biophys J 99 (2):675–684
11. Gimber N et al (2015) Diffusional spread and confinement of newly exocytosed synaptic vesicle proteins. Nat Commun 6:8392
12. Hua Y et al (2011) A readily retrievable pool of synaptic vesicles. Nat Neurosci 14(7):833–839
13. Willig KI et al (2006) STED microscopy reveals that synaptotagmin remains clustered after synaptic vesicle exocytosis. Nature 440 (7086):935–939
14. Hu Y, Qu L, Schikorski T (2008) Mean synaptic vesicle size varies among individual excitatory hippocampal synapses. Synapse 62 (12):953–957
15. Lemke EA, Klingauf J (2005) Single synaptic vesicle tracking in individual hippocampal boutons at rest and during synaptic activity. J Neurosci 25(47):11034–11044
16. Westphal V et al (2008) Video-rate far-field optical nanoscopy dissects synaptic vesicle movement. Science 320(5873):246–249
17. Hoopmann P et al (2010) Endosomal sorting of readily releasable synaptic vesicles. Proc Natl Acad Sci U S A 107(44):19055–19060
18. Lehmann M et al (2015) Multicolor caged dSTORM resolves the ultrastructure of synaptic vesicles in the brain. Angew Chem Int Ed Engl 54(45):13230–13235
19. Maschi D, Klyachko VA (2017) Spatiotemporal regulation of synaptic vesicle fusion sites in central synapses. Neuron 94(1):65–73.e3
20. Peng A et al (2012) Differential motion dynamics of synaptic vesicles undergoing spontaneous and activity-evoked endocytosis. Neuron 73(6):1108–1115
21. Joensuu M et al (2016) Subdiffractional tracking of internalized molecules reveals heterogeneous motion states of synaptic vesicles. J Cell Biol 215(2):277–292
22. Joensuu M et al (2017) Visualizing endocytic recycling and trafficking in live neurons by subdiffractional tracking of internalized molecules. Nat Protoc 12(12):2590–2622
23. Miesenbock G, De Angelis DA, Rothman JE (1998) Visualizing secretion and synaptic transmission with pH-sensitive green fluorescent proteins. Nature 394(6689):192–195
24. Royle SJ et al (2008) Imaging phluorin-based probes at hippocampal synapses. Methods Mol Biol 457:293–303
25. Villarreal S, Lee SH, Wu LG (2017) Measuring synaptic vesicle endocytosis in cultured hippocampal neurons. J Vis Exp (127)
26. Fiolka R (2016) Clearer view for TIRF and oblique illumination microscopy. Opt Express 24(26):29556–29567
27. Giannone G et al (2010) Dynamic superresolution imaging of endogenous proteins on living cells at ultra-high density. Biophys J 99 (4):1303–1310
28. Rothbauer U et al (2006) Targeting and tracing antigens in live cells with fluorescent nanobodies. Nat Methods 3(11):887–889
29. Harper CB et al (2016) Botulinum neurotoxin type-A enters a non-recycling pool of synaptic vesicles. Sci Rep 6:19654
30. Nair D et al (2013) Super-resolution imaging reveals that AMPA receptors inside synapses are dynamically organized in nanodomains regulated by PSD95. J Neurosci 33 (32):13204–13224
31. Giannone G et al (2013) High-content super-resolution imaging of live cell by uPAINT. Methods Mol Biol 950:95–110
32. Kechkar A et al (2013) Real-time analysis and visualization for single-molecule based super-resolution microscopy. PLoS One 8(4):e62918
33. Tinevez JY et al (2017) TrackMate: an open and extensible platform for single-particle tracking. Methods 115:80–90
34. Kubala MH et al (2010) Structural and thermodynamic analysis of the GFP:GFP-nanobody complex. Protein Sci 19 (12):2389–2401
35. Durisic N et al (2014) Single-molecule evaluation of fluorescent protein photoactivation efficiency using an in vivo nanotemplate. Nat Methods 11(2):156–162
36. Craig AM, Banker G (1994) Neuronal polarity. Annu Rev Neurosci 17:267–310
37. Kaech S, Banker G (2006) Culturing hippocampal neurons. Nat Protoc 1(5):2406–2415
38. Chenouard N et al (2014) Objective comparison of particle tracking methods. Nat Methods 11(3):281–289
39. Izeddin I et al (2012) Wavelet analysis for single molecule localization microscopy. Opt Express 20(3):2081–2095

40. Racine V et al. (2006) Multiple-target tracking of 3D fluorescent objects based on simulated annealing. In: 2006 3rd IEEE international symposium on biomedical imaging: macro to nano, vol 1–3, pp 1020–1023
41. Kerr RA et al (2008) Fast Monte Carlo simulation methods for biological reaction-diffusion systems in solution and on surfaces. SIAM J Sci Comput 30(6):3126
42. Wilhelm BG et al (2014) Composition of isolated synaptic boutons reveals the amounts of vesicle trafficking proteins. Science 344 (6187):1023–1028
43. Mutch SA et al (2011) Protein quantification at the single vesicle level reveals that a subset of synaptic vesicle proteins are trafficked with high precision. J Neurosci 31(4):1461–1470
44. Takamori S et al (2006) Molecular anatomy of a trafficking organelle. Cell 127(4):831–846
45. Meunier FA et al (2010) Sustained synaptic-vesicle recycling by bulk endocytosis contributes to the maintenance of high-rate neurotransmitter release stimulated by glycerotoxin. J Cell Sci 123(Pt 7):1131–1140
46. Nwabuisi-Heath E, LaDu MJ, Yu C (2012) Simultaneous analysis of dendritic spine density, morphology and excitatory glutamate receptors during neuron maturation in vitro by quantitative immunocytochemistry. J Neurosci Methods 207(2):137–147
47. Grillo FW et al (2013) Increased axonal bouton dynamics in the aging mouse cortex. Proc Natl Acad Sci U S A 110(16):E1514–E1523
48. Truckenbrodt S, Rizzoli SO (2015) Synaptic vesicle pools: classical and emerging roles. In: Mochida S (ed) Presynaptic terminals. Springer, Tokyo
49. Iwabuchi S et al (2014) Examination of synaptic vesicle recycling using FM dyes during evoked, spontaneous, and miniature synaptic activities. J Vis Exp (85)
50. Raingo J et al (2012) VAMP4 directs synaptic vesicles to a pool that selectively maintains asynchronous neurotransmission. Nat Neurosci 15(5):738–745
51. Hua Z et al (2011) v-SNARE composition distinguishes synaptic vesicle pools. Neuron 71(3):474–487
52. Pan PY, Marrs J, Ryan TA (2015) Vesicular glutamate transporter 1 orchestrates recruitment of other synaptic vesicle cargo proteins during synaptic vesicle recycling. J Biol Chem 290(37):22593–22601
53. Kwon SE, Chapman ER (2011) Synaptophysin regulates the kinetics of synaptic vesicle endocytosis in central neurons. Neuron 70 (5):847–854
54. Granseth B et al (2006) Clathrin-mediated endocytosis is the dominant mechanism of vesicle retrieval at hippocampal synapses. Neuron 51(6):773–786
55. Voglmaier SM et al (2006) Distinct endocytic pathways control the rate and extent of synaptic vesicle protein recycling. Neuron 51(1):71–84
56. Ramirez DM et al (2012) Vti1a identifies a vesicle pool that preferentially recycles at rest and maintains spontaneous neurotransmission. Neuron 73(1):121–134
57. Harper CB et al (2011) Dynamin inhibition blocks botulinum neurotoxin type A endocytosis in neurons and delays botulism. J Biol Chem 286(41):35966–35976
58. Pelkmans L et al (2004) Caveolin-stabilized membrane domains as multifunctional transport and sorting devices in endocytic membrane traffic. Cell 118(6):767–780
59. Haas BL et al (2015) Single-molecule tracking in live Vibrio cholerae reveals that ToxR recruits the membrane-bound virulence regulator TcpP to the toxT promoter. Mol Microbiol 96(1):4–13
60. Jiang M, Chen G (2006) High Ca2+-phosphate transfection efficiency in low-density neuronal cultures. Nat Protoc 1 (2):695–700
61. Zeitelhofer M et al (2007) High-efficiency transfection of mammalian neurons via nucleofection. Nat Protoc 2(7):1692–1704
62. Belevich I et al (2016) Microscopy image browser: a platform for segmentation and analysis of multidimensional datasets. PLoS Biol 14 (1):e1002340
63. Sudhof TC, Rothman JE (2009) Membrane fusion: grappling with SNARE and SM proteins. Science 323(5913):474–477
64. Chanaday NL, Kavalali ET (2018) Presynaptic origins of distinct modes of neurotransmitter release. Curr Opin Neurobiol 51:119–126
65. Maritzen T, Haucke V (2018) Coupling of exocytosis and endocytosis at the presynaptic active zone. Neurosci Res 127:45–52
66. Heller JP, Rusakov DA (2017) The nanoworld of the tripartite synapse: insights from super-resolution microscopy. Front Cell Neurosci 11:374
67. Monnier N et al (2015) Inferring transient particle transport dynamics in live cells. Nat Methods 12(9):838–840
68. Persson F et al (2013) Extracting intracellular diffusive states and transition rates from single-molecule tracking data. Nat Methods 10 (3):265–269

Chapter 6

Approaching Protein-Protein Interactions in Membranes Using Single-Particle Tracking and Packing Coefficient Analysis

Marianne Renner and Antoine Triller

Abstract

Molecules diffuse randomly in the plane of the plasma membrane due to membrane fluidity. However, interactions with other molecules may introduce noticeable changes in diffusion behavior such as a slowdown or immobilization by formation of complexes. As a consequence, the analysis of transitions between diffusive behaviors can provide effective rate constants of molecular interactions. Tracking the positions of labeled single molecules, i.e., using single-particle tracking (SPT), is particularly pertinent for this kind of studies. We provide here a step-by-step protocol for SPT experiments and analyses needed to estimate the effective rate constants of molecular interactions. Classical SPT data analyses using the mean square displacement provide the average diffusive behavior, thus precluding the analysis of transitions. We propose an alternative approach to overcome this problem, namely the packing coefficient (*Pc*) analysis. We illustrate the application of this method to the interactions of neurotransmitter receptors with their scaffolding proteins.

Key words Single-particle tracking, Protein-protein interactions, Lateral diffusion, Confinement, Diffusion-capture, Scaffolding interactions

1 Introduction

Membrane molecules exhibit complex lateral diffusion behaviors that are the consequence of the local heterogeneity of the membrane and the interactions established with other molecules. The formation of membrane domains enriched with given membrane proteins results from the interactions of the latter with other molecules which transiently impede their lateral diffusion. The kinetics of these protein-protein interactions in the cell cannot be measured using classical bulk biochemistry approaches since they tend to favor strong interactions and the experimental conditions are different from the physical constraints in cells [1]. This is particularly important in case of molecular interactions in cell membranes,

Nobuhiko Yamamoto and Yasushi Okada (eds.), *Single Molecule Microscopy in Neurobiology*, Neuromethods, vol. 154, https://doi.org/10.1007/978-1-0716-0532-5_6,

which is a badly mixed system (diffusion of receptors in 2D, obstacles to diffusion, etc.) involving a small number of molecules.

The analysis of transitions between different diffusive sequences (e.g., Brownian-like to confined and vice versa) may give access to the effective rate constants of molecular interactions. The experimental strategy is to follow the same molecule over a long period of time in order to observe transitions of diffusion behavior. SPT provides the high pointing accuracy needed to localize molecules in sub-micrometer membrane domains. Long trajectories of individual molecules can be obtained using stable tracers such as quantum dots (QD [2]).

We propose a method for the analysis of QD-based SPT data in order to estimate the effective rate constants of an interaction between a diffusing receptor and immobile membrane-associated proteins. The analytical approach employs what we named the *Pc* analysis [3]. *Pc* parameter allows the detection and quantification of receptor transient confinement sequences (i.e., their frequency and duration) likely to correspond to interactions with their scaffolding protein(s) [4–6]. It is then possible to derivate their effective kinetic rates even if in case of weak and transient interactions. Actually, short-lived interactions are difficult to detect with classical bulk methods (such as co-immunoprecipitation, mass spectrometry, or isotitration calorimetry [1]) although they are important factors being regulated during cellular regulations such as synaptic plasticity.

We focus here on the immobilization (stabilization) of receptors for neurotransmitters on the postsynaptic side of neuronal synapses. More precisely, we take the example of inhibitory glycinergic synapses [4, 7]. Briefly, glycine receptors (GlyR) interact with the scaffolding protein Gephyrin [8, 9]. In young neurons, the co-transfection of Gephyrin with a modified myc-tagged α1 subunit of GlyR (α1βgb, containing the binding sequence to Gephyrin) leads to the formation of receptor-Gephyrin clusters at the plasma membrane in the absence of presynaptic terminals [4, 7]. This is a convenient system to analyze the interactions that stop GlyR diffusion on top of postsynaptic Gephyrin scaffolds.

We provide here a step-by-step protocol to obtain SPT data compatible with *Pc* analysis and we illustrate its application in a model of neuronal synapse formation (*see* **Note 1**). The workflow is shown in Fig. 1. Advantages and limitations of *Pc* analysis are listed in Table 1.

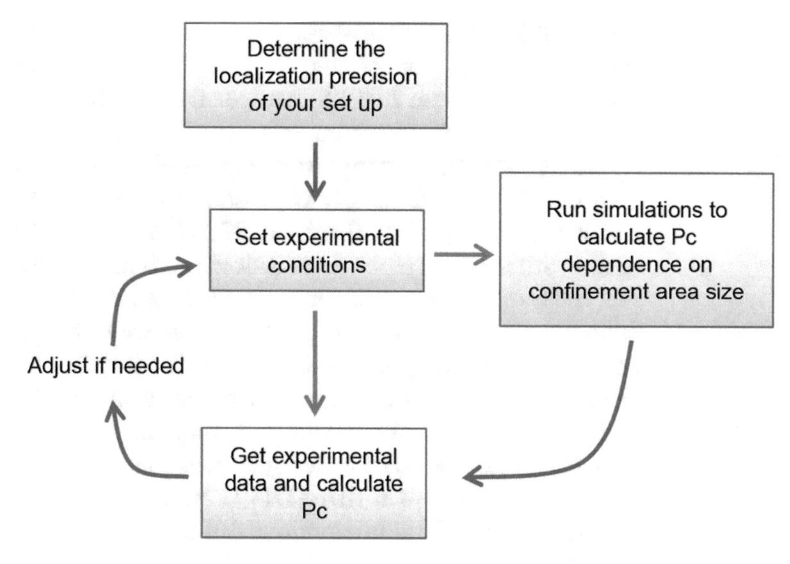

Fig. 1 Workflow of the method

Table 1
Strength and weaknesses of *Pc* analysis, compared to *MSD* analysis and confinement index calculation [22]

	MSD	Confinement index	*Pc*
Discrimination between directed, free, anomalous, and confined diffusion	Yes	No	No
Spatial and temporal localization of confinement period along the trajectory	No	Yes[a]	Yes
Independence from diffusion values (i.e., slow or fast molecules)	No	No	Yes
Determination of confinement surface area	Yes[b]	No	Yes

Notes: (a) Only when *D* is substantially different between free diffusion and confinement period [3]. (b) Only when the confinement period lasts during the whole trajectory [3]

2 Materials and Equipment

2.1 Neuronal Cell Culture and Transfection

Monolayer cell cultures are particularly suitable for SPT due to the absence of imaging issues such as light scattering and out-of-focus fluorescence. Nevertheless, experiments could be done on slices [10]. The application presented here uses neuronal primary cultures from rat hippocampus (references in [11]).

The day before, coat coverslips with 80 μg/ml poly-D,L-ornithine (Sigma Aldrich, Lyon, France). On the day of the culture, prepare fresh media (dissection medium: 20 mM HEPES, HBSS 1× in water; attachment medium: glutamine 2 mM, horse serum 10%, sodium pyruvate 1 mM in MEM; culture medium: glutamine

2 mM, B27 1× in Neurobasal medium) and sterile PBS. Keep all the media cold. Dissect the embryos (E18–19, Sprague-Dawley rat) and place them in PBS, on ice. Dissect the brains and place them in dissection medium, on ice. Dissect several hippocampi (typically two hippocampi for one 12-well plate) in dissection medium. Proceed as rapidly as possible to reduce cell death. Trypsinize hippocampi for 10 min at 37 °C and gently dissociate the tissue with a pipet (pipet up and down ~20 times). After counting the cells, dilute the solution of dissociated cells in the required volume of attachment medium (1 ml per well, for 12-well plates) and plate at a density of ~6–10 × 10^4 cells/cm^2. 2–4 h later, change the medium to culture medium (1 ml per well for a 12-well plate). Cultures can be maintained for 3–4 weeks in the incubator (37 °C, 5% CO_2). In this kind of mixed glia-neuronal cultures, synaptogenesis starts at 4–5 days in vitro (DIV) and it is completed by 21 DIV.

Receptor-scaffold interactions can be studied on artificial postsynaptic scaffolds before synapse formation [7]. To do so, transfect neurons at 2 DIV with Venus-tagged Gephyrin (Ve::Ge) together with the glycine receptor chimera α1βgb (for descriptions of the constructs, *see* [7]). Artificial postsynaptic scaffolds are formed in 24 h. Transfections can be done with, for example, Lipofectamine 2000 (Invitrogen, Cergy-Pontoise, France). Transfection efficiency in primary neuronal cultures using this method is low (~7 or less transfected cells/cm^2, 10–20 transfected neurons per coverslip) but the amount of transfected cells is sufficient for SPT acquisitions.

2.2 Reagents

2.2.1 Culture Reagents and Media

- Poly-ornithine (Sigma, ref. P3655) stock solution (1 mg/ml in water).
- D-PBS (10×, Ref. 14200-067 Gibco, Fisher Scientific, Illkirch, France).
- HEPES (Ref. 15630-56 Gibco, Fisher Scientific, Illkirch, France).
- HBSS 10× without calcium, magnesium, phenol red (10×) (Ref. 14185052 Gibco, Fisher Scientific, Illkirch, France).
- MEM without glutamine (Ref. 11550556 Gibco, Fisher Scientific, Illkirch, France).
- Sodium pyruvate (Ref. 11360-039 Gibco, Fisher Scientific, Illkirch, France).
- Glutamine (Ref. 25030-024, Fisher Scientific, Illkirch, France).
- Horse serum (Ref. 16050122 Gibco, Fisher Scientific, Illkirch, France).
- Neurobasal medium (Ref. 21103049 Gibco, Fisher Scientific, Illkirch, France).
- B27 (Ref. 17504-044, Fisher Scientific, Illkirch, France).
- Trypsin (2.5%), no phenol red (Ref. 15090046 Gibco, Fisher Scientific, Illkirch, France).

2.2.2 Preparation of Pre-coupled QD

- Goat anti-rabbit *F(ab′)*2-tagged QDs emitting at 655 nm (Ref. Q11422MP Invitrogen, Fisher Scientific, Illkirch, France).
- Primary antibody: anti-c-MYC rabbit monoclonal clone Y69 (Cat. No. 790-4628, Roche Diagnostics, Meylan, France).
- Casein 10× (Ref. SP5020, Vector Labs, CA, USA).

2.2.3 Media for Imaging (Neuronal Cultures): MEMv

- Phenol red-free MEM (Ref. 51200-046 Gibco, Fisher Scientific, Illkirch, France).
- 33 mM Glucose (Ref. G5767, Sigma Aldrich, Merck, Darmstadt, Germany).
- 20 mM HEPES (Ref. 15630-56 Gibco, Fisher Scientific, Illkirch, France).
- 2 mM Glutamine (Ref. 25030-024, Fisher Scientific, Illkirch, France).
- 1 mM Sodium pyruvate (Ref. 11360-039 Gibco, Fisher Scientific, Illkirch, France).
- 1× B27 (Ref. 17504-044, Fisher Scientific, Illkirch, France).

2.3 Equipment

- IX70 inverted microscope (Olympus France, France) equipped with a 60× objective (NA 1.45; Olympus France) and a heating chamber (Chamlide CU-501, Live Cell Instrument, Seoul, Korea).
- Xenon lamp (75 W, XCite Series 120Q, Lumen Dynamics, Excelitas Technologies, Wiesbaden, Germany).
- Filters to be used: for QD: FF01-460/60-25, FF510-DiO1-25x36, FF01-655/15-25, Semrock, IDEX Health & Science, New York, USA. For GFP: HQ500/20, HQ535/30m; Chroma Technology, Olching, Germany.
- EMCCD camera (Cascade 512BFT, Roper Scientific, Evry, France).
- Personal computer (Dell Precision T1700, 8 GB RAM, TX, USA) equipped with Matlab software (Matlab 2015b, Mathworks, Natick, USA).

3 Methods

3.1 Cell Labeling and Imaging

Prepare pre-coupled QDs and media ideally on the day of the experiment. These solutions may be kept for a couple of days.

3.1.1 Preparation of Pre-coupled QD (Stock Solution)

Incubate 1 μl of *F(ab′)*2-tagged QDs with the primary antibody (ideally at a stoichiometry of less than one antibody molecule per QD) for 30 min in PBS (final volume 9 μl) at room temperature

(RT) with gentle agitation. Add casein 10× (1 μl) to block unspecific binding of antibodies and continue the incubation at RT for 15 min (*see* **Note 2**).

3.1.2 Preparation of Imaging Medium (MEMv)

Dilute the reagents in phenol red-free MEM and heat the solution to 37 °C. Avoid freezing and thawing B27 and glutamine reagents; prepare single-use aliquots to be kept at −20 °C. This medium can be replaced by any other suitable for the cells being studied (*see* **Note 3**).

3.1.3 Cell Labeling

Before starting the experiment, be sure that the heating system of the imaging setup is equilibrated at 37 °C (e.g., inside the environmental chamber). Incubate cells with pre-coupled QDs (start with a stock solution diluted to 1:600–1:1000 in MEMv) for 5 min at 37 °C. Rinse twice with 1 ml MEMv. The concentration of pre-coupled QDs must be adjusted to be able to detect and track correctly as many QDs as possible (*see* next section).

3.2 Imaging and Acquisition Parameters

3.2.1 Evaluation of Localization Precision

Coat a coverslip with a drop of QDs (dilute a QD solution at 1:50,000 or more). Once dried, the QDs will be immobilized; then acquire sets of frames at the same frequency and illumination conditions as those for cell preparations. Figure 2a illustrates the sparsity of QD labeling ideally needed to correctly detect and track single molecules.

Set the labeling conditions and image acquisition to detect individual QDs with an optimum signal-to-noise ratio (Fig. 2a). Choose maximal linear EM (electron multiplying) gain on the EMCCD camera and set the acquisition frequency for an optimal trade-off between getting enough photons for an optimal detection and the best possible temporal resolution (*see* **Note 4**).

Detect and localize single immobile QDs with a tracking software (*see* Subheading 3.3) to analyze the distribution of positions. The localization precision corresponds to the standard deviation of the positions [12] and should be of ~10–25 nm (in the illustrated example, it was 21 nm for QDs).

Free detection and tracking software are available, e.g., MTT (implemented in Matlab or Octave [13]) and TrackMate (for Fiji [14]). In our experiments, we have used a homemade software (SPTrack_v5) in Matlab (MathWorks, Natick, MA), available upon request. In the abovementioned software, detection and tracking approaches are based on a least-squared fit to a two-dimensional Gaussian point spread function in order to determine the coordinates of each QD.

The quality of labeling and detection can be assessed by analyzing the resulting fluorescent spots (blinking, size, intensity, and background, Fig. 2b, c). Individual QDs are identified by their blinking behavior ([2], Fig. 2b). The presence of multiple QDs in single spots can be detected using an intensity histogram: the spots

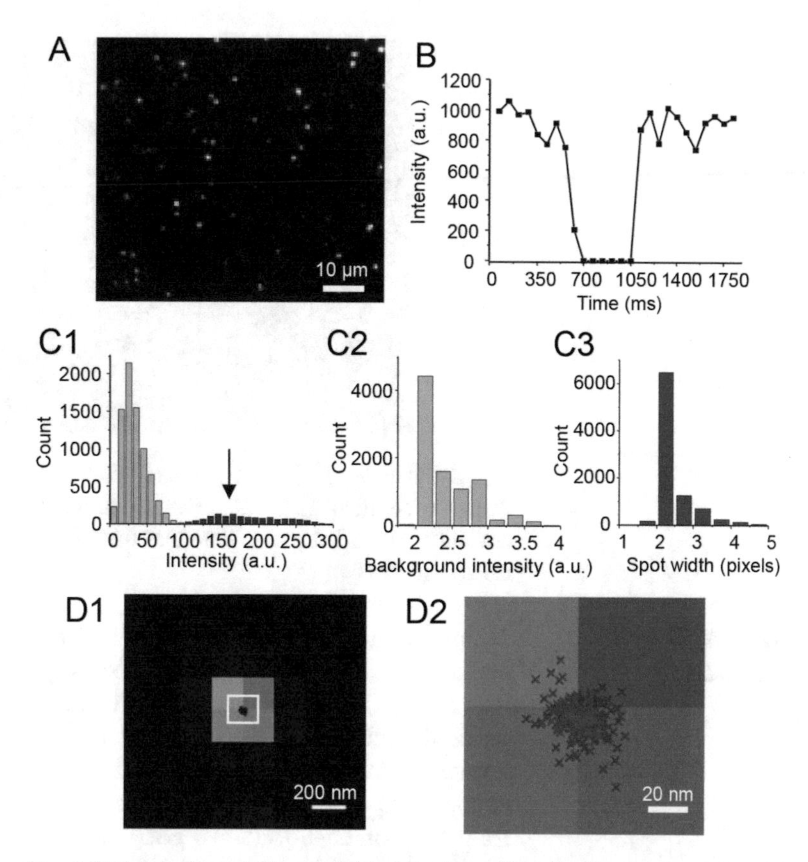

Fig. 2 SPT detection of QDs and determination of the localization precision. (**a**) Example of a recording field showing the fluorescence of QDs dried (immobile) on a coverslip. (**b**) Fluorescence intensity upon time for one QD. Note the blinking period: the light emission is intermittent. (**c**) Histograms for the characterization of QD spots during one recording session: QD intensity (**c1**); background intensity around each QD (**c2**); width of QD spots (**c3**). In (**c1**), the second peak of intensity (arrow) denotes the presence of spots containing more than one QDs. (**d**) Successive detected positions of a single immobile QD along time, overlaid on its fluorescence image (**d1**); (**d2**) higher magnification of the region indicated by the white frame in (**d1**)

with twice or more the value of intensity of most of the signals contain multiple QDs (Fig. 2c1). These spots are discarded (intensity threshold) to obtain individual QD trajectories. From the values of background noise, one can estimate the signal-to-noise ratio (Fig. 2c2). The optimal width (at half maximum intensity) for a spot of one single QD is 2–3 times the width of a pixel, following Nyquist's rule of sampling (Fig. 2c3, *see* **Note 5**). The best magnification to use is such that the pixel size is in the range of the standard deviation of the point spread function (optimal condition for single-molecule localization). If localization precision is higher than ~10–25 nm (Fig. 2d), check the illumination parameters to maximize the signal-to-noise ratio.

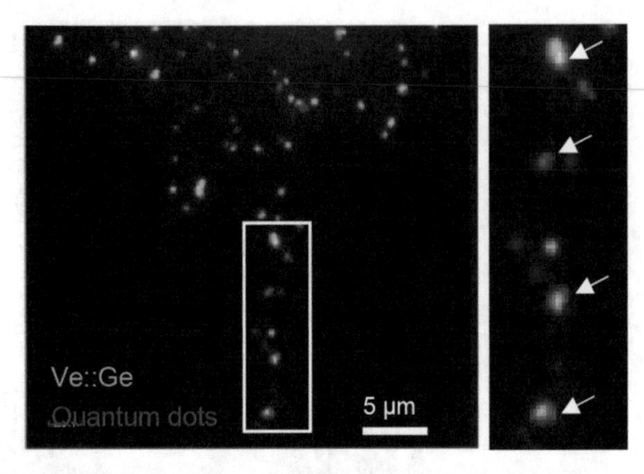

Fig. 3 Detection of QDs on neurons. Image of Ve::Ge (green) and QD (red) fluorescence. Right panel: Magnification of the region delimitated by the white rectangle on the left panel. Note that most QDs are associated with Gephyrin clusters (arrows)

3.2.2 Acquisition of SPT Data on Cells

Maintain samples at a stable physiological temperature (37 °C) to avoid metabolic and membrane viscosity changes. The total duration of the QD stream image acquisition has to be adjusted to each particular interaction duration (*see* below). Ideally, the time interval between images has to maximize the localization precision without losing temporal accuracy (*see* **Note 4**). Typically, an acquisition frequency of 12–30 Hz is a good starting point. Here, in the case of GlyR [7], we have acquired images at 13.3 Hz.

If there are other fluorescent labels in your sample which are excited at the same wavelength than QDs, image them first in order to avoid their bleaching (here, Ve::Ge, Fig. 3). Phototoxicity may be a problem and it has to be controlled. In case of neurons, limit the recording session to 1 h. When cells undergo oxidative stress the autofluorescence of mitochondria increases due to the accumulation of NADH [15], reducing the localization precision.

3.3 Tracking and Analysis of Diffusion

In most tracking routines, the selected spots (those considered as single emitters by their blinking, intensity, and size criteria) in consecutive frames (time points) are likely to be connected to form a trajectory if they are within a distance coherent with the expected diffusion coefficient D (Fig. 4a). When QDs blink, trajectories are interrupted. As a consequence, the initial tracking procedure may yield more than one trajectory per QD. A second tracking procedure has to be executed to reconnect all the trajectories belonging to the same QD.

If the interactions are studied in a defined membrane surface area, trajectories can be chopped to analyze separately portions in and out of these domains. In the proposed example, trajectories are defined as "IN" (in synapses) if they co-localize with Ve::Ge fluorescent clusters (Fig. 4b).

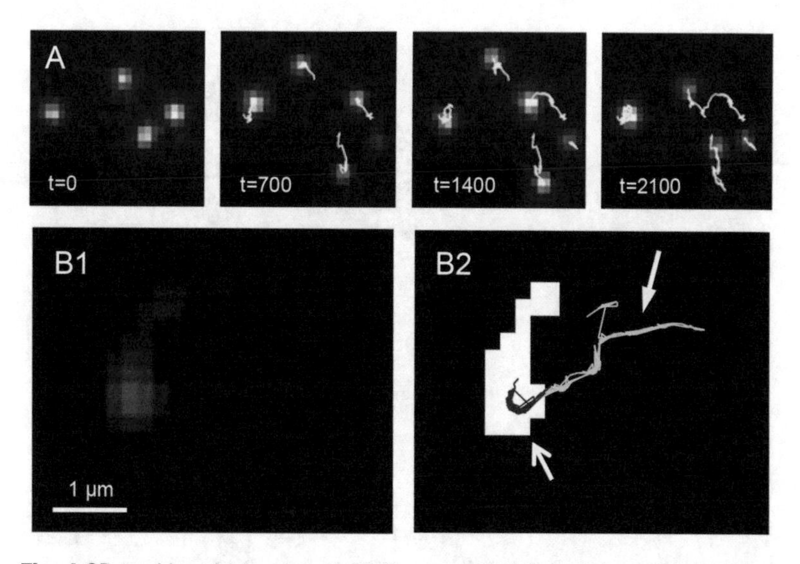

Fig. 4 QD tracking. (**a**) Images of QD fluorescence taken at the indicated times (ms) on the same recording field, overlaid with the trajectories reconstructed by the tracking software (solid line). (**b**) Example of a Ve::Ge fluorescence cluster before (**b1**) and after (**b2**) binarization. In (**b2**), note a 35-s-long trajectory (solid line) entering a Gephyrin cluster (in: dark gray; out: light gray)

3.4 Analysis of Diffusion and Calculation of Effective Rate Constants

The mean square displacement (*MSD*) is a simple and popular way to characterize diffusion [16]. It allows the calculation of a diffusion coefficient (initial slope of the *MSD* vs. time interval plot) and the characterization the global diffusion behavior. In case of Brownian motion, $MSD = 4D\tau$ (in two dimensions) where D is the diffusion coefficient and τ the time interval. Anomalous diffusion is characterized by a nonlinear $MSD = 4D\tau^{\alpha}$ with $\alpha < 1$. In case of confined diffusion the *MSD* curve is asymptotic toward a value related to the surface of the confinement area [16]. Unfortunately *MSD* analysis hides information when a molecule switches between different diffusive behaviors, e.g., immobilized vs. diffusing [3]. The classical *MSD* analysis cannot reveal these switches unless it is derived from portions of trajectory using a sliding time window on the trajectory. However, short trajectories produce noisy and less reliable *MSD* [17, 18]. To overcome this drawback we have developed another parameter, the packing coefficient [3], that can easily be implemented.

3.4.1 Packing Coefficient

A sliding window with a given number of time points (Fig. 5a, b) allows the calculation of the packing coefficient (Pc) at each time point i:

$$Pc_i = \sum_{i}^{i+n-1} \frac{(x_{i+1} - x_i)^2 + (y_{i+1} - y_i)^2}{S_i^2}$$

where x_i and y_i are the coordinates at time i; $x_i + 1$ and $y_i + 1$ are the coordinates at time $i + 1$; n is the length of the time window (in the

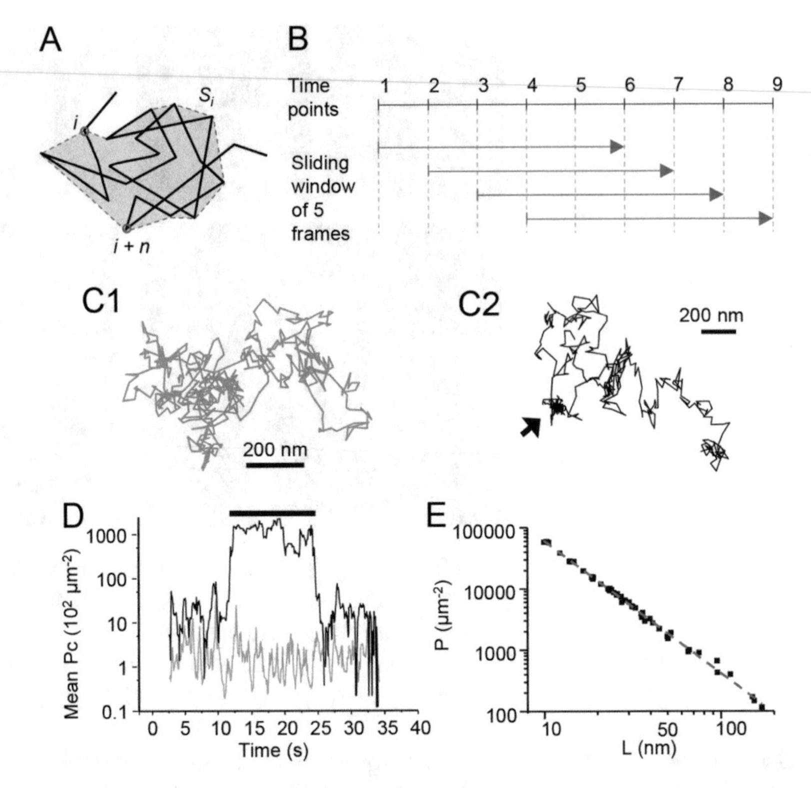

Fig. 5 Determination of *Pc*. (**a**) Schematic representation illustrating the principle of *Pc* calculation at a given time point *i*. S_i is the convex hull of the trajectory segment i–$i + n$. *Pc* is the ratio between the sum of the square displacements between successive time points i to $i + n$ and the squared surface area S_i^2 (convex hull of the trajectory segment i to $i + n$). (**b**) Calculation of *Pc* on a trajectory using a sliding window (here, five frames). *Pc* of time point 1 is calculated on the trajectory segment between points 1 and 6. For point 2, the trajectory segment goes from points 2 to 7, etc. (**c**) Examples of simulated random walk without (**c1**) or with a confinement period (arrow in **c2**). (**d**) *Pc* values of the trajectories in (**c**) (**c1** in gray, **c2** in black). The horizontal line in black indicates the time span of the confinement period. (**e**) Power law fit (dotted line) matching the mean *Pc* vs. the diameter of the confinement area *L* on simulated trajectories

example presented here, $n = 30$ time points); and S_i is the surface area of the convex hull of the trajectory segment between time points i and $i + n$. S_i is calculated using the *convhull* function of Matlab. The length of the sliding window is a trade-off: it has to be short enough to detect short stabilization events but long enough to detect confinement and to decrease statistical associated uncertainty [3]. For the experimental data presented here, we set this value to 30 time points. The blinking behavior of QDs introduces additional uncertainty in the calculation of *Pc*. Therefore, discard *Pc* values if the dark state of a QD lasted 20% or more of the corresponding sliding window length.

Pc scales inversely to the size of the confinement area *L* (Fig. 5c–e) following a power law that can be assessed by fitting *Pc* values obtained from simulated trajectories [3] (Fig. 5e). The procedure is explained in Subheading 3.4.2. Next, determine a threshold value of *Pc* (Pc_{thresh}) that corresponds to a given confinement area L_{thresh}. If the interaction studied results in the immobilization of the molecule, L_{thresh} is the localization precision (*see* **Note 6**). Periods with *Pc* above Pc_{thresh} correspond to the time in which the molecule is confined in an area smaller than L_{thresh}.

Brownian diffusion can temporarily mimic confinement due to random fluctuations of the length of the displacements; thus trajectories may display periods that look like confinement to the eye (Fig. 5c). To improve the detection of confinement periods, it is advisable to set a time threshold (t_{thresh}) that reduces spurious detections due to statistical fluctuations of *Pc* along the trajectory [3]. Therefore, a confinement period corresponds to a sequence that lasts more than t_{thresh}, in which $Pc > P_{\text{thresh}}$. P_{thresh} is set to the P95 or P99 percentile of the *Pc* distribution of simulated random walks. These simulations have to match the experimental data in acquisition frequency, localization precision, and length of trajectories. The duration time threshold t_{thresh} is then chosen by applying P_{thresh} to random walk trajectories and extracting the P95 or P99 of the distribution of durations. Simulations are explained in the next section.

3.4.2 Determination of Thresholds to Detect Transitions

– *Monte Carlo simulations*

Generate simulated random walks and confined trajectories matching the experimental data in terms of acquisition frequency, localization precision, and length (Fig. 5c) as in Renner et al. [3, 19]. Briefly, the *x* and *y* components of the *i*th displacement step in the trajectory are randomly selected from two independent normal distributions with the mean of zero and the variance equal to $2D_{sim}\,\Delta t$. The noise introduced in SPT trajectories by the limited accuracy of localization is simulated by adding a distance to *x* and *y*. Choose this distance randomly from an independent normal distribution with the mean of zero and a variance that corresponds to the localization precision. Here the distance is calculated independently for *x* and *y* at each time point. Periods of confinement are imposed by inserting a boundary at a given time point during the simulation and removing it at a later prescribed time (Fig. 5c2). During this period, positions are forced to stay within a circle of the selected diameter *L* and reflecting border. Calculate the mean *Pc* for different values of *L* in order to plot *Pc* vs. *L* (Fig. 5e) and fit the resulting data (linear fit of log *L* vs. log *Pc*). With our data,

$$\log L \cong 3.2 - 0.46 \log Pc$$

From this relationship it is possible to set the threshold Pc_{thresh} that corresponds to a given confinement area size.

– *Pc threshold to detect immobilization periods*

Since the localization precision in SPT acquisitions on cells may be impaired by background noise not present in the preparation of dried QDs, we advise to set a Pc_{thresh} to twice the localization precision measured in the dried preparation. In our case, $Pc_{thresh} = 5600\ \mu m^{-2}$ (corresponding to a confinement area size with a diameter of ~42 nm) and t_{thresh} of 0.375 s (5 time points).

In the illustrated example, QD-bound α1βgb trajectories are not confined or they display transient immobilization (stabilization) periods (Fig. 6a). QD-α1βgb diffusion is reduced on top of Ve::Ge clusters (Fig. 6b, synaptic domain). The median diffusion coefficient D decreases from 1.85 to $1.46 \times 10^{-2}\ \mu m^2/s$ outside and within clusters, respectively. However, the population of trajectories on top of clusters is heterogeneous, since receptors can still diffuse when they are not stabilized by the scaffold. Trajectories within synaptic domains were then sorted into "stabilized" (those undergoing at least one period of stabilization) and "not stabilized" subgroups. Stabilized trajectories (62.84 ± 3.55% of the trajectories) display a median D of $3.28 \times 10^{-4}\ \mu m^2/s$, whereas the median D of non-stabilized trajectories is ~100-fold higher ($3.11 \times 10^{-2}\ \mu m^2/s$) (Fig. 6b).

3.4.3 Calculation of Effective Rate Constants

Assuming that confinement periods arise from interaction with a scaffolding protein, the effective forward binding rate k_{on} is defined as the frequency of the binding events. The effective backward binding rate k_{off} is defined as the reciprocal of the mean duration of the stabilization periods [3]. More specifically, in our experiments QD-α1βgb undergoes 4.66 ± 0.20 stabilization events per minute, thus indicating an effective $k_{on} = 7.76 \times 10^{-2}\ s^{-1}$. The distribution duration of the stabilization events is broad, with a mean of 5.48 ± 8.64 (mean ± SD that corresponds to a k_{off} of $0.18\ s^{-1}$). These data confirm previous studies showing that Gephyrin scaffolds act as shallow energy traps for GlyR, despite the strong affinity for Gephyrin of these receptors [5, 9, 20].

4 Notes

1. Another SPT protocol is presented by Hiroko Bannai in Chap. 7.
2. Pre-coupling is not suitable for all antibodies. Check the level of nonspecific labeling using non-coupled QDs incubated only with casein (in the absence of primary antibody). If you do not get a specific labeling, an alternative is to use streptavidin-

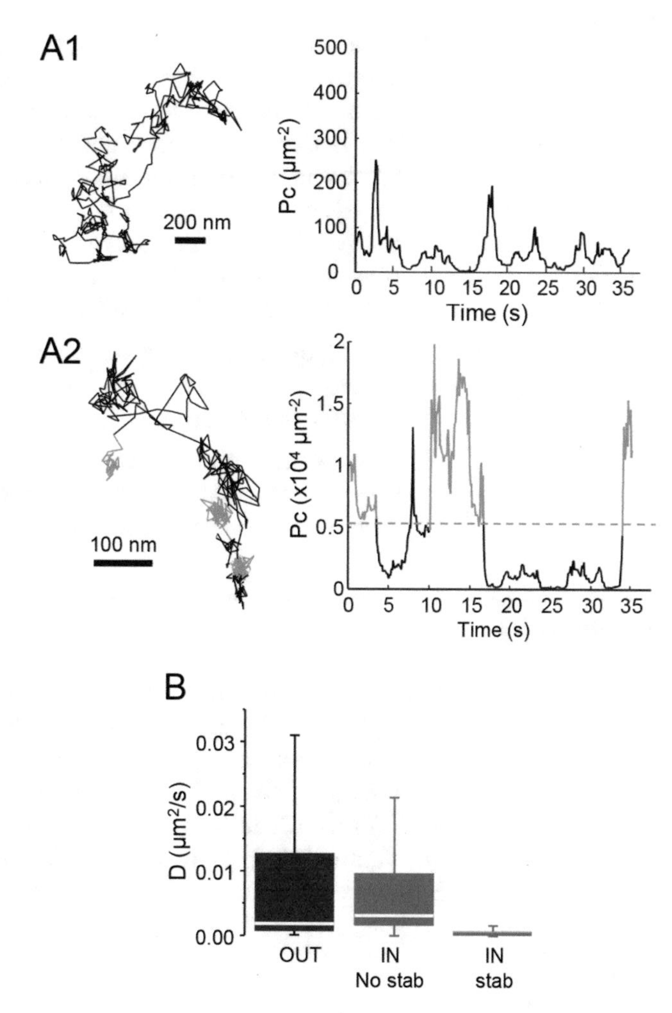

Fig. 6 Interaction of GlyR-binding sequence (α1βgb) with Gephyrin clusters during the recording session. (**a**) Examples of QD-α1βgb trajectories without (**a1**) or with stabilization periods (**a2**) with their *Pc* vs. time plots. (**a2**) Trajectory alternating between stabilization periods (gray) and non-confined diffusion (black). (**b**) Distribution of diffusion coefficients (*D*; median, box: 25–75% IQR, whiskers: 5–95%) for QD-α1βgb trajectories outside (OUT) or on top of clusters being not stabilized (IN no stab) or stabilized (IN stab)

coupled QDs and a sequential labeling of cells with primary antibody, a biotinylated secondary antibody, and streptavidin-coupled QDs [21].

3. In case of neurons, the composition of the culture media (in particular the concentration of ions that determine the membrane resting potential) may alter their activity. The medium presented here is appropriate for standard experiments. It can be modified as needed; the only requirements are that it has to be nonfluorescent (i.e., without phenol red) and must contain a buffer system.

4. When analyzing interactions, the acquisition frequency and the length of recordings depend on the duration of the interactions. Higher frequency of acquisition allows better tracking and better detection of short-lasting interactions, but it also results in lower localization precision. With respect to the length of recordings, in our case ~5% of the events lasted for the whole recording session (35.175 s), hence introducing a bias in k_{off} measurements. Longer acquisitions, however, increase the possibility of light-induced toxicity.
5. The Nyquist rule states that for sufficient sampling, a minimum of two data points (pixels with fluorescent signal) are needed per unit of resolution (spot).
6. Due to the limited localization precision in SPT, immobilization is translated into confinement in an area which size is within the localization precision [12].

Acknowledgments

This work was supported by the Agence Nationale de la Recherche "Synaptune" (Programme blanc, ANR-12-BSV4-0019-01), the ERC advanced research grant "PlasltInhib," the program "Investissements d'Avenir" (ANR-10-LABX-54 MEMOLIFE and ANR-11-IDEX-0001-02 PSL Research University), and the Institut National de la Santé et de la Recherche Médicale (INSERM). Authors declare no conflict of interest.

References

1. Kasai RS, Kusumi A (2014) Single-molecule imaging revealed dynamic GPCR dimerization. Curr Opin Cell Biol 27:78–86
2. Pinaud F, Clarke S, Sittner A, Dahan M (2010) Probing cellular events, one quantum dot at a time. Nat Methods 7:275–285
3. Renner M, Wang L, Levi S, Hennekinne L, Triller A (2017) A simple and powerful analysis of lateral subdiffusion using single particle tracking. Biophys J 113:2452–2463
4. Ehrensperger MV, Hanus C, Vannier C, Triller A, Dahan M (2007) Multiple association states between glycine receptors and gephyrin identified by SPT analysis. Biophys J 92:3706–3718
5. Dahan M, Levi S, Luccardini C, Rostaing P, Riveau B et al (2003) Diffusion dynamics of glycine receptors revealed by single-quantum dot tracking. Science 302:442–445
6. Specht CG, Grunewald N, Pascual O, Rostgaard N, Schwarz G et al (2011) Regulation of glycine receptor diffusion properties and gephyrin interactions by protein kinase C. EMBO J 30:3842–3853
7. Renner M, Choquet D, Triller A (2009) Control of the postsynaptic membrane viscosity. J Neurosci 29:2926–2937
8. Kneussel M, Loebrich S (2007) Trafficking and synaptic anchoring of ionotropic inhibitory neurotransmitter receptors. Biol Cell 99:297–309
9. Grunewald N, Jan A, Salvatico C, Kress V, Renner M et al (2018) Sequences flanking the gephyrin-binding site of GlyRbeta tune receptor stabilization at synapses. eNeuro 5. https://doi.org/10.1523/ENEURO.0042-17.2018
10. Biermann B, Sokoll S, Klueva J, Missler M, Wiegert JS et al (2014) Imaging of molecular surface dynamics in brain slices using single-particle tracking. Nat Commun 5:3024
11. Renner M, Schweizer C, Bannai H, Triller A, Levi S (2012) Diffusion barriers constrain receptors at synapses. PLoS One 7:e43032

12. Manzo C, Garcia-Parajo MF (2015) A review of progress in single particle tracking: from methods to biophysical insights. Rep Prog Phys 78:124601
13. Serge A, Bertaux N, Rigneault H, Marguet D (2008) Dynamic multiple-target tracing to probe spatiotemporal cartography of cell membranes. Nat Methods 5:687–694
14. Tinevez JY, Perry N, Schindelin J, Hoopes GM, Reynolds GD et al (2017) TrackMate: an open and extensible platform for single-particle tracking. Methods 115:80–90
15. Bartolome F, Abramov AY (2015) Measurement of mitochondrial NADH and FAD autofluorescence in live cells. Methods Mol Biol 1264:263–270
16. Saxton MJ, Jacobson K (1997) Single-particle tracking: applications to membrane dynamics. Annu Rev Biophys Biomol Struct 26:373–399
17. Michalet X (2010) Mean square displacement analysis of single-particle trajectories with localization error: Brownian motion in an isotropic medium. Phys Rev E Stat Nonlinear Soft Matter Phys 82:041914
18. Michalet X, Berglund AJ (2012) Optimal diffusion coefficient estimation in single-particle tracking. Phys Rev E Stat Nonlinear Soft Matter Phys 85:061916
19. Renner M, Domanov Y, Sandrin F, Izeddin I, Bassereau P et al (2011) Lateral diffusion on tubular membranes: quantification of measurements bias. PLoS One 6:e25731
20. Masson JB, Dionne P, Salvatico C, Renner M, Specht CG et al (2014) Mapping the energy and diffusion landscapes of membrane proteins at the cell surface using high-density single-molecule imaging and Bayesian inference: application to the multiscale dynamics of glycine receptors in the neuronal membrane. Biophys J 106:74–83
21. Cantaut-Belarif Y, Antri M, Pizzarelli R, Colasse S, Vaccari I et al (2017) Microglia control the glycinergic but not the GABAergic synapses via prostaglandin E2 in the spinal cord. J Cell Biol 216:2979–2989
22. Simson R, Sheets ED, Jacobson K (1995) Detection of temporary lateral confinement of membrane proteins using single-particle tracking analysis. Biophys J 69:989–993

Chapter 7

Synaptic Function and Neuropathological Disease Revealed by Quantum Dot-Single-Particle Tracking

Hiroko Bannai, Takafumi Inoue, Matsumi Hirose, Fumihiro Niwa, and Katsuhiko Mikoshiba

Abstract

Quantum dot-single-particle tracking (QD-SPT) is a super-resolution imaging technique that uses semiconductor nanocrystal quantum dots as fluorescent probes and is a powerful tool for analyzing protein and lipid behavior in the plasma membrane. Recent QD-SPT experiments have provided critical insight into the mechanism and physiological relevance of membrane self-organization in neurons and astrocytes in the brain. The mobility of some membrane molecules may become abnormal in cellular models of epilepsy and Alzheimer's disease. Based on these findings, we propose that the behavior of membrane molecules reflects the condition of neurons in pathological disease states. In this chapter, we describe the latest, simple QD-SPT technique, which is feasible with epifluorescence microscopy and dissociated cell cultures.

Key words Membrane molecules, Lateral diffusion, Quantum dot, Single-particle tracking, Neuron, Astrocyte, Dissociated culture, Synapse, Receptor, Ion channel, Lipids

1 Introduction

Synapses are specialized structures critical for the formation and retrieval of memory. Efficient synaptic transmission is enabled by high-density localization of neurotransmitter receptors on the postsynaptic membrane, while synaptic components such as receptors and scaffold proteins are continuously replenished [1]. The integrity of postsynaptic structures and ultimately synaptic transmission is determined by a complex dynamic equilibrium that includes balance between protein synthesis and degradation as well as balance between endocytosis and exocytosis to yield a certain number of proteins or receptors at the plasma membrane [2, 3]. Another important factor is the lateral diffusion of neurotransmitter receptors. The plasma membrane has a mosaic structure composed of two layers of phospholipids (i.e., a lipid bilayer) in which proteins and other substances such as cholesterol are embedded. At physiological temperatures, phospholipids remain in a fluid state [4],

Nobuhiko Yamamoto and Yasushi Okada (eds.), *Single Molecule Microscopy in Neurobiology*, Neuromethods, vol. 154, https://doi.org/10.1007/978-1-0716-0532-5_7,

allowing transmembrane proteins including postsynaptic receptors to diffuse freely by Brownian movement within the viscous two-dimensional phospholipid matrix [5]. The question of how receptors overcome free diffusion to form the postsynaptic density (PSD) has become a fundamental question in the field of neuroscience.

To address this question, the single-molecule imaging technique has emerged as a powerful tool to visualize the dynamics of endogenous membrane molecules. The largest advantage of single-molecule imaging is that it avoids ensemble averaging of multiple molecules. Unique behaviors that take place within subpopulations of molecules and their transient behavior can be captured by imaging at single-molecule resolution. Single-molecule imaging techniques include single-particle tracking (SPT) using particles such as beads or nanoparticles as a label, and single-fluorophore tracking (SFT) in which target molecules are labeled with organic fluorophores, or with genetically encoded fluorescent proteins. Quantum dot-single-particle tracking (QD-SPT) is a hybrid of the SFT and SPT techniques in which quantum dots (QDs), fluorescent nanocrystals that are composed of semiconductor materials, are used as "fluorescent particles" to label the molecule of interest. Fluorescence from a single QD is much brighter and more photostable, allowing single-molecule imaging with a good signal-to-noise ratio for a longer period compared with SFT using chemical dyes and fluorescent proteins. QD has broad absorption spectra and an emission wavelength that is determined by the crystal size (2–9.5 nm in diameter) [6]. QDs for biological applications are conjugated with functional biomolecules such as streptavidin, protein A, antibodies, and biotin through which QD can bind to target molecules such as membrane proteins [7, 8] (Fig. 1a). Phospholipids such as 1,2-dioleoyl-sn-glycero-3-phosphoethanolamine (DOPE) [9] and GM1 ganglioside [10] can also be labeled with streptavidin-biotin (Fig. 1b). The hydrodynamic radius of commercially available streptavidin-conjugated QDs is 4–7 nm [11]; therefore the diameter of a functionalized QD is estimated to be less than 20 nm, which is smaller than the size of the synaptic cleft.

QD-SPT has contributed to addressing questions in the field of neuroscience, such as those regarding synaptic structure and molecular mechanisms underlying learning and memory both in the excitatory synapse [12–18] and in the inhibitory synapse [19–28]. QD-SPT may also highlight abnormal membrane protein dynamics associated with disease such as epilepsy and Alzheimer's disease [29–32]. Although QD-SPT is a powerful and feasible technique as described above, a few limitations should also be considered. QDs targeted to neurotransmitter receptors through the antibodies can reach in the synaptic cleft [15, 19]. However, the QD-antibody-receptor complex penetrates less into the perisynaptic area and shows more suppressed diffusion at narrow synaptic

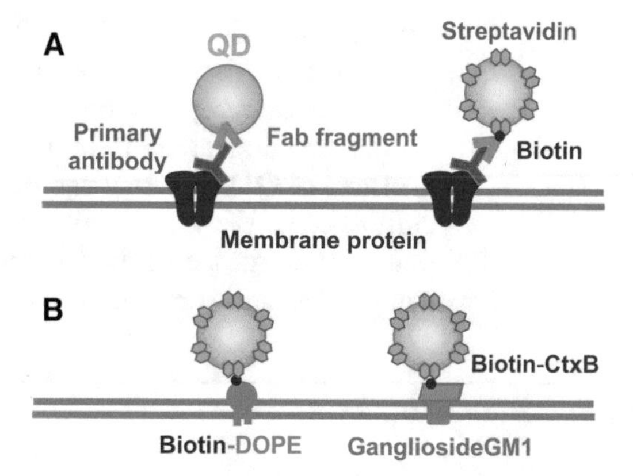

Fig. 1 Strategies to label membrane molecules with QDs. (**a**) Two methods for labeling membrane proteins with QDs. In the first method, the primary antibody against the target protein is incubated with QDs conjugated with the secondary antibodies, and this QD-antibody complex is targeted toward the proteins of interest (left) [7]. The other method uses streptavidin-coated QDs [8] (right). The membrane protein is recognized by the primary antibody, which in turn is recognized by a biotinylated Fab fragment of a secondary antibody. A streptavidin-coated QD then binds to the biotin on the molecule-antibody complex. In this protocol, we describe the second method. (**b**) Methods to label lipids with QDs. The phospholipids (e.g., DOPE) conjugated with biotin are incorporated into the plasma membrane, and streptavidin-coated QDs are bound to the biotin-phospholipid complex (left) [9]. Endogenous GM1-ganglioside is targeted with streptavidin-coated QDs through biotinylated cholera toxin subunit B (CtxB) [10]

clefts compared with receptors tagged with organic dyes [7, 12], suggesting that the hydrodynamic volume of the QD-antibody complex affects the diffusion and penetration properties of the target molecules at highly confined areas such as synaptic clefts. Recently, new QD-SPT using small QDs (sQDs) with thin and stable ligand layer has been developed [33]. This technique revealed that the localization and behavior of sQD-conjugated AMPA receptors (AMPARs) were comparable to fluorescent dyes and were significantly different from those observed using larger conventional QDs: larger population sQDs were targeted to AMPARs in postsynaptic cleft, and the sQD-AMPAR complex was confined in "nanodomains" in postsynaptic density with lifetime longer than 15 min [33, 34]. It is important to note that although sQDs currently need to be produced by the researchers who require it, one can regulate their size. Reducing the size of QDs will greatly improve its accessibility to synaptic clefts and will significantly contribute to overcome its drawbacks. Another drawback of the QD-SPT technique is that the oligomerization status of the target molecule cannot be determined, except when the diffusion property is highly dependent on its oligomerization states [35], due to the labeling ratio that is kept as low as possible to resolve single QDs. Yet, QD-SPT remains the most powerful technique to track receptor movement for long periods to visualize the transition of receptors between synapses [36] using a relatively simple imaging system. In this section, we describe a standard protocol of QD-SPT that can be performed easily using dissociated culture cells. We will describe here the preparation of the primary neuronal culture, and the labeling protocol of membrane proteins using QDs, and the analysis of QD-SPT data.

2 Materials

2.1 Reagents and Equipment for Primary Cell Culture

- CO_2 incubator.
- 2% (V/V) PEI stock solution (50×): 2 ml of 50% poly(ethyleneimine) (PEI) solution (Sigma-Aldrich, St. Louis, MI; #P3143), 48 ml sterile distilled water; sterile by filtration (pore size of 0.22 μm). Stock PEI can be stored at −20 °C. Preparing small aliquots of 250–750 μl is recommended.
- 200 mM L-Glutamine stock (100×): 1.46 g L-Glutamine and 50 ml sterile distilled water. Sterile by filtration (0.22 μm). Commercially available (e.g., Thermo Fisher, Waltham, MA; #25030081). The solution can be stored at −20 °C for at least 1 year. Preparing small aliquots of 250–750 μl is recommended.
- 1 M N-2-hydroxyethylpiperazine-N-2-ethane sulfonic acid (HEPES) (pH 7.2–7.5): 11.9 g HEPES and 50 ml sterile distilled water. Adjust pH to 7.2–7.5 with NaOH. Sterile filtration (0.22 μm pore). The solution can be stored at 4 °C for 2 years. Commercially available (e.g., Thermo Fisher #15630080).
- 100 mM Sodium pyruvate stock (100×): e.g., Thermo Fisher #11360070. Aliquots (10 ml) can be stored at −20 °C. After thawing, the solution can be maintained at 4 °C for 2 months.
- B-27™ supplement (50×): B-27™ supplement (Thermo Fisher #17504044). This can be replaced by B-27™ plus supplement (Thermo Fisher #A3582801) or MACS NeuroBrew-21 (Miltenyi Biotec, Bergisch Gladbach, Germany; #130-093-566).
- Penicillin-streptomycin solution: Penicillin 10,000 U/ml and streptomycin 10,000 μg/ml (e.g., Thermo Fisher #15140122).
- Dissection medium (49 ml HBSS and 1 ml of 1 M HEPES).
- Cell plating medium (for 4 × 12-well plates): 48 ml Minimum essential medium (MEM, Thermo Fisher #11090-081) supplemented with 1 ml B-27™ supplement, 500 μl of L-glutamine stock (final concentration: 2 mM), 500 μl of sodium pyruvate stock (1 mM), and 25 μl penicillin-streptomycin solution (penicillin 5 U/ml, streptomycin 5 μg/ml). The concentration of the penicillin-streptomycin solution is critical for cell survival.

- Maintenance medium for 4 × 12-well plates: 48.5 ml Neurobasal™-A medium (Thermo Fisher #10888022) supplemented with 1 ml B-27™, 500 μl of L-glutamine stock (final concentration: 2 mM), and 25 μl penicillin-streptomycin solution (penicillin 5 U/ml, streptomycin 5 μg/ml). Neurobasal™ plus medium can be used instead of Neurobasal™-A medium.
- Thickness 1, 18 mm diameter circular coverslips (Karl Hecht "Assistent" GmbH, Altnau, Switzerland; #41001118 and Matsunami Glass IND LTD, Osaka, Japan; #C018001).
- 12-Well multiwell culture plates with low-evaporation rid (Falcon®, Lincoln Park, NJ; #353043).

2.2 Reagents and Equipment for QD Labeling

- 667 mM D-Glucose in MEM (20×): 6 g D-Glucose in 50 ml MEM without phenol red and glutamine (Thermo Fisher #51200038). Sterile filtration (0.22 μm pore). The solution can be stored at 4 °C for at least 6 months when handled under sterile conditions.
- Stock QD-binding buffer (4×): 2.0 g Sodium tetraborate decahydrate (grade ACS reagent 99.5–105.0%); 4.885 g boric acid (grade for molecular biology approx. 99%); and 400 ml distilled water. After adjusting the pH to 8.0 ± 0.2 with 1 M HCl or 1 M NaOH, adjust volume to 500 ml with distilled water. This buffer can be stored at room temperature.
- QD-binding buffer (1×): 5 ml of 4× QD-binding buffer supplemented; 2% (w/v) bovine serum albumin (Sigma-Aldrich #A7030) supplemented with 0.05% (w/v) sodium azide. Sterilized by filtration. This buffer can be stored at 4 °C for several months.
- Primary antibody against extracellular domain of the target protein. Examples are shown in Table 1.
- Fab fragment of secondary antibody conjugated with biotin: e.g., Jackson ImmunoResearch, West Grove, PA; #115-067-003 (mouse); #111-067-003(rabbit), and #112-067-003(rat). These antibodies can be stored at −20 °C for at least 2 years, after adding an equal amount of autoclaved glycerol (1:1 dilution) to prevent freeze/thaw damage. Preparing aliquots of 10–20 μl is recommended for longer storage.
- Qdot® streptavidin conjugate 1 μM solution: Multiple colors are available (Thermo Fisher, Qdot®565: Q10131MP; Qdot®605: #Q10101MP; Qdot®625: #A10196; Qdot®655: #Q10121MP; and Qdot®705: #Q10161MP). A 1 μM QD solution can be stored at 4 °C for at least 2 years. Avoid freezing and centrifugation. We recommend Qdot®655, as a first choice for these QDots, because it provides the best signal-to-noise ratio and

Table 1
Recommended antibodies and incubation time for QD-SPT in mouse hippocampal neurons

	Primary antibody				Secondary antibody			
Target protein	Supplier	Catalog number	Conc. (μg/ml)	Incubation time (min)	Supplier	Catalog number	Conc. (μg/ml)	Incubation time (min)
GluA1	M	MAB2263	2.5	3	J	115-067-003	10 μg/ml	3
NMDAR1	A	AGC-001	12	5				
$GABA_AR$	Custom made [26]		8	5				
TRPA1	A	ACC-037	8	5				
Aquaporin11	A	AQP-011	8	5				
mGluR5	A	AGC-007	9.4	5				
GABAB receptor	A	AGB-001	6.4	5	J	111-067-003	10 μg/ml	5
P2Y receptors	A	APR-021	12	5				
GABA transporter type 1 (GAT1)	A	AGT-001	6.4	5				
Neural cell adhesion molecule (NCaM)	A	ANR-041	6.4	5				
Neuroligin1	A	ANR-111	3.2	5				

M Millipore, *A* Alomone Labs, *J* Jackson ImmunoResearch

facilitates analysis. The combination of Qdot®605, Qdot®655, and Qdot®705 allows for multicolor imaging on the same sample by using appropriate filter sets [37].

- 100 nM Intermediate QD dilution: Dilute 1 μM solution Qdot® streptavidin conjugates with QD-binding buffer (10×). This intermediate dilution is stable for 1 week at 4 °C.
- 1.43 M Sucrose solution: 4.9 g Sucrose and 10 ml distilled water; stable for at least 1 year at 4 °C.
- Imaging medium (50 ml): 44.5 ml MEM without phenol red (Thermo Fisher #51200038); 1 ml 1 M HEPES (final 20 mM), 2.5 ml 660 mM D-glucose in MEM (33.3 mM), 500 μl L-glutamine stock (2 mM), 500 μl sodium pyruvate stock

(1 mM), and 1 ml B-27™ supplement. The imaging medium should be fresh or stored for no more than 2 days.

- Heating block (37 °C) for cell labeling.
- A vacuum pump, an aspirator.

2.3 Image Acquisition and Processing (Fig. 2a)

1. Inverted fluorescence microscope (e.g., Olympus Tokyo, Japan; IX73; Nikon Instech Co., Ltd., Tokyo, Japan; Eclipse Ti) equipped with a Plan-Apochromat objective lens (NA ≥1.1, NA ≥1.4 is highly recommended, 60× or 100×) and appropriate filter sets for the wavelength of QDs. As an excitation filter, a blue excitation filter with a broad excitation passband range (450/70 nm, e.g., Semrock, Rochester, NY; BrightLine® single-band band-pass filter #FF01-450/70-25) is highly recommended. Choose a narrow band-pass filter (EM) and dichroic beam splitter (DM) according to each QD

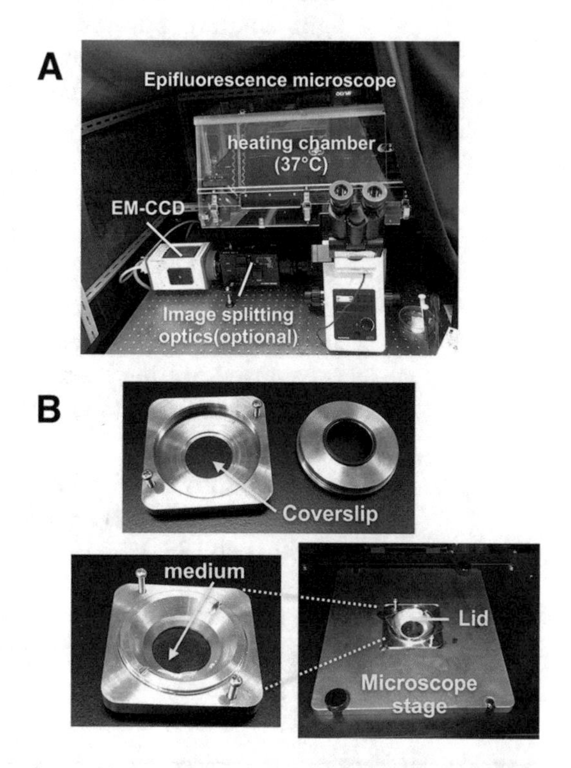

Fig. 2 Imaging system and recording chamber for QD-SPT. (**a**) Inverted epifluorescence microscope equipped with a chamber to keep the microscope at 37 °C. For the detection of QD signals, an EM-CCD camera is recommended. Image-splitting optics (optional) allow multicolor QD-SPT. (**b**) A recording chamber allowing the observation of live cells with high numerical aperture objective lenses, while keeping cells in a good condition in the imaging medium. Coverslips are maintained in the middle of the bottom part and sandwich the silicon O-ring installed in the top part. The glass lid should be placed on the microscope to avoid evaporation of the imaging medium

emission wavelength (e.g., Semrock, Qdot®565: EM 567/15 nm #FF01-567/15-25, DM #FF552-Di02-25x36; Qdot®605: 605/15 nm #FF01-605/15-25, DM #FF593-Di03-25x36; Qdot®625: 625/15 nm #FF01-625/15-25, DM #FF605-Di02-25x36; Qdot®655: 655/15 nm #FF01-655/15-25, DM #FF640-FDi01-25x36; Qdot®705: 711/25 nm #FF01-711/25-25, DM #FF700-Di01-25x36).

2. Detectors: EM-CCD camera (e.g., Hamamatsu Photonics, Hamamatsu, Japan; ImagEM; Andor, Belfast, UK; iXon). A spinning disk confocal unit (e.g., Yokogawa, Tokyo, Japan; CSU W-1) is available.
3. Illumination: The following light sources are available, mercury lamp (100 W), xenon lamp (75 W), light-emitting diode (LED) illumination system (e.g., CoolLED Ltd., Andover, UK; precisExcite; Thorlabs Inc., Newton, NJ; 4-Wavelength LED Source; Lumencor, Inc., Beaverton, OR; SPECTRA X light engine). For the spinning disk confocal system, both lasers (488 nm, 470 nm) and the SPECTRA X light engine are confirmed to be suitable for QD imaging.
4. Recording chamber: Use a chamber that allows recording in the imaging medium (e.g., Elveflow, Paris, France; Life Imaging Service Ludin chamber, Fig. 2b). A glass-bottom dish is also usable, but drift should be minimized by placing a weight on the top or by employing other means.
5. Microscope heating system: A heating system to maintain cells at 37 °C during the imaging. To avoid drift caused by thermal expansion, heating systems covering the entire microscope itself (e.g., Tokai Hit, Fujinomiya, Japan; Thermobox) are recommended.
6. PC and software for image acquisition (e.g., MetaMorph® from Molecular Devices, Sunnyvale, CA; Micromanager; TI Workbench [38]).

2.4 Data Analysis

1. Mac computers (Apple) running on Mac OS 10.6 or above.
2. Image analysis Software "TI Workbench" [38] installed on the Mac computer.

3 Methods

3.1 Preparation of Primary Neuronal Culture on the Coverslip

This protocol is appropriate for freshly prepared primary neuron-astrocyte mixed culture from E18-P1 mouse and rat hippocampus and cortex, and frozen cells from E18-20 rat cortex, and optimized for QD-SPT experiments.

3.1.1 <Day 0> Preparation of PEI-Coated Coverslip

PEI coating is highly recommended for QD-SPT experiments, as it allows tight cell attachment to the coverslips without preventing the development of neurons and astrocytes. Other coating methods (e.g., poly-ornithine, poly L-lysine, laminin coating) are also available if necessary; however, the cells should be resistant to multiple washing steps.

1. Place an 18-mm-diameter glass coverslip in each well of a 12-well plate.
2. Dilute the 2% PEI solution stock to prepare the 0.04% PEI solution with sterilized water.
3. Put 1 ml of 0.04% PEI solution in each well. Ensure that there are no bubbles underneath the coverslip.
4. Incubate the plates in a CO_2 incubator overnight.

3.1.2 <Day 1> Plating of Primary Cells

5. Wash the coated coverslips with 1 ml of sterilized water three times. Remove the PEI solution with an aspirator, put 1 ml sterilized water in the well, and then shake the 12-well plate so that the PEI solution between the coverslip and the plate can be washed out thoroughly. The remaining PEI is toxic for cells. Ensure that the water is aspirated completely after the final wash.
6. Dry and sterilize the coverslips inside the hood with UV light for at least 15 min. The PEI-coated dish can be stored at 4 °C for up to 2 months. Illuminate with UV light for 15 min before use.
7. Put 5 ml sterile distilled water in the space between wells. This prevents evaporation of the culture medium.
8. Prepare fresh primary culture cells according to appropriate protocols [39, 40], or use frozen cells.
9. Dilute cells with the cell plating medium at the following densities as appropriate. Freshly prepared rat hippocampal and cortical cells: 1.4×10^5 cells/ml. Freshly prepared mouse hippocampal and cortical cells: 2.5×10^5 cells/ml. Frozen rat cells $3.0–5.0 \times 10^5$ cells/ml (should be optimized depending on the survival rate). Plate 1 ml of cell suspension in each well of the PEI-coated coverslips.
10. Maintain cells in the 37 °C CO_2 incubator for 2–3 days.

3.1.3 <Days 2–3> Medium Change and Maintenance

11. Pre-warm the maintenance medium to 37 °C.
12. Remove culture medium using an aspirator, and gently add 1 ml of the maintenance medium. Do not let the cells dry out.
13. Maintain cells in the 37 °C CO_2 incubator until the imaging experiments. The culture can be maintained for at least 4 weeks without any further treatment.

3.1.4 <Days 3–14> (Optional) Expression of Marker Proteins

To highlight the morphology of neurons and astrocytes or to label the postsynaptic structures, transfection of plasmid DNA or infection of adeno-associated virus (AAV) coding marker fluorescent proteins may be necessary. The timing of transfection should be optimized depending on the purpose of the experiments. Purified AAV can be directly introduced in the culture medium 1–2 week (s) prior to the imaging experiments, at appropriate titer. Here, we describe a plasmid DNA transfection protocol using the Lipofectamine 3000 transfection reagent (Thermo Fisher #L3000-001). Cultures at 3–5 days in vitro (DIV) have good transfection efficacy for neurons, and those at 5–7 DIV are appropriate for transfection to astrocytes.

1. Label two tubes, one for "plasmid DNA" and the other for "Lipofectamine 3000."
2. Put 50 μl of OPTIMEM (Thermo Fisher #31985-062)/ coverslip in each tube.
3. Add 0.5 μg of plasmid DNA/coverslip and P3000 reagent in the plasmid DNA tube.
4. Add 1 μl of Lipofectamine 3000/coverslip in the Lipofectamine 3000 tube.
5. Vortex both tubes for 1–2 s.
6. Add the Lipofectamine 3000/OPTIMEM solution into the DNA/OPTIMEM solution; mix by pipetting gently and incubate the mixture (100 μl/coverslip) for 5 min at room temperature.
7. Apply the mixture to the cells droplet by droplet.
8. Incubate the cells in the incubator for 2–3 days, until the marker proteins are expressed.

3.2 QD Labeling

The labeling of membrane proteins includes (1) primary antibody targeting, (2) biotinylated secondary antibody targeting, and (3) streptavidin-conjugated QD targeting steps, which are followed by washing steps to remove unconjugated antibodies and QDs (Fig. 3).

1. Prepare a moist chamber for staining. Put a Parafilm (approximately 5 × 5 cm) on a heating block set at 37 °C and use a moist paper towel or facial tissue (wet paper) to form a ring around 9 cm in diameter. Put a 10-cm-diameter culture dish on the heating block as a lid (Fig. 4a). Alternatively, label cells in a 37 °C CO_2 incubator using the double-layer moist chamber, which is composed of the "inner chamber" (10-cm-diameter culture dish) and the "outer chamber" (15-cm-diameter culture dish with wet paper) (Fig. 4b). In the inner chamber (10-cm-diameter culture dish), put a 5 × 5 cm Parafilm and wet paper.

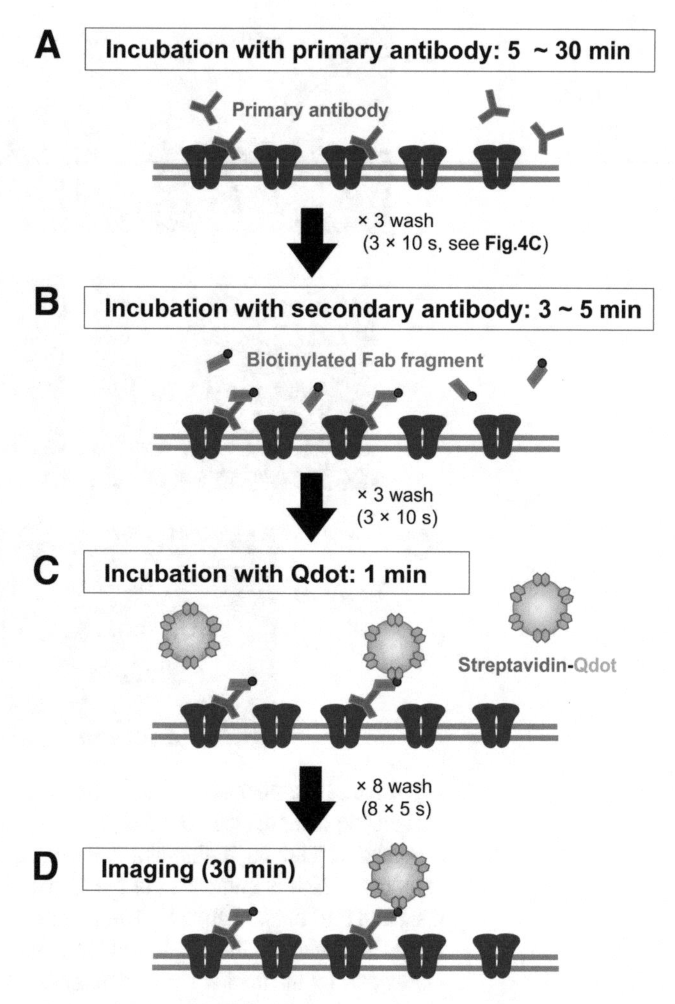

Fig. 3 Flow diagram of this protocol. The procedure is described in detail in Subheading 3.2

2. Transfer a coverslip from the culture dish to the center of the moist chamber on the Parafilm sheet. Parafilm prevents the liquid from spreading and allows staining at a small volume (100 μl for an 18 mm glass coverslip). Maintain the cell side up.
3. Wash cells briefly by sucking the medium with a vacuum pump from one edge of the coverslip and then by immediately adding 250 μl of imaging medium from the other edge (Fig. 4c). Make sure that the coverslip is not dried out.
4. Incubate cells with the primary antibody solution (Fig. 3a). The antibody concentration and the incubation time (3–30 min) should be optimized for each antibody and culture condition (*see* **Note 1**). When cells are incubated for >10 min, incubation in the CO_2 incubator using the double-layer moist chamber (Fig. 4b) is recommended.

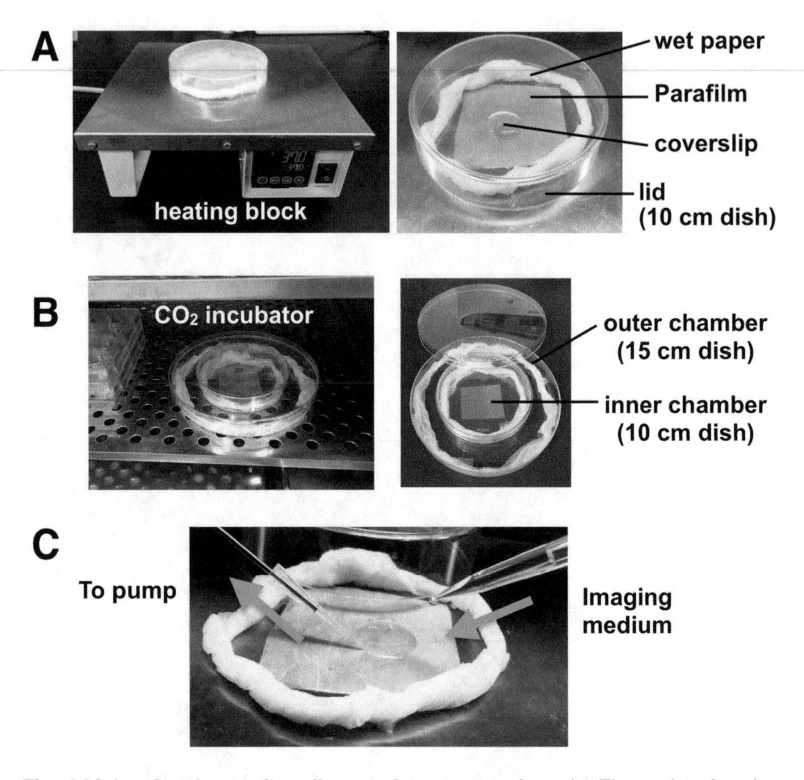

Fig. 4 Moist chamber and medium exchange procedure. (**a**) The moist chamber for labeling comprises a Petri dish lid over a wet facial tissue (wet paper) and a coverslip placed on a Parafilm on a heating block. (**b**) Double-layer moist chamber, which is composed of the "inner chamber" (10-cm-diameter culture dish) and the "outer chamber" (15-cm-diameter culture dish with wet paper), for longer incubation in the CO_2 incubator. In the inner chamber, put a Parafilm and wet paper. (**c**) Incubation medium is removed with a sucking pipette connected to a vacuum pump. Then, imaging medium is added rapidly and gently on the coverslip. One wash requires 10 s

5. Wash cells three times and incubate them with biotinylated secondary Fab antibody (10 μg/ml) for 3–5 min (Fig. 3b). The concentration of Fab antibody and the incubation time can be modified if necessary.
6. Wash cells three times. Prepare 1 nM QD-streptavidin conjugate in 1× QD-binding buffer supplemented with 215 mM sucrose. The sucrose increases the osmolarity of the QD-binding buffer to a physiological level (300 mOsm). The final QD mixture should be prepared just before use. Vortex mix the 100 nM intermediate QD solution just before final dilution to avoid aggregation.
7. Incubate cells with the 1 nM QD-streptavidin conjugate mixture for 1 min (Fig. 3c).

8. Wash cells approximately eight times. Wash cells thoroughly after QD incubation to remove unbound QDs.
9. (Optional) To stain the presynaptic terminal, incubate cells with the imaging medium containing 1 μM FM 4-64 and 40 mM KCl for 15 s, and wash the cells three times.

3.3 QD-SPT Data Acquisition (Fig. 3d)

1. Mount the coverslip in the recording chamber, add some imaging medium (approximately 400 μl for 18 mm coverslips) in the chamber, and set it on the microscope stage. Place the lid.
2. Take a snapshot of the marker proteins (e.g., GFP, YFP-tagged synaptic proteins) and save the image when necessary (Fig. 5a, left). One limitation of this step is that the puncta, highlighted by the fluorescent synaptic marker, appear larger than the actual size of the postsynaptic terminal, due to their diffraction limits. Nevertheless, the synaptic area can be defined by thresholding the synaptic marker images after applying a median filter to remove noise. Alternatively, the synaptic area can be defined by processing the synaptic marker images with a 2D object segmentation tool using wavelet transform [41] (Fig. 5a, middle).
3. Change the optical filter sets for QD and carry out real-time or time-lapse imaging in the same field of view (Fig. 5a, right). If labeling is successful, significantly larger numbers of QDs should be observed compared with cells stained without the primary antibody (Fig. 5b). Fifty QDs in the subregion of 250 × 250 pixels at Binning 1 (corresponds to 34 μm × 34 μm in our system) is the ideal labeling density. The data acquisition frequency and recording period should be optimized depending on the molecules of interest and the purpose of the experiments (*see* **Note 2**). Test for transmembrane proteins via continuous recording at 13 Hz (75 ms exposure) and for phospholipids at 33 Hz (30 ms exposure) in the first trial. To achieve fast recording rate, the excitation light needs to be continuously illuminated during the recording. However, continuous illumination beyond 60 s is not recommended, as longer exposure to the excitation light causes photodynamic damage to the cell, photobleaching, and frequent blinking of QDs. To track QDs for longer than 60 s, a time-lapse recording with 100–500 ms interval and 75 ms exposure will help in avoiding phototoxicity. Simultaneous multicolor imaging (e.g., EGFP and QD605, 625, 655; QD605 and QD705) can be achieved using an emission splitter, for example, Dual-View® (Optical Insights, Tucson, AZ) or W-View Gemini (Hamamatsu Photonics).

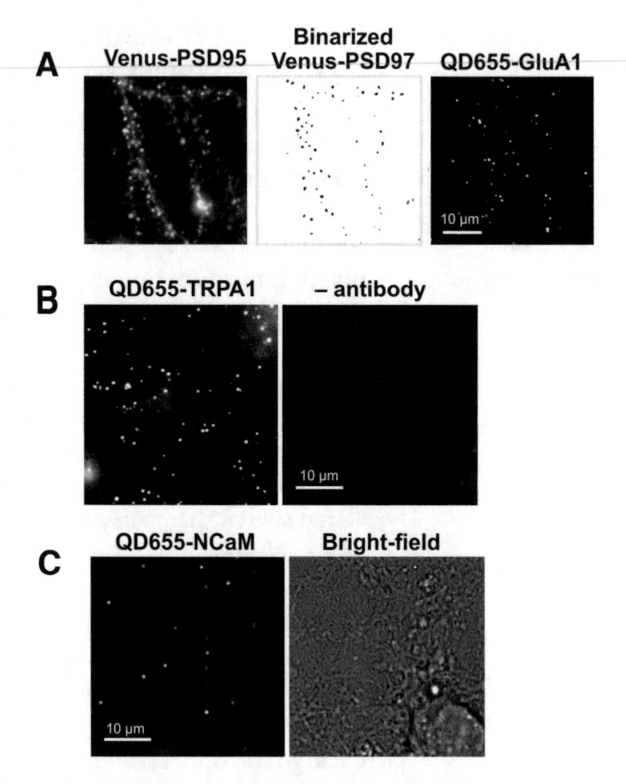

Fig. 5 Examples of neurons labeled with QDs and marker proteins on mouse hippocampal neurons (16–27 days in vitro). (**a**) An example of a mouse hippocampal neuron expressing Venus-PSD95 by AAV infection. Glutamate receptor A1 (GluA1) on the same field of view is labeled with QD655 (courtesy of M. Kanatani and A. Takashima, Gakushuin University, Japan). The Venus signal and QD655 signals can be separated using appropriate filter sets. For the analysis of diffusion in the subcellular compartments such as synapses (*see* Subheading 3.4), marker protein images should be binarized by using appropriate algorithms. (**b**) Labeling of neurons with QD655 with (left) and without (right) anti-TRPA1 antibody. Confirm that only few QDs are observed when the primary antibody is omitted. (**c**) Labeling of neural cell adhesion molecule (NCaM) with QD655 (left) and bright-field microscopy image in the same field of view (right)

4. Acquire a bright-field microscopy image in the same field of view as the QDs, if necessary (Fig. 5c).
5. Locate other cells and continue imaging for up to 30 min (*see* **Note 3**).

3.4 QD-SPT Data Analysis

Analysis of QD-SPT data consists of two main steps. The first step is to determine the location of QDs from the fluorescent spots. The second step is to assemble QD trajectories by linking detected spots presumably from the same QD. These analyses can be performed through custom-made software, the commercial software G-track (G-Angstrom Inc.), or ImageJ or Fiji plug-ins such as the "particle tracker" [42] or "TrackMate" [43]. In this protocol, we describe the QD-SPT data analysis using the "QDot tracker" function in the custom-made software "TI Workbench" [24, 38]. QDs intrinsically have intermittency of fluorescence emission, termed "blinking" (Fig. 6a, left). This blinking behavior provides a criterion to identify single QDs. Moreover, QDs that are internalized by endocytosis do not show blinking (Fig. 6a, right). Therefore, only QDs with blinking should be included in the analysis. Here, we will describe the reconstitution of QD trajectories, calculation of diffusion parameters such as mean-square displacement, diffusion coefficient, and confinement size. Analysis of other parameters is described in Chapter 6.

1. In the current version of TI Workbench, fluorescent spots of QDs can be detected by cross-correlation of the image with a Gaussian model with a point spread function [44] (Fig. 6b). A least-squares Gaussian fit is applied (around the local maximum above a threshold) to determine the center of each spot with a spatial accuracy of 5–10 nm (depending on the signal-to-noise ratio). For Gaussian fit and creating the trajectory, users should input the following parameters: threshold, max frame gaps for blinking, initial diffusion coefficient (D), pixel width, frame interval (ms), QD emission wavelength, and numerical aperture of the objective lens (Fig. 6b).
2. QD trajectories are assembled automatically by linking the centers of fluorescent spots which are judged as the same QD from frame to frame [45, 46]. The association criterion is based on the assumption of free Brownian diffusion and takes short blinking events into account (Fig. 6c). After completion of the process, a manual association step is performed, in which QD trajectories of maximal length are assembled from smaller fragments separated by blinking events (Fig. 6d). Possible errors of the program can be reduced at this step. Then, trajectories are subjected to further analyses. The coordinates (x, y, t) of each trajectory can be output into text format or Microsoft Excel format.
3. The following parameters are calculated from the trajectories composed of the sequential images of QD: mean square displacement (MSD), diffusion coefficient (D), transition between compartments, and dwell time within a compartment.

3.4.1 Mean Square Displacement

Physical parameters can be extracted from each trajectory ($x(t)$, $y(t)$) by computing the MSD [47], determined by the following formula:

$$\mathrm{MSD}(n\tau) = \frac{1}{N-n}\sum_{i=1}^{N-n}\left[(x((i+n)\tau) - x(i\tau))^2 + (y((i+n)\tau) - y(i\tau))^2\right]$$

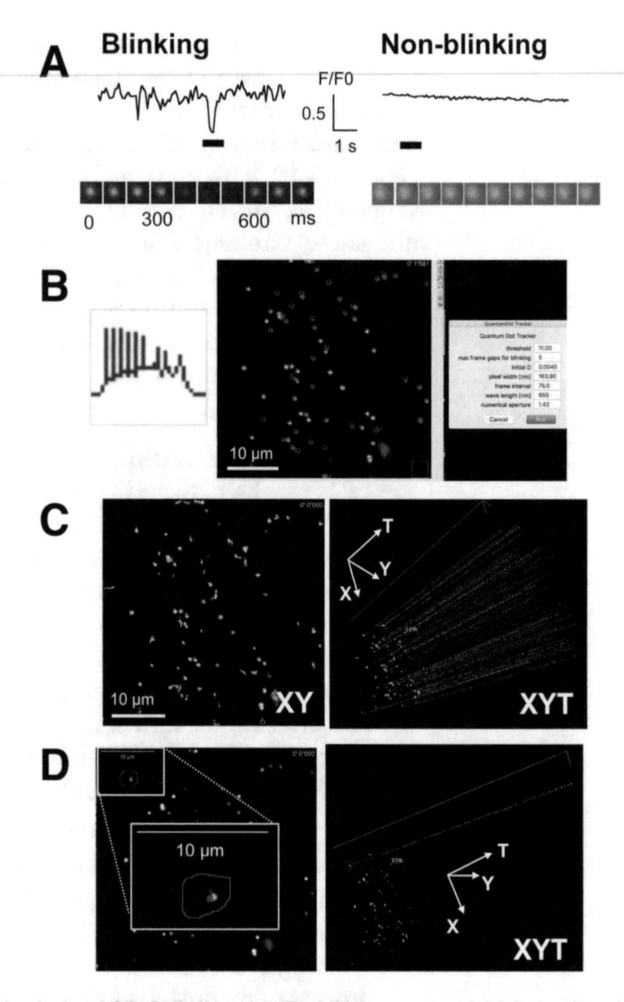

Fig. 6 Analysis of QD-SPT data. (**a**) (Top) Time course of QD intensity of blinking (left) and non-blinking (right) QDs. (Bottom) QD images at the timing indicated by the black bar in the time course. (**b**) Determination of the centers of QDs using the "QDot tracker" function in the custom-made software "TI Workbench" [38] (left icon). QDs are detected by cross-correlating the fluorescent spots in the image with a Gaussian model of the point spread function (right, red circles). (**c**) QD center determined by tracking. Left: *XY* image of tracking results superimposed on the original QD movie. Right: Tracking results in a 3D space (*XYT*). At this stage, the centers of fluorescent spots which are judged as the same QD from frame to frame are automatically linked by the program to produce short pieces of trajectories. (**d**) Assembly of the QD trajectory by linking small pieces of trajectories. The short trajectories included in the area surrounded by red circles are incorporated into one trajectory. Error spots derived from noise should be removed at this step

where τ is the acquisition time and N the total number of frames. Different types of motion can be distinguished from the time dependence of the MSD [47]. For simple two-dimensional Brownian motion, the MSD-$n\tau$ plot is linear with a slope of $4D$ where

D is the diffusion constant. If the MSD-$n\tau$ plot tends toward a plateau, the diffusion is classified as confined. If an additional directed motion with velocity v is present, MSD-$n\tau$ is on average equal to $4Dn\tau + v^2n^2\tau^2$.

3.4.2 Diffusion Coefficient

The diffusion coefficient (D) is determined by fitting the initial few points of the MSD-$n\tau$ curve with $\text{MSD}(n\tau) = 4Dn\tau + b$ (Fig. 7a, b, blue lines). This fit is generally used because it determines D independently of the type of motion [48]. The detection limit of D can be estimated by the analysis of QD videos that are immobilized on the coverslips (*see* **Note 4**). In our setup, 95% of fixed QDs are less than $D < 0.0004$; therefore, we define the QDs with $D < 0.0004$ as "immobile" (Fig. 7c).

3.4.3 Confinement Area

For molecules undergoing diffusion within a limited area (Fig. 7b), the size of the domain in which diffusion is confined can be estimated by fitting the MSD-$n\tau$ plot with the following equation [22, 48]:

$$\text{MSD}(n\tau) = \frac{L^2}{3}\left(1 - \exp\left(-\frac{12Dn\tau}{L^2}\right)\right)$$

where L (μm) is the side length of the square in which QD diffusion is restricted. L^2 (μm^2) represents the confined area.

3.4.4 Transition Between Compartments and Dwell Time Within a Compartment

Given that compartments such as synapses can be labeled with fluorescent markers (Fig. 5a), overlaying the image of a compartment marker with the QD trajectory allows the analysis of diffusion parameters within each compartment (Fig. 7d), and visualization of transitions of a QD-labeled molecule between compartments (Fig. 7e). The number of transitions between two compartments and the dwell time within each compartment can be estimated from the output data of the trajectory (Fig. 7e, right).

3.5 Examples of QD-SPT Studies on Synaptic Mechanisms and Neurological Disorders

For decades, QD-SPT has contributed to revealing the synaptic structure and the molecular mechanisms underlying learning and memory, as described in Subheading 1. Figure 8a is an example showing the involvement of intracellular Ca^{2+} signaling in the regulation of the GABAergic synapse in rat hippocampal neurons [28]. Loss of Ca^{2+} release by knockout by type 1 IP_3 receptor (IP_3R1KO), which is a Ca^{2+} channel on the intracellular Ca^{2+} store endoplasmic reticulum (ER), resulted in change in several diffusion parameters of $GABA_A$Rs (Fig. 8a). In IP_3RKO neurons, D increased both inside and outside the synapse. The dwell time in the synapse was decreased and the synaptic confinement size increased in IP_3RKO neurons. These results indicate that Ca^{2+} release from the ER is required for the stabilization of GABAergic synapses.

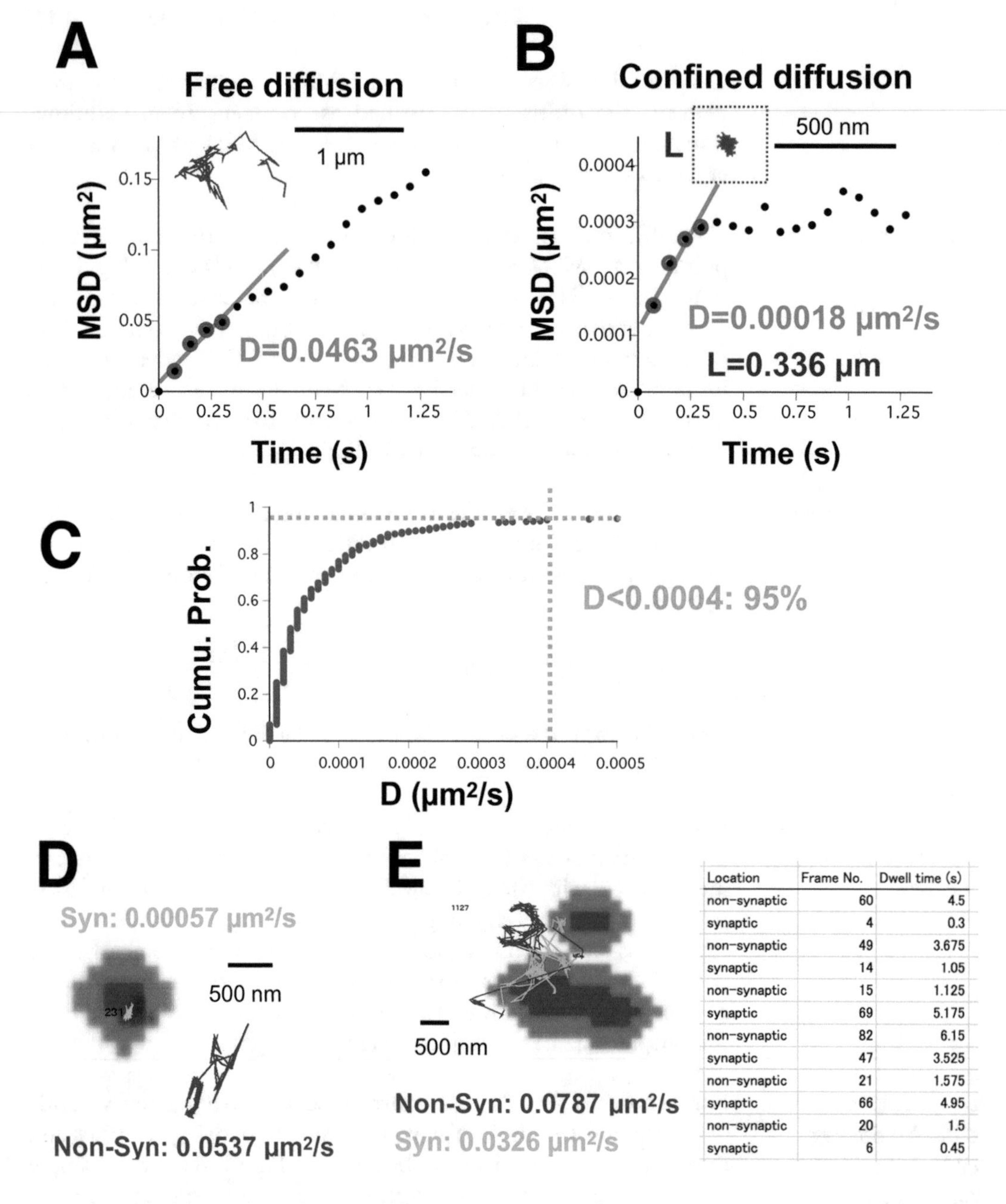

Location	Frame No.	Dwell time (s)
non-synaptic	60	4.5
synaptic	4	0.3
non-synaptic	49	3.675
synaptic	14	1.05
non-synaptic	15	1.125
synaptic	69	5.175
non-synaptic	82	6.15
synaptic	47	3.525
non-synaptic	21	1.575
synaptic	66	4.95
non-synaptic	20	1.5
synaptic	6	0.45

Fig. 7 Analysis of QD-SPT trajectories. (**a**, **b**) Examples of MSD–time plot of QD trajectories (7.5 s, inset) with free (**a**) and confined (**b**) diffusion of QD655-GluA1. Note that the *y*-axis in (**b**) represents smaller MSD than in (**a**). The slope of the linear fit (blue line) corresponds to the diffusion coefficient (*D*). The brown square in (**b**) indicates the area in which the diffusion is confined, and *L* indicates the side length of the square. (**c**) Cumulative plot for the *D* of QD655 fixed on the coverslip. In our imaging system, 95% QDs showed $D < 0.0004\ \mu m^2/s$. (**d**) Example of a QD655-GluA1 trajectory in the synaptic area (Syn: green) and non-synaptic area (Non-Syn: blue). Synapses are labeled with Venus-PSD95. Binarized Venus-PSD95 signals are indicated by purple, and the perisynaptic area (2 pixels neighboring the PSD95 signal), which is often regarded as synaptic compartment, is indicated by magenta. (**e**) Example of a QD655-GluA1 trajectory entering (green) and leaving (blue) synaptic areas. *D* can be analyzed in each compartment. The table indicates the TI Workbench output showing the timing of transition and the dwell time in each compartment for this trajectory (**a**–**d**: courtesy of M. Kanatani and A Takashima, Gakushuin University, Japan)

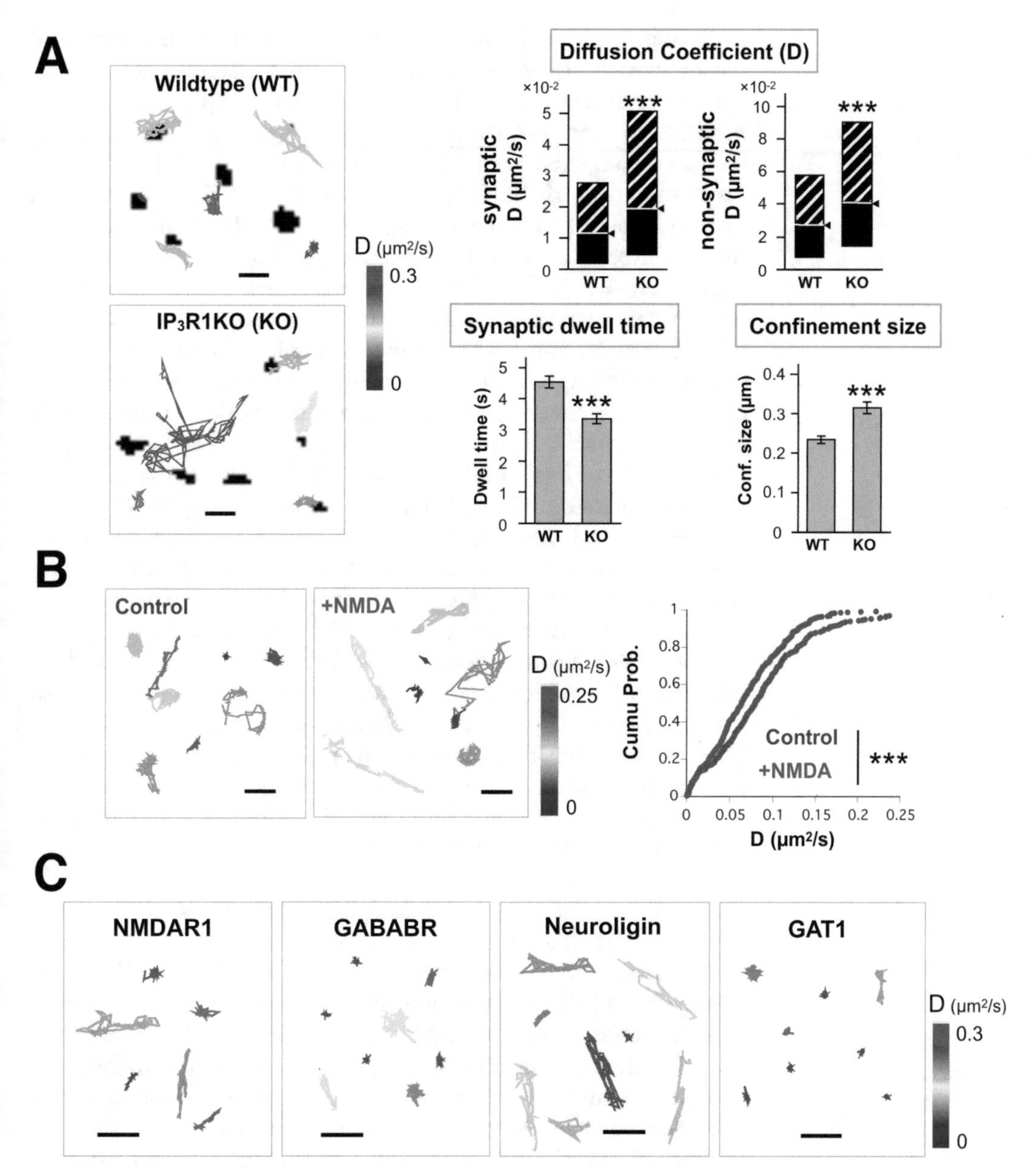

Fig. 8 Examples of QD-SPT experiments. (**a**) Comparison of $GABA_AR$ diffusion in mouse hippocampal neurons from wild-type (WT) and IP_3R1KO neurons (KO). QD trajectories in WT and KO reconstructed from recording sequences (15.2 s) are overlaid with FM4-64 signals (black) in order to identify synapses. The level of the D ($\mu m^2/s$) is indicated by pseudocolor. Quantifications of median diffusion coefficients (median $D \pm$ 25–75% interquartile range IQR) for synaptic and non-synaptic receptors, the mean ($\pm$SEM) synaptic dwell time, and the size of confinement domains (mean $\pm$ SEM) highlight increased diffusion dynamics and reduced synaptic confinement. (**b**) $GABA_AR$ diffusion dynamics in rat cortical neurons (21–27 days in vitro) in the absence (control) and presence (+NMDA) of 50 μM NMDA and 5 μM glycine. The pseudocolor presentation of QD trajectories and the cumulative plot of D indicate that NMDA stimulation increases the $GABA_AR$ D, as has been reported for hippocampal neurons [24, 26, 49]. Number of QDs analyzed for control: 433; for +NMDA: 298. $^{***}p = 0.0003$, Mann-Whitney U-test. (**c**) Pseudocolor presentation of QD-SPT trajectories for NMDA receptor type 1 (NMDAR1), GABA B receptor (GABABR), neuroligin, and GABA transporter type 1 (GAT1) in mouse hippocampal neurons (21–27 days in vitro). Scale bars for trajectories: 1 μm

Additionally, a growing body of evidence shows that neuronal receptor dynamics become abnormal in disease states [32]. In neurons with excess neuronal activity mimicking epilepsy, diffusion motility of the $GABA_AR$ in hippocampal neurons markedly increased compared to that in normal cells [24, 26, 49]. Moreover, point mutation of the $GABA_AR$ γ2 subunit at K289M, which is seen in human generalized epilepsy with febrile seizure plus syndrome, increases $GABA_AR$ diffusion evoked by high temperatures, suggesting a role for altered $GABA_AR$ membrane diffusion in human febrile seizures [30]. Cortical neurons incubated with 50 μM NMDA and 5 μM glycine to mimic epilepsy also showed increased $GABA_AR$ diffusion (Fig. 8b). Abnormalities in molecular dynamics have also been reported in in vitro models of Alzheimer's disease; β-amyloid oligomer aggregation reduces metabotropic glutamate receptor 5 (mGluR5) mobility and accelerates the abnormal clustering of mGluRs in both neurons and astrocytes [29, 31]. Additionally, mGluR diffusion was altered in another glial cell model mimicking the pathology of Alzheimer's disease; the overexpression of mGluR5 led to dysfunction of the diffusion barrier between the astrocytic soma and its processes [9]. Altered mGluR5 dynamics have also been reported in other neurological disorders. mGluR5 was significantly more mobile at synapses in hippocampal neurons lacking fragile X mental retardation 1 gene (Fmr1), which is a causal gene for fragile X syndrome [50]. Taken together, altered molecular diffusion dynamics are now considered as a pathological phenotype at the cellular level and provide insight into the molecular mechanism of neurological disorders.

3.6 Conclusion

QD-SPT is a super-resolution imaging technique to visualize the dynamics of endogenous membrane molecules with a conventional and relatively simple imaging system. Since the first QD-SPT study was reported [19], various molecular mechanisms supporting synaptic structures and the regulatory mechanism of synapses have been highlighted. Additionally, QD-SPT has been employed to reveal that abnormality in membrane molecular diffusion dynamics is associated with disease states of neurons and glial cells. The analysis of membrane molecular dynamics will be an effective approach for elucidating the pathogenesis and establishing the diagnosis of neurological disorders. We believe that it is important to investigate the dynamics of various membrane molecules in disease-model neurons, and are presently attempting to increase the molecular species that can be examined with QD-SPT (Fig. 8c and Table 1). As three-dimensional tracking of QD movement is possible [51], the uses of QD-SPT can also be expanded to monitor various 3D biological phenomena, such as the kiss-and-run release of single synaptic vesicles [52], endocytotic and exocytotic processes of lectin [53], and real-time monitoring of an influenza virus infection [54].

4 Notes

1. The QD labeling conditions have to be determined by the user and depend on the abundance of the target molecule, the affinity of the primary antibody, the maturation stage of the cells, and other factors. Both polyclonal and monoclonal antibodies can be used. Antibodies should be tested first with live immunostaining to examine the accessibility to the epitope. Minimize the concentrations of primary antibody to prevent cross-linking of molecules: the lower the concentration of primary antibodies, the lower the possibility of cross-linking. The density of QDs should also be optimized to avoid overlap of QD trajectories. If QDs are specifically targeted to the primary antibody, the quantity of QDs will be greatly reduced when primary antibody incubation is omitted, as shown in Fig. 5b. We recommend clamping the secondary antibody concentration and the incubation time as described in this protocol (Table 1), and identifying the best primary antibody concentration to obtain good QD labeling.
2. Bright fluorescence of QD allows fast recording at 1.75 kHz of the sampling frequency, with mercury lump and an EMCCD camera [55]. Although QDs are more resistant to photobleach compared with other fluorescent dyes and fluorescent proteins, they indeed photobleach under long-time strong excitation light; therefore, there is trade-off between the frame rate and the recording duration. The optimal sampling rate and duration should be determined according to the purpose of the experiments and the performance of the detector. Faster frame rate is recommended for molecules undergoing fast lateral diffusion. To obtain diffusion parameters such as diffusion coefficient of membrane proteins, we record 50–100 consecutive frames at a frame rate of 13 Hz.
3. Live imaging should not exceed 30 min at 37 °C for cell survival and should be reduced depending on the turnover of the targeted molecules. Imaging at room temperature may decrease the endocytosis rate of the targeted molecules. The proportion of endocytosed QDs can be estimated by acid wash, which breaks the epitope-ligand interaction and allows release of QDs located in the extracellular space. The protocol of acid wash is as follows:
   (a) Store 10 ml of imaging medium on ice and add 70 μl of 37% HCl to it. The pH should be at approximately 2.0 (pH 2 imaging medium).
   (b) Place the recording chamber after the QD-SPT recording on ice for 2 min.

(c) Replace the medium to pH 2 imaging medium and incubate for 2 min. Continuously agitate the chamber during incubation.
(d) Wash cells once with the pH 2 imaging medium.
(e) Wash cells three times with normal imaging medium at room temperature.
(f) Observe the cells under the same condition as of the QD-SPT. The remaining QD signals are estimated as "endocytosed QDs."

4. The analysis of movies of immobilized QD provides the detection limit of *D*. Measure the *D* of immobilized QDs on the coverslips using the following procedure:
   (a) Dilute biotinylated Fab fragments at 10 μg/ml with PBS adjusting the pH to 8–9.
   (b) Incubate a coverslip with the above solution overnight, at 4 °C.
   (c) Wash the coverslip with PBS (normal) three times.
   (d) Incubate the coverslip for 1 min with approximately 1 nM Qdot in QD-binding buffer, at 37 °C.
   (e) Wash the coverslip more than five times with PBS (normal).
   (f) Record QD movies under the same condition as of the QD-SPT.
   (g) Analyze the QD movies to calculate the *D* and determine the threshold for immobilized QDs (Fig. 7c).

Acknowledgments

Grant support: JST/PRESTO (grant number JPMJPR15F8, Japan); JSPS/KAKENHI (grant numbers JP18H05414, JP17H05710, JP16K07316), Takeda Foundation. The authors would like to deeply thank Maxime Dahan, Sabine Lévi, Misa Arizono, and Antoine Triller for establishing this protocol. We thank Misa Kanatani and Akihiko Takashima for providing QD655-GluA1 data for this chapter.

References

1. Choquet D, Triller A (2013) The dynamic synapse. Neuron 80:691–703
2. Bredt DS, Nicoll RA (2003) AMPA receptor trafficking at excitatory synapses. Neuron 40:361–379
3. Jacob TC, Moss SJ, Jurd R (2008) GABA (A) receptor trafficking and its role in the dynamic modulation of neuronal inhibition. Nat Rev Neurosci 9:331–343
4. Singer SJ, Nicolson GL (1972) The fluid mosaic model of the structure of cell membranes. Science 175:720–731
5. Triller A, Choquet D (2005) Surface trafficking of receptors between synaptic and extrasynaptic

membranes: and yet they do move! Trends Neurosci 28:133–139

6. Michalet X, Pinaud FF, Bentolila LA, Tsay JM, Doose S, Li JJ, Sundaresan G, Wu AM, Gambhir SS, Weiss S (2005) Quantum dots for live cells, in vivo imaging, and diagnostics. Science 307:538–544
7. Groc L, Lafourcade M, Heine M, Renner M, Racine V, Sibarita JB, Lounis B, Choquet D, Cognet L (2007) Surface trafficking of neurotransmitter receptor: comparison between single-molecule/quantum dot strategies. J Neurosci 27:12433–12437
8. Bannai H, Levi S, Schweizer C, Dahan M, Triller A (2006) Imaging the lateral diffusion of membrane molecules with quantum dots. Nat Protoc 1:2628–2634
9. Arizono M, Bannai H, Nakamura K, Niwa F, Enomoto M, Matsu-Ura T, Miyamoto A, Sherwood MW, Nakamura T, Mikoshiba K (2012) Receptor-selective diffusion barrier enhances sensitivity of astrocytic processes to metabotropic glutamate receptor stimulation. Sci Signal 5:ra27
10. Renner M, Choquet D, Triller A (2009) Control of the postsynaptic membrane viscosity. J Neurosci 29:2926–2937
11. Swift JL, Cramb DT (2008) Nanoparticles as fluorescence labels: is size all that matters? Biophys J 95:865–876
12. Groc L, Heine M, Cognet L, Brickley K, Stephenson FA, Lounis B, Choquet D (2004) Differential activity-dependent regulation of the lateral mobilities of AMPA and NMDA receptors. Nat Neurosci 7:695–696
13. Bats C, Groc L, Choquet D (2007) The interaction between Stargazin and PSD-95 regulates AMPA receptor surface trafficking. Neuron 53:719–734
14. Ehlers MD, Heine M, Groc L, Lee MC, Choquet D (2007) Diffusional trapping of GluR1 AMPA receptors by input-specific synaptic activity. Neuron 54:447–460
15. Heine M, Groc L, Frischknecht R, Beique JC, Lounis B, Rumbaugh G, Huganir RL, Cognet L, Choquet D (2008) Surface mobility of postsynaptic AMPARs tunes synaptic transmission. Science 320:201–205
16. Groc L, Heine M, Cousins SL, Stephenson FA, Lounis B, Cognet L, Choquet D (2006) NMDA receptor surface mobility depends on NR2A-2B subunits. Proc Natl Acad Sci U S A 103:18769–18774
17. Groc L, Choquet D, Stephenson FA, Verrier D, Manzoni OJ, Chavis P (2007) NMDA receptor surface trafficking and synaptic subunit composition are developmentally regulated by the extracellular matrix protein Reelin. J Neurosci 27:10165–10175
18. Dupuis JP, Ladepeche L, Seth H, Bard L, Varela J, Mikasova L, Bouchet D, Rogemond V, Honnorat J, Hanse E, Groc L (2014) Surface dynamics of GluN2B-NMDA receptors controls plasticity of maturing glutamate synapses. EMBO J 33:842–861
19. Dahan M, Lévi S, Luccardini C, Rostaing P, Riveau B, Triller A (2003) Diffusion dynamics of glycine receptors revealed by single-quantum dot tracking. Science 302:442–445
20. Hanus C, Ehrensperger MV, Triller A (2006) Activity-dependent movements of postsynaptic scaffolds at inhibitory synapses. J Neurosci 26:4586–4595
21. Ehrensperger MV, Hanus C, Vannier C, Triller A, Dahan M (2007) Multiple association states between glycine receptors and gephyrin identified by SPT analysis. Biophys J 92:3706–3718
22. Charrier C, Ehrensperger MV, Dahan M, Levi S, Triller A (2006) Cytoskeleton regulation of glycine receptor number at synapses and diffusion in the plasma membrane. J Neurosci 26:8502–8511
23. Levi S, Schweizer C, Bannai H, Pascual O, Charrier C, Triller A (2008) Homeostatic regulation of synaptic GlyR numbers driven by lateral diffusion. Neuron 59:261–273
24. Bannai H, Levi S, Schweizer C, Inoue T, Launey T, Racine V, Sibarita JB, Mikoshiba K, Triller A (2009) Activity-dependent tuning of inhibitory neurotransmission based on GABAAR diffusion dynamics. Neuron 62:670–682
25. Marsden KC, Shemesh A, Bayer KU, Carroll RC (2010) Selective translocation of Ca2+/calmodulin protein kinase IIalpha (CaMKIIalpha) to inhibitory synapses. Proc Natl Acad Sci U S A 107:20559–20564
26. Niwa F, Bannai H, Arizono M, Fukatsu K, Triller A, Mikoshiba K (2012) Gephyrin-independent GABA(A)R mobility and clustering during plasticity. PLoS One 7:e36148
27. Petrini M, Ravasenga T, Hausrat TJ, Iurilli G, Olcese U, Racine V, Sibarita JB, Jacob TC, Moss SJ, Benfenati F, Medini P, Kneussel M, Barberis A (2014) Synaptic recruitment of gephyrin regulates surface GABAA receptor dynamics for the expression of inhibitory LTP. Nat Commun 5:3921
28. Bannai H, Niwa F, Sherwood MW, Shrivastava AN, Arizono M, Miyamoto A, Sugiura K, Levi S, Triller A, Mikoshiba K (2015)

Bidirectional control of synaptic GABAAR clustering by glutamate and calcium. Cell Rep 13:2768–2780

29. Renner M, Lacor PN, Velasco PT, Xu J, Contractor A, Klein WL, Triller A (2010) Deleterious effects of amyloid beta oligomers acting as an extracellular scaffold for mGluR5. Neuron 66:739–754
30. Bouthour W, Leroy F, Emmanuelli C, Carnaud M, Dahan M, Poncer JC, Levi S (2012) A human mutation in Gabrg2 associated with generalized epilepsy alters the membrane dynamics of GABAA receptors. Cereb Cortex 22:1542–1553
31. Shrivastava N, Kowalewski JM, Renner M, Bousset L, Koulakoff A, Melki R, Giaume C, Triller A (2013) Beta-amyloid and ATP-induced diffusional trapping of astrocyte and neuronal metabotropic glutamate type-5 receptors. Glia 61:1673–1686
32. Shrivastava N, Aperia A, Melki R, Triller A (2017) Physico-pathologic mechanisms involved in neurodegeneration: misfolded protein-plasma membrane interactions. Neuron 95:33–50
33. Cai E, Ge P, Lee SH, Jeyifous O, Wang Y, Liu Y, Wilson KM, Lim SJ, Baird MA, Stone JE, Lee KY, Davidson MW, Chung HJ, Schulten K, Smith AM, Green WN, Selvin PR (2014) Stable small quantum dots for synaptic receptor tracking on live neurons. Angew Chem Int Ed Engl 53:12484–12488
34. Lee SH, Jin C, Cai E, Ge P, Ishitsuka Y, Teng KW, de Thomaz AA, Nall D, Baday M, Jeyifous O, Demonte D, Dundas CM, Park S, Delgado JY, Green WN, Selvin PR (2017) Super-resolution imaging of synaptic and extra-synaptic AMPA receptors with different-sized fluorescent probes. Elife 6:e27744
35. Chung I, Akita R, Vandlen R, Toomre D, Schlessinger J, Mellman I (2010) Spatial control of EGF receptor activation by reversible dimerization on living cells. Nature 464:783–787
36. de Luca E, Ravasenga T, Petrini EM, Polenghi A, Nieus T, Guazzi S, Barberis A (2017) Inter-synaptic lateral diffusion of GABAA receptors shapes inhibitory synaptic currents. Neuron 95:63–69.e65
37. Clausen MP, Arnspang EC, Ballou B, Bear JE, Lagerholm BC (2014) Simultaneous multi-species tracking in live cells with quantum dot conjugates. PLoS One 9:e97671
38. Inoue T (2018) TI Workbench, an integrated software package for electrophysiology and imaging. Microscopy (Oxf) 67:129–143
39. K. Goslin, H. Asmussen, G. Banker, in Culturing nerve cells, G. Banker, K. Goslin. (MIT Press, Cambridge, 1998), pp. 339–370
40. Arizono M, Bannai H, Mikoshiba K (2014) Imaging mGluR5 dynamics in astrocytes using quantum dots. Curr Protoc Neurosci 66:2.21.1–2.21.18
41. Racine V, Sachse M, Salamero J, Fraisier V, Trubuil A, Sibarita JB (2007) Visualization and quantification of vesicle trafficking on a three-dimensional cytoskeleton network in living cells. J Microsc 225:214–228
42. Sbalzarini F, Koumoutsakos P (2005) Feature point tracking and trajectory analysis for video imaging in cell biology. J Struct Biol 151:182–195
43. Tinevez Y, Perry N, Schindelin J, Hoopes GM, Reynolds GD, Laplantine E, Bednarek SY, Shorte SL, Eliceiri KW (2017) TrackMate: an open and extensible platform for single-particle tracking. Methods 115:80–90
44. Stallinga S, Rieger B (2010) Accuracy of the Gaussian point spread function model in 2D localization microscopy. Opt Express 18:24461–24476
45. Bonneau S, Cohen L, Dahan M (2004) A multiple target approach for single quantum dot tracking. In: Proceedings of the IEEE international symposium on biological imaging, p 664
46. Bonneau S, Dahan M, Cohen LD (2005) Single quantum dot tracking based on perceptual grouping using minimal paths in a spatiotemporal volume. IEEE Trans Image Process 14:1384–1395
47. Saxton MJ, Jacobson K (1997) Single-particle tracking: applications to membrane dynamics. Annu Rev Biophys Biomol Struct 26:373–399
48. Kusumi A, Sako Y, Yamamoto M (1993) Confined lateral diffusion of membrane receptors as studied by single particle tracking (nanovid microscopy). Effects of calcium-induced differentiation in cultured epithelial cells. Biophys J 65:2021–2040
49. Muir J, Arancibia-Carcamo IL, MacAskill AF, Smith KR, Griffin LD, Kittler JT (2010) NMDA receptors regulate GABAA receptor lateral mobility and clustering at inhibitory synapses through serine 327 on the gamma2 subunit. Proc Natl Acad Sci U S A 107:16679–16684
50. Aloisi E, Le Corf K, Dupuis J, Zhang P, Ginger M, Labrousse V, Spatuzza M, Georg Haberl M, Costa L, Shigemoto R, Tappe-Theodor A, Drago F, Vincenzo Piazza P, Mulle C, Groc L, Ciranna L, Catania MV, Frick A (2017) Altered surface mGluR5 dynamics provoke synaptic NMDAR

dysfunction and cognitive defects in Fmr1 knockout mice. Nat Commun 8:1103

51. Wang Y, Fruhwirth G, Cai E, Ng T, Selvin PR (2013) 3D super-resolution imaging with blinking quantum dots. Nano Lett 13:5233–5241
52. Zhang Q, Li Y, Tsien RW (2009) The dynamic control of kiss-and-run and vesicular reuse probed with single nanoparticles. Science 323:1448–1453
53. Liu SL, Zhang ZL, Sun EZ, Peng J, Xie M, Tian ZQ, Lin Y, Pang DW (2011) Visualizing the endocytic and exocytic processes of wheat germ agglutinin by quantum dot-based single-particle tracking. Biomaterials 32:7616–7624
54. Liu SL, Tian ZQ, Zhang ZL, Wu QM, Zhao HS, Ren B, Pang DW (2012) High-efficiency dual labeling of influenza virus for single-virus imaging. Biomaterials 33:7828–7833
55. Clausen P, Lagerholm BC (2013) Visualization of plasma membrane compartmentalization by high-speed quantum dot tracking. Nano Lett 13:2332–2337

Chapter 8

Multipolarization Dark-Field Imaging of Single Endosomes in Microfluidic Neuronal Culture for Simultaneous Orientation and Displacement Tracking

Luke Kaplan and Bianxiao Cui

Abstract

Robust axonal transport of endosomes is critical for neuronal function. Increasingly it is appreciated that single axonal endosomes are transported by teams of molecular motors. How these motors are regulated remains poorly understood. In part, this is due to the high variability between individual endosomes. Conventional imaging approaches also only track the position of endosomes, effectively measuring one variable in a team of several motors. Here we describe how to image transport of individual axonal endosomes using multipolarization dark-field microscopy to measure both the position and the orientation of axonal endosomes. This is accomplished in a high-throughput manner by combining the imaging approach with microfluidic cell culture.

Key words Axonal transport, Gold nanorods, Polarization, Orientation, Dark field, Imaging, Microfluidic, DRG neurons, Molecular motors, Dynein, Single-particle tracking

1 Introduction

Neurons are one of the largest cell types in metazoans, and their axons can reach lengths of over a meter [1]. At the same time, they are generally long-lived and postmitotic. These factors make active intracellular transport in neurons for signaling [2, 3] and degradative pathways [4, 5] of paramount importance. Unsurprisingly, many neurodegenerative diseases are associated with defects in intracellular transport [6–9].

An important class of cargos in axonal transport is endosomal cargos, which include signaling endosomes and degradative cargos such as autophagosomes and lysosomes. These endosomes can be quite heterogenous, ranging in size from 50 to 150 nm in diameter [10, 11]. Their transport is regulated by a variety of molecules, both on the endosome, itself [12, 13], and on the cytoskeleton [14–16]. These regulators are upstream of the molecular motors, dynein and kinesin, which physically power the transport of

Nobuhiko Yamamoto and Yasushi Okada (eds.), *Single Molecule Microscopy in Neurobiology*, Neuromethods, vol. 154,
https://doi.org/10.1007/978-1-0716-0532-5_8,

endosomes in the retrograde (toward the cell body) and anterograde (away from the cell body) directions, respectively.

Recent work has demonstrated that most endosomes have both kinesin and dynein molecules on them [17–19]. Attempts to model such multimotor transport based on in vitro-determined single-molecule properties of these motors predict frequent tugs-of-war, with motors on a single endosome pulling in opposite directions. Endosomal transport in the axon, however, seems to proceed with long unidirectional runs, suggesting that the mere presence of a motor on an endosome does not ensure that it is active [20]. Consequently, the heterogeneity and spatiotemporal dynamics of axonal transport call for high-resolution single-endosome imaging approaches.

Nanoparticle-based imaging approaches can fill this need, providing bright optical probes with high photostability. Delivery of a nanoparticle to the lumen of an endosome can proceed by the endogenous endocytosis pathways and transport is executed by the native cellular motors. As such, nanoparticles can function as biologically active cargos in axonal transport, causing minimal perturbation to the endogenous machinery [10].

Nanoparticle-based imaging is also compatible with compartmentalized microfluidic culture. This technique is based on culturing neurons close to microchannels which are large enough to allow neurite outgrowth but too small to allow the cell bodies to pass through (Fig. 2) [21, 22]. The result is three fluidically separate compartments containing the cell bodies, the axon termini, and the mid-axon.

This approach has three important advantages over conventional mass cell culture. *First*, in mass culture it can be hard to identify which axon belongs to which cell body. It is therefore often difficult or impossible to determine which direction is anterograde and which is retrograde. In compartmentalized cell culture, all the axons are unidirectionally aligned making anterograde vs. retrograde determination unambiguous. *Second*, since all the cell bodies are in one compartment and the other compartment only contains axonal termini, the investigator can selectively add nanoparticles to either compartment, resulting in predominantly retrograde (nanoparticles added to the axonal compartment) or anterograde (nanoparticles added to the cell body compartment) nanoparticle-endosome flux in the channels. *Finally*, since the microchannels are fluidically isolated, nanoparticles incubated in the other compartments can only get into the microchannels by active intracellular transport. This results in minimal background of nanoparticles nonspecifically adsorbed to the surface or bound to the outside of the plasma membrane.

Gold nanoparticles are particularly attractive candidates for axonal transport studies as they are extremely bright and photostable due to localized plasmon resonances that are in the optical frequency range. Furthermore, gold nanoparticles are readily conjugated to biological ligands. Particularly in scattering-based approaches, several groups have demonstrated their utility, obtaining nanometer-level and sub-millisecond spatial and temporal resolutions in live imaging [23, 24]. This level of high spatiotemporal precision can be important for the study of endosomal transport as molecular motor step sizes are often <10 nm and occur on the millisecond timescale [23].

1.1 Detecting Rotational Motion of Axonal Endosomes Using Gold Nanorods

Molecular motors can be attached at different locations on the endosome [25] and as such exert torques resulting in rotations. Active motors also act as tethers to the microtubule, restricting thermally driven rotation which dominates in untethered endosomes [26, 27]. Consequently, if motors fully detach from microtubules translational motion will pause (as is the case for mitochondria [28]), but the cargo will be free to rotate. Rotational motion can therefore be an important measure of motor activity, but is difficult to measure with conventional optical approaches.

Rod-shaped gold nanoparticles, specifically, scatter light anisotropically, making them useful as optical probes for rotation. Detection of their orientation can be done by a variety of means [29–33], but one of the simplest relies on measuring the polarization of the light they scatter in dark-field microscopy. Polarization can be measured by splitting the image acquired by the microscope into different component polarizations of light. These different polarization "channels" are analogous to different color "channels" in multicolor fluorescence microscopy. Importantly, gold nanorods also enjoy all of the excellent photophysical properties described in the prior section and as such can be used simultaneously as rotational and translational probes.

2 Materials and Equipment

- Dissociated embryonic DRG neurons:
  - HBSS.
  - Trypsin + 0.25% EDTA (Thermo Fisher 25200056).
  - Cell culture media:

    Neurobasal medium (Thermo Fisher A3582901).

    GlutaMAX (Thermo Fisher 35050061).

    Nerve growth factor (Cedarlane Labs CLMCNET-001.25, final concentration 50 ng/mL).

    Penicillin-streptomycin (Thermo Fisher 15070063, final concentration 100 U/mL).
  - Plating media:

    Dulbecco's modified Eagle's medium (Thermo Fisher 11995073).

Fetal bovine serum (10% final concentration).

Nerve growth factor (50 ng/mL).

Penicillin-streptomycin (100 U/mL).

- CO_2-independent medium (Thermo Fisher 18045-088).
- B-27 supplement (Thermo Fisher A3582801, sold at 50× concentration).
- Ara-C (Sigma-Aldrich C1768, final concentration 1 μM).

- Microscope and detection optics:
  - Non-polarizing cube beam splitter.
  - Polarizing cube beam splitter (Thorlabs PBS252).
  - High-numerical-aperture (NA) condenser (NA = 1.5).
  - Adjustable NA objective (Nikon CFI Plan Fluor 60xS Oil).
  - Inverted microscope (Nikon Eclipse Ti-U).
  - Adjustable rectangular aperture.
  - Linear polarizer.
  - Relay lenses (Thorlabs LBF254-200-A).
  - Knife-edge mirror, regular, polarization-preserving mirrors.
  - sCMOS camera (PCO.edge 5.5 or similar).
  - Dark-field stop.
- Rotational mount (Thorlabs LCRM2).
- CTAB-coated gold nanorods (Nanopartz A12-25-CTAB-DI-25).
- Thiol-PEG-biotin 5 kDa molecular weight (Nanocs PG2-BNTH-5K).
- Biotinylated wheat germ agglutinin (WGA) (Vector Labs B-1025).
- Streptavidin (1 mg/mL).
- Phosphate-buffered saline (PBS).
- Microfluidic:
  - Si master.
  - Trimethylchlorosilane.
  - Polydimethylsiloxane (PDMS) and cross-linker.
  - 70 °C Oven.
  - Alconox detergent.
- Cell culture:
  - Poly-L-lysine (PLL) 150–300 kDa (Sigma-Aldrich P1399-500MG).
  - #1.5 Coverslips (VWR 470019-010).
  - 5% CO_2 cell culture incubator.

2.1 Assembling the Microscope

The schematic for the polarization-resolved imaging setup is shown in Fig. 1. Generally, it is a 4*f* imaging setup, so called because the ultimate image plane is relayed four focal lengths after the image plane immediately outside the body of the microscope. This design facilitates placement of polarization optics and is flexible enough to toggle the system between different applications.

Alignment of the condenser and objective can be done by closing the field aperture of the condenser all the way and adjusting the NA on the objective to its maximum. A coverslip with writing on it can be placed on the microscope as something that is easy to find focus on through the eyepiece. Once the writing is in focus, adjust the height of the condenser until the edges of the field stop are in focus. Now the position of the condenser relative to the objective can be adjusted with the setscrews until it is centered. This can be done without the dark-field stop in place in order to allow more light to pass through the imaging optics, making alignment easier.

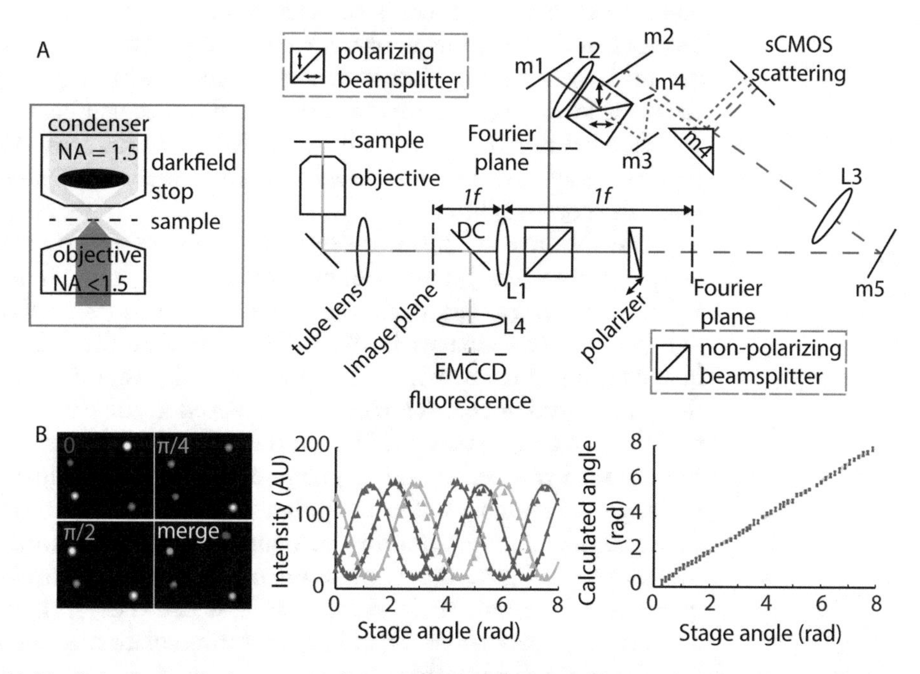

Fig. 1 Multipolarization optical setup. (**a**) Schematic of excitation (left) shows white light passing through the condenser and focused onto the sample. Nanorods selectively scatter red light which can be collected by the objective and relayed by the imaging optics (right). Conjugate image and Fourier planes are indicated with dashed lines. Lower wavelength fluorescent light (dashed green line) can be split off from red scattered light. Red scattered light from the nanorods can then be split into component polarizations (indicated by double-headed arrows) by the *4f* imaging setup. Different polarizations indicated by different dashed red lines. Scattered image paths can be recombined by the knife-edge mirror (m4) and imaged onto the sCMOS camera. (**b**) Example of four nanorods adsorbed to a glass coverslip (left). As the stage is rotated the intensities of a nanorod in each polarization channel oscillate sinusoidally (middle; triangles are measured intensity, solid lines are fit to the data). These intensities can then be converted to an angle (right)

The imaging optics are aligned as follows. With the same sample on the microscope, the image plane outside the microscope can be found by placing a screen and finding focus. This is the location where the rectangular aperture must be placed. The size of the aperture may be adjusted to roughly two times the magnified width of the microchannels. For a 60× objective and 20 μm channels, this is 1.2 mm. Next, place the first relay lens one focal length away from the aperture, ensuring that it is aligned with the optical axis.

Immediately after the first relay lens, a non-polarizing beam splitter is placed on a magnetic mount, again ensuring that the reflected light stays parallel to the surface of the optical table. By placing the first relay lens before the beam splitter, the number of optical elements common to all polarization channels is maximized. Each split path can now be aligned independently, starting with the reflected path.

The reflected path will be further split into s- and p-polarized light by a polarizing beam splitter. Prior to doing so, care must be taken to ensure that there is sufficient room to recombine all the images on the same image plane, roughly three focal lengths after the initial relay lens. Note that since there are multiple optical elements in the image path, the effective distance will be less than the nominal distance listed as the focal length of the lens. As with the first relay lens, the next lens is placed immediately before the polarizing beam splitter.

The mounts for the mirrors in the path of the p-polarized image (transmitted) must be minimal so as not to block the s-polarized image. Adjust the angles of the mirrors so that the light of the two images is parallel and so that the s-polarized image light just barely misses the D-mirror (Fig. 1a, m3) reflecting the p-polarized image. A screen can be placed in the path to check that the two images come in focus at the same plane. If they do not, the distance traveled by the p-polarized light can be adjusted by moving the first mirror after the polarizing beam splitter further or closer to the cube. Using only the s- and p-polarized channels it is possible to estimate the orientation of the nanorod within one quadrant of the unit circle. Adding at least one more polarization helps break this degeneracy and allows for calculation of the absolute angle of the nanorod.

The third image path can be aligned, traveling in the opposite direction from the s-polarized image path. A piece of paper can be used to ensure that the s-polarized image and unpolarized image overlap exactly. To further confirm this, irises can be centered on both paths out of the non-polarizing beam splitter and the light from image path 1 should reflect back onto the center of the iris in image path 2 and vice versa. At this point, the polarizer for image path 1 may be placed after the non-polarizing beam splitter without affecting alignment.

The knife-edge mirror (m4) can be placed to combine all three image paths. The position of the mirror should be adjusted such that all image paths are completely reflected 90° toward the final position of the camera. Ideally, this is accomplished by mounting the knife-edge mirror on an *x–y* translational stage. The *y*-axis of the stage moves the knife-edge mirror closer or further from the camera while the *x*-axis can help ensure that the optical path lengths of the three channels are equal.

Finally, the camera can be placed at the final image plane. With just transmitted light, the rectangular aperture should appear in sharp focus on the camera. The sample can also be imaged and should appear in simultaneous sharp focus (parfocal) on all three image channels as well as in the microscope eyepiece.

Rotational dynamics of the nanorods can be quite fast (on the several millisecond timescale). Additionally, the range of scattered light intensity can be large given that it is dependent on the orientation of the gold nanorod. For this reason, it is helpful to use an sCMOS camera with fast readout rate and high dynamic range over an EMCCD.

2.2 Calibrating the Polarized Dark-Field Microscope

In order to verify that the multipolarization dark-field imaging setup is functioning properly, a sample of gold nanorods should be imaged, rotating the sample through known angles on a rotational mount. This must be done in dark field, so the dark-field stop can be placed into the condenser and the condenser apertures can be fully opened at this point. A dilute sample of gold nanorods in PBS with 10 mM magnesium chloride should be prepared and plated on a bare glass coverslip. This can be sandwiched with a thin polydimethylsiloxane (PDMS) spacer and another glass coverslip in order to mount the sample on the microscope.

With the NA of the objective is set to its minimum, the sample can be brought to focus in the eyepiece. The nanorods should appear as bright, red diffraction-limited spots on the surface. The NA of the objective can be slowly increased until transmitted light begins to be captured by the objective. The condenser height should simultaneously be optimized to maximize signal to noise.

The sample is now properly mounted and can be imaged by the camera. On the camera, the images in the three channels should appear fairly different from one another (Fig. 1b) as nanorods will appear bright in some channels but undetectable in others. The stage should be rotated through at least one full revolution, capturing images of the sample every 10° or so. The stage may need to be moved in order to maintain the same nanorods in the field of view. The intensities of the nanorods should show a $\sin^2$ dependence on stage angle, offset in phase for each of the polarization channels (Fig. 1b).

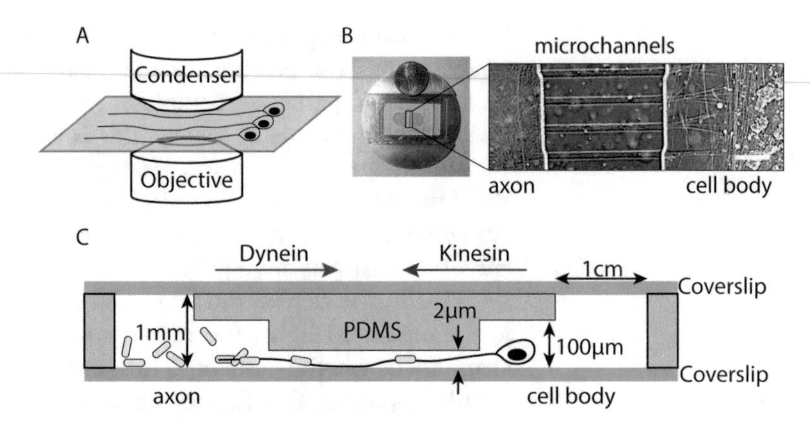

Fig. 2 Microfluidic culture of neurons for multipolarization imaging. Neurons plated in the cell body chamber of a microfluidic culture device will grow out aligned neurites through the microchannels to be imaged in dark field (**a**). The microchannels are small enough to not allow cell bodies (**b**) as is evident in bright field (scale bar = 100 μm). A schematic of a side view of the microfluidic culture device (**c**) shows that the PDMS must be kept thin enough such that neurites remain within the working distance of the condenser

2.3 Microfluidic Neuronal Culture

The fabrication of silicon master for the microfluidic culture platform is thoroughly described elsewhere [34, 35]. Custom masters can also be purchased commercially. The design features 20 μm wide, 2 μm tall channels that are at least 100 μm long. The channels connect areas that are 100 μm high on each side (Fig. 2).

Once a master is available, PDMS devices can be made by mixing degassed silicone elastomer and curing agent at a 10:1 w/w ratio. Prior to pouring the PDMS onto the master for replica molding, the master should be pretreated by vapor coating the surface with trichloromethylsilane (TCMS) to facilitate removal of the PDMS. This is accomplished by placing the master in a sealed box with an open bottle of TCMS for at least 5 min.

Two important considerations in making the PDMS device are keeping the devices flat during curing and keeping the thickness of the PDMS to roughly 1 mm. This thickness can be achieved by pouring a minimum amount of mixed PDMS/curing agent onto the silicone wafer and tilting it until the whole surface is covered. For condenser-type dark field as presented here, the excitation light must pass through the PDMS prior to reaching the neurons (Fig. 2). Therefore, it is essential that the thickness of the PDMS poured onto the master be less than the working distance of the condenser. A thinner device also reduces the total scattering of the excitation light, thereby improving image quality and maximizing the useable NA of the objective.

The wafers can now be placed into a desiccator under vacuum until the PDMS is fully degassed, at which point they can be transferred to an oven at 70 °C overnight to cure. Once cured, the axon and cell body chambers can be cut into the PDMS with a sharp razor, keeping one chamber larger than the other. Cut PDMS devices can be washed several hours in Alconox solution, rinsed with ddH_2O, dried, and autoclaved.

2.4 Neuronal Cell Culture

To maximize outgrowth of axons through the channels, coverslips should be cleaned with Alconox, thoroughly rinsed with ddH_2O, and dried under UV light to sterilize. One day before plating neurons, coverslips are coated with poly-L-lysine (PLL) overnight at 37 °C, then rinsed again with sterile ddH_2O, and allowed to fully dry prior to assembly. The PDMS should be carefully placed onto the clean coverslips and gentle pressure can be applied to ensure a proper seal between the PDMS and the glass. As these devices are quite thin, excess pressure over the channels, themselves, should be avoided as this can collapse the channels and disrupt the PLL coating underneath.

The microfluidic culture device is set up as depicted in Fig. 2. Dorsal root ganglion (DRG) neurons from embryonic day-18 (E18) Sprague-Dawley rats can be prepared by standard methods [36]. Briefly, the embryos are removed from the mother, decapitated, and the bodes are immediately placed in ice-cold HBSS. One embryo should yield enough neurons for 2–4 microfluidic culture devices. In a clean petri dish on a bed of crushed ice, an embryo can be laid dorsal side up and the spinal cord is removed using fine tweezers. Afterwards, the DRGs are visible as small white balls on each side of the spinal column. They can be removed with tweezers and placed in a fresh tube of HBSS. Once dissected, the HBSS can be aspirated off and replaced with 5 mL of trypsin + EDTA and incubated at 37 °C for 30 min. Every 10 min, the DRG suspension should be mixed by inverting the tube. After 30 min, 5 mL of plating media is added and the suspension is triturated with a 1 mL pipette until no clumps of tissue are visible.

The cells can be concentrated by spinning down and resuspending in plating media. The final cell density immediately prior to plating should be at least $10 * 10^6$ cells/mL. If the investigator wishes to eventually incubate nanoparticles on the cell body side, lower cell densities tend to result in higher flux of particles through the channels. Higher densities can inhibit penetration of the nanoparticles close to the microchannels. If axonal side particle incubation is intended, higher cell densities should be used on the order of $25 * 10^6$ cells/mL. This ensures a high density of axons crossing through the microchannels of the PDMS and higher flux.

Immediately before plating cells, 20 μL of plating media should be added to the axon chamber. Proper PLL coating of the coverslip will result in fast uptake of the liquid into the 100 μm groove and microchannels by capillary action. At this point, 4 μL of cell suspension is added to the cell body side and allowed to attach for 5 min with the coverslip tilted, taking care not to let the liquid dry out. Finally, several hundred microliters of maintenance media can be added to the cell body chamber plus enough to fill the axon chamber. Due to the thinness of the PDMS, extra care should be taken to avoid allowing mixing of the liquid between the cell body and axon compartments as this will impede nanoparticle incubation.

After 24 h, the media is replaced with maintenance media supplemented with arabinofuranosyl cytidine (Ara-C) to suppress the growth of proliferative glial cells. After an additional 24 h, the Ara-C-containing maintenance media should be replaced with fresh maintenance media without Ara-C once every 48 h.

2.5 Gold Nanorod Preparation

Gold nanorods, particularly with non-neutral zeta potentials, are stable for several months in solutions containing cetyl trimethylammonium bromide (CTAB) concentrations above the critical micelle concentration (1 mM). CTAB is incompatible with live-cell experiments and is first replaced with thiolated polyethylene glycol (PEG). This is done by incubating 5 mL of gold nanorods with 5 mL ddH_2O and 5 mg of SH-PEG-biotin, rocking overnight at room temperature. This solution can be kept for up to a week at 4 °C.

On the day of the transport experiment, PEGylated nanorods can be conjugated to streptavidin. This conjugate can further be conjugated to biotinylated wheat germ agglutinin to drive endocytosis of the nanoparticles by the neurons. The conjugate is less stable than either the CTAB-coated nanorods or the PEGylated nanorods alone and will begin to aggregate after a day. Therefore it should be prepared fresh.

For the conjugation, 1.5 mL of PEGylated nanorods are mixed with Tween-20 up to 0.1% Tween-20, vortexed, and centrifuged at 3000 × *g* for 10 min at room temperature. As much supernatant as possible should be aspirated from the pellet, which should be quite soft. The pellet is resuspended in a solution up to a final concentration of 0.2 mg/mL streptavidin + 0.05% Tween-20 in ddH_2O. The concentration of the nanorods at this point should be roughly 4 nM with saturating streptavidin, which is allowed to bind for 30 min at room temperature with occasional mixing.

Wheat germ agglutinin (WGA) is a lectin that can bind glycosylated proteins on neuronal plasma membranes. It has been shown to robustly transport both anterogradely and retrogradely and therefore can be used as a model cargo for endosome transport in neurons [37]. Once nanorods are coated in streptavidin, biotinylated WGA can be added to the solution up to a final concentration of 30 μM of WGA. Insufficient amounts of biotin-WGA can lead to cross-linking of multiple nanorods. The solution should be allowed to conjugate for 30 additional minutes at room temperature with occasional mixing. Excess WGA should then be removed by diluting the conjugate up to 2 mL in ddH_2O + 0.05% Tween-20, and then centrifuging at 3000 × *g* for 10 min. As much supernatant as possible should be aspirated and the pellet can be resuspended in PBS up to a final volume of 50 μL. The pellet should still be easily resuspended with flaky, difficult-to-resuspend pellets indicative of nanorod aggregation. This is enough conjugate for either two cell-body incubations or four axon-side incubations.

2.6 Imaging Transport

WGA-conjugated gold nanorods can be incubated with either cell bodies or axonal termini of cultured neurons which have grown axons all the way through the microchannels. This will result in predominantly anterograde or retrograde flux in the microchannels, respectively. Incubation is done by removing the cell culture media from the desired chamber and replacing it with enough WGA-conjugated gold nanorods to cover the surface. The opposite chamber should be kept full with cell culture media to prevent diffusion of gold nanorods into the channels.

WGA-conjugated gold nanorods are incubated with the neurons for 30 min in the cell culture incubator, and then replaced with warm maintenance media for an additional 1 h for axon-side incubation. The incubation for cell-body-side incubation is 2 h in order to allow time for sufficient flux in the microchannels.

At the end of the incubation the maintenance media is replaced with warm CO_2-independent medium. The meniscus of the liquid should be higher than the height of the PDMS to avoid introduction of air bubbles by placing a coverslip on top of the PDMS. The coverslip sandwich can be placed on the microscope with the channels oriented in the direction of the aperture of the imaging path.

Both cells and nanorods should be easily observable through the eyepiece. A good WGA-nanorod conjugation should result in dense coverage of the plasma membrane with WGA-conjugated nanorods in the chamber where particles were incubated (Fig. 3a). In this example, WGA-nanorods were added to the

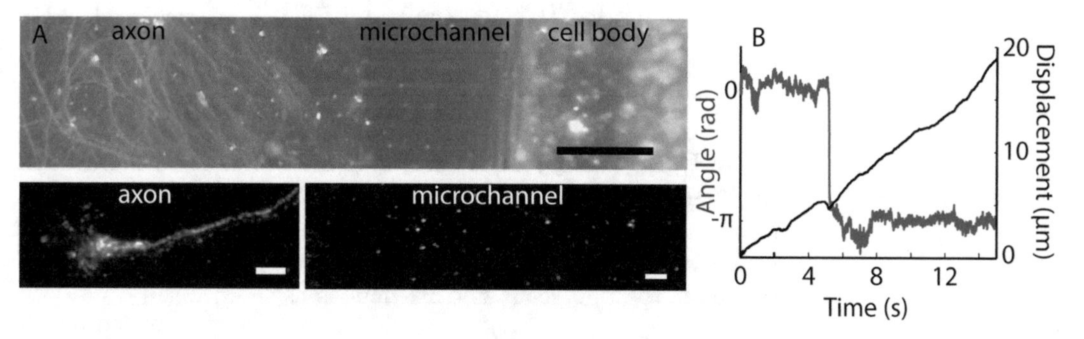

Fig. 3 Selective incubation with nanorods. WGA-conjugated nanorods added to the axon side of the microfluidic culture device make the axon chamber appear red even after washing out free nanorods (**a**). The microchannels and cell body compartment appear clear except for nanorods that have been endocytosed and transported (scale bars 100 and 5 μm for top and bottom images, respectively). (**b**) Transporting nanorods can be tracked in multipolarization dark-field microscopy to calculate both angle and displacement of the endosome carrying them. In this instance, a rotational motion clearly coincides with a reversal in direction

axon chamber making it appear red. Axonal termini are densely stained by the nanorods, but only transporting nanorods are observed in the microchannel. A microchannel with high transport flux can also be found through the eyepiece. Once an adequate channel is found, the microscope light path can be switched to the camera, allowing imaging of transport.

With nanorods of the size indicated here, the scattering cross section is quite high and therefore exposure time of the camera can be set around 1 ms while maintaining high localization precision [27, 38]. The rate-limiting parameter for imaging is the readout time of the camera being used, which is set by the number of rows of pixels being read out. Faster rotations of endocytosed nanorods are on the order of 10 ms [27]; thus the imaging region of interest should be set small enough to maintain a sampling interval of 10 ms.

2.7 Analysis of Transport Data

A variety of methods are available for robust single-particle tracking for time-lapse imaging [39–40]. For tracking rotational motion, because gold nanorods do not photobleach or blink, the particle should be detectable in every frame. The intensities in the different polarization channels will vary, however. It is essential, therefore, to calculate a 2D registration map to allow calculation of the position of the particle position across all polarization channels. This can be done by imaging spherical gold particles which will appear in all polarization channels.

The azimuthal (φ) and polar (θ) angles of the nanorod can be calculated by the following expressions for three-channel imaging [41]:

$$\varphi = \frac{1}{2} \tan^{-1} \left\{ \frac{I_{45} - \frac{I_0 + I_{90}}{2}}{\frac{I_0 - I_{90}}{2}} \right\} \tag{1}$$

$$I_{\text{tot}} = \frac{1}{2A} \left[I_0 \left(1 - \frac{B}{C \cos 2\varphi} \right) + I_{90} \left(1 + \frac{B}{C \cos 2\varphi} \right) \right] \tag{2}$$

$$\theta = \sin^{-1} \left[\sqrt{\frac{I_0 - I_{90}}{2 I_{\text{tot}} C \cos 2\varphi}} \right] \tag{3}$$

where I_0, I_{45}, and I_{90} are the measured intensities of the nanorod in the 0°, 45°, and 90° polarization channels. I_{tot} is the calculated total intensity emitted by the nanorod (only a fraction of which is collected by the objective). The quantities, A, B, and C are a function of the microscope, itself, and are defined as

$$A = \frac{1}{6} - \frac{1}{4} \cos \alpha + \frac{1}{12} \cos^3 \alpha \tag{4}$$

$$B = \frac{1}{8} \cos \alpha (1 - \cos^2 \alpha) \tag{5}$$

$$C = \frac{7}{48} - \frac{1}{16}\cos\alpha - \frac{1}{16}\cos^2\alpha - \frac{1}{48}\cos^3\alpha \quad (6)$$

$$\alpha = \sin^{-1}\frac{\text{N.A.}}{n} \quad (7)$$

with n being the refractive index of the immersion oil. While in principle it is possible to calculate the polar angle of the nanorod, in practice this quantity is much more sensitive to errors and is much more dependent on high NA optics than the azimuthal angle. Figure 3b shows an example of a transporting nanorod in this imaging setup.

3 Discussion

Conventional fluorescence microscopy has been an enormously useful tool for studying the transport of endosomes in axons. It is increasingly clear, however, that this transport is the product of several molecular motors that can be differentially regulated in space and time. As such, conventional fluorescence microscopy, tracking translational motion alone, cannot capture their dynamics. Here we describe a relatively simple approach for monitoring rotational motion of endosomes in addition to their translational motion. The two kinds of motion can provide complementary information describing axonal transport [26, 27].

Importantly, gold nanorods scatter light at specific wavelengths, with higher aspect nanorods scattering more red-shifted wavelengths [42]. By building the imaging optics in a 4*f* configuration, there is ample space for addition of a fluorescence imaging module which can be spectrally separated from the fast rotational measurements. This allows the investigator to observe how endosome transport dynamics correlate with fluorescently tagged modulators of transport such as Rab7 [5]. The 4*f* imaging setup also allows room for additional filters, such as phase masks or cylindrical lenses for resolving axial motions [43, 44].

References

1. Craig A, Banker G (1994) Neuronal polarity. Annu Rev Neurosci 17:267–310
2. Chowdary PD, Che DL, Cui B (2012) Neurotrophin signaling via long-distance axonal transport. Annu Rev Phys Chem 63:571–594
3. Howe CL, Mobley WC (2004) Signaling endosome hypothesis: a cellular mechanism for long distance communication. J Neurobiol 58:207–216
4. Maday S, Wallace KE, Holzbaur ELF (2012) Autophagosomes initiate distally and mature during transport toward the cell soma in primary neurons. J Cell Biol 196:407–417
5. Zhang K et al (2013) Defective axonal transport of Rab7 GTPase results in dysregulated trophic signaling. J Neurosci 33:7451–7462
6. Goldstein LSB (2012) Axonal transport and neurodegenerative disease: can we see the elephant? Prog Neurobiol 99:186–190
7. Her L-S, Goldstein LSB (2008) Enhanced sensitivity of striatal neurons to axonal transport defects induced by mutant huntingtin. J Neurosci 28:13662–13672

8. Florenzano F (2012) Localization of axonal motor molecules machinery in neurodegenerative disorders. Int J Mol Sci 13:5195–5206
9. Brady ST, Morfini GA (2017) Regulation of motor proteins, axonal transport deficits and adult-onset neurodegenerative diseases. Neurobiol Dis 105:273–282
10. Cui B et al (2007) One at a time, live tracking of NGF axonal transport using quantum dots. Proc Natl Acad Sci U S A 104:13666–13671
11. Hendricks AG et al (2010) Motor coordination via a tug-of-war mechanism drives bidirectional vesicle transport. Curr Biol 20:697–702
12. Caviston JP, Holzbaur ELF (2009) Huntingtin as an essential integrator of intracellular vesicular trafficking. Trends Cell Biol 19:147–155
13. Deinhardt K et al (2006) Rab5 and Rab7 control endocytic sorting along the axonal retrograde transport pathway. Neuron 52:293–305
14. Dixit R, Ross JL, Goldman YE, Holzbaur ELF (2008) Differential regulation of dynein and kinesin motor proteins by tau. Science 319:1086–1089
15. Nakata T, Niwa S, Okada Y, Perez F, Hirokawa N (2011) Preferential binding of a kinesin-1 motor to GTP-tubulin-rich microtubules underlies polarized vesicle transport. J Cell Biol 194:245–255
16. Janke C, Kneussel M (2010) Tubulin post-translational modifications: encoding functions on the neuronal microtubule cytoskeleton. Trends Neurosci 33:362–372
17. Shubeita GT et al (2008) Consequences of motor copy number on the intracellular transport of kinesin-1-driven lipid droplets. Cell 135:1098–1107
18. Lu Q, Li J, Zhang M (2014) Cargo recognition and cargo-mediated regulation of unconventional myosins. Acc Chem Res 47:3061. https://doi.org/10.1021/ar500216z
19. Welte MA, Gross SP, Postner M, Block SM, Wieschaus EF (1998) Developmental regulation of vesicle transport in Drosophila embryos: forces and kinetics. Cell 92:547–557
20. Chowdary PD, Che DL, Zhang K, Cui B (2015) Retrograde NGF axonal transport—motor coordination in the unidirectional motility regime. Biophys J 108:2691–2703
21. Campenot RB (1977) Local control of neurite development by nerve growth factor. Proc Natl Acad Sci U S A 74:4516–4519
22. Taylor AM et al (2003) Microfluidic multicompartment device for neuroscience research. Langmuir 19:1551–1556
23. Nan X, Sims PA, Xie XS (2008) Organelle tracking in a living cell with microsecond time resolution and nanometer spatial precision. Chemphyschem 9:707–712
24. Ortega Arroyo J, Cole D, Kukura P (2016) Interferometric scattering microscopy and its combination with single-molecule fluorescence imaging. Nat Protoc 11:617–633
25. Soppina V, Rai AK, Ramaiya AJ, Barak P, Mallik R (2009) Tug-of-war between dissimilar teams of microtubule motors regulates transport and fission of endosomes. Proc Natl Acad Sci U S A 106:19381–19386
26. Gu Y et al (2012) Rotational dynamics of cargos at pauses during axonal transport. Nat Commun 3:1030
27. Kaplan L, Ierokomos A, Chowdary P, Bryant Z, Cui B (2018) Rotation of endosomes demonstrates coordination of molecular motors during axonal transport. Sci Adv 4: e1602170
28. Wang X et al (2011) PINK1 and Parkin target Miro for phosphorylation and degradation to arrest mitochondrial motility. Cell 147:893–906
29. Wang G, Sun W, Luo Y, Fang N (2010) Resolving rotational motions of nano-objects in engineered environments and live cells with gold nanorods and differential interference contrast microscopy. J Am Chem Soc 132:16417–16422
30. Li T et al (2012) Three-dimensional orientation sensors by defocused imaging of gold nanorods through an ordinary wide-field microscope. ACS Nano 6:1268–1277
31. Xiao L, Qiao Y, He Y, Yeung ES (2010) Three dimensional orientational imaging of nanoparticles with dark-field microscopy. Anal Chem 82:5268–5274
32. Sönnichsen C, Alivisatos AP (2005) Gold nanorods as novel nonbleaching plasmon-based orientation sensors for polarized single-particle microscopy. Nano Lett 5:301–304
33. Boyer D, Tamarat P, Maali A, Lounis B, Orrit M (2002) Photothermal imaging of nanometer-sized metal particles among scatterers. Science 297:1160–1163
34. Zhang K et al (2010) Single-molecule imaging of NGF axonal transport in microfluidic devices. Lab Chip 10:2566–2573
35. Park JW, Vahidi B, Taylor AM, Rhee SW, Jeon NL (2006) Microfluidic culture platform for neuroscience research. Nat Protoc 1:2128–2136
36. Liu R, Lin G, Xu H (2013) An efficient method for dorsal root ganglia neurons purification with a one-time anti-mitotic reagent treatment. PLoS One 8:e60558. https://doi.org/10.1371/journal.pone.0060558

37. Dumas, M., Schwab, M.E. & Thoenen, H. (1979). Retrograde axonal transport of specific macromolecules as a tool for characterizing nerve terminal membranes. J. Neurobiol., 10: 179–197. https://doi.org/10.1002/neu.480100207
38. Chowdary PD, Kaplan L, Che DL, Cui B (2018) Dynamic clustering of dyneins on axonal endosomes: evidence from high-speed dark-field imaging. Biophys J 115:230–241
39. Zhang K, Osakada Y, Xie W, Cui B (2011) Automated image analysis for tracking cargo transport in axons. Microsc Res Tech 74:605–613
40. Jaqaman K et al (2008) Robust single-particle tracking in live-cell time-lapse sequences. Nat Methods 5:695–702
41. Fourkas JT (2001) Rapid determination of the three-dimensional orientation of single molecules. Opt Lett 26:211–213
42. Yu Y, Chang S, Lee C, Wang CRC (1997) Gold nanorods: electrochemical synthesis and optical properties. J Phys Chem B 101:6661–6664
43. Lee H-LD, Sahl SJ, Lew MD, Moerner WE (2012) The double-helix microscope super-resolves extended biological structures by localizing single blinking molecules in three dimensions with nanoscale precision. Appl Phys Lett 100:153701–1537013
44. Kao HP, Verkman AS (1994) Tracking of single fluorescent particles in three dimensions: use of cylindrical optics to encode particle position. Biophys J 67:1291–1300

Chapter 9

Practical Guidelines for Two-Color SMLM of Synaptic Proteins in Cultured Neurons

Xiaojuan Yang and Christian G. Specht

Abstract

The application of single-molecule localization microscopy (SMLM) to the study of synaptic proteins has shown that the postsynaptic density (PSD) is organized heterogeneously in subsynaptic domains (SSDs) that are thought to play important roles in neurotransmission and synaptic plasticity. However, the dense packing of neurotransmitter receptors and scaffold proteins at synapses, together with the small total number of target molecules, makes SMLM of synaptic components particularly challenging. Here, we discuss the technical difficulties of SMLM imaging that are specific to synapses. We present a method for dual-color direct stochastic optical reconstruction microscopy (dSTORM) of two inhibitory synaptic proteins, the glycine receptor (GlyR) and the scaffold protein gephyrin (GPHN), highlighting strategic choices and practical solutions for imaging quality control. Our aim is to provide biologists with guidelines for the implementation of two-color dSTORM imaging of synaptic proteins from sample preparation to data analysis.

Key words Super-resolution imaging, Single-molecule localization microscopy (SMLM), Two-color direct stochastic optical reconstruction microscopy (dSTORM), Synaptic proteins, Subsynaptic domain (SSD), Trans-synaptic nanocolumn, Correlation analysis

1 Introduction

1.1 Overview of SMLM Techniques

In conventional optical microscopy, the lateral resolution is limited by the wavelength (λ) and the numerical aperture (NA) of the objective. The minimal distance to differentiate two points in the image is several hundred nanometers due to the diffraction limit ($d \approx 0.61\ \lambda/\mathrm{NA}$). Conventional fluorescence microscopy therefore cannot gain detailed information on structures below the diffraction limit such as synapses, which have a size of only a few hundred nanometers. The advance of super-resolution fluorescence microscopy allows researchers to probe structural details on the nanometer scale. Single-molecule localization microscopy (SMLM) takes advantage of the blinking behavior of fluorophores (Fig. 1a) to calculate the precise position of individual molecules. During

Nobuhiko Yamamoto and Yasushi Okada (eds.), *Single Molecule Microscopy in Neurobiology*, Neuromethods, vol. 154,
https://doi.org/10.1007/978-1-0716-0532-5_9,

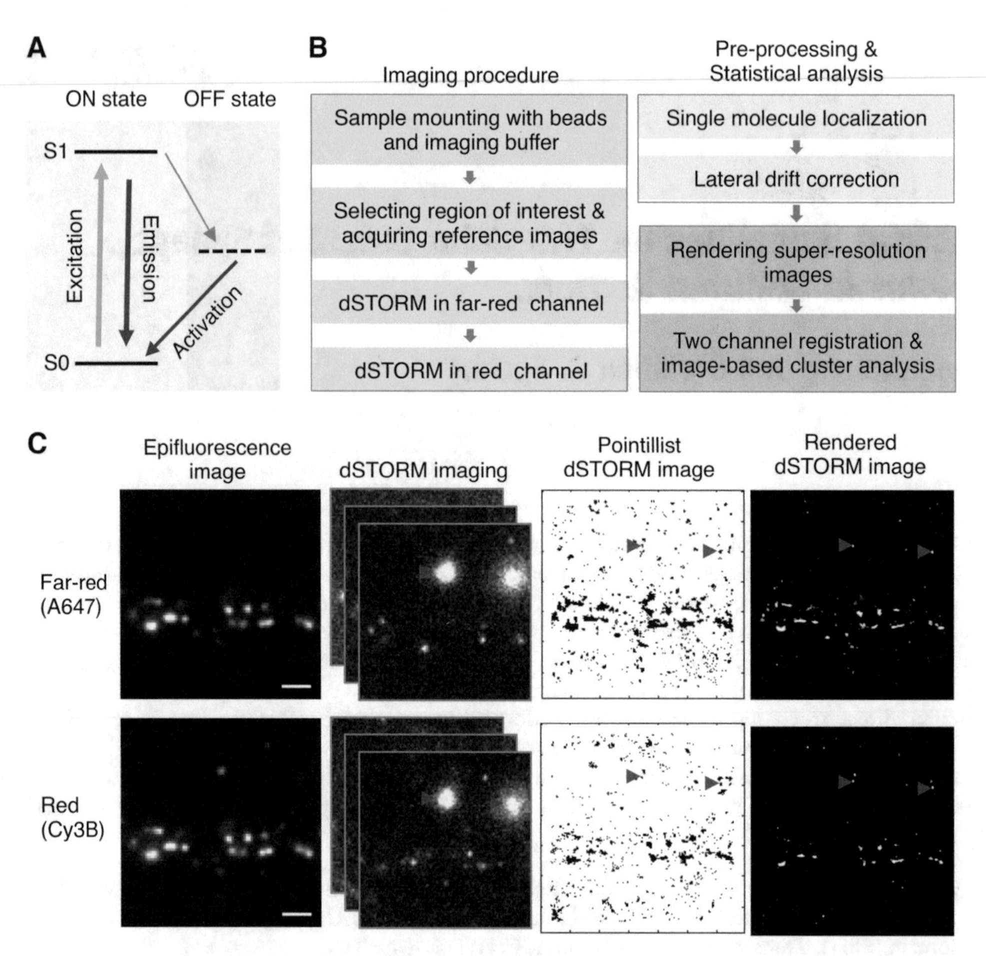

Fig. 1 Sequential two-color dSTORM imaging of synaptic proteins. (**a**) Simplified Jablonski diagram, showing the switching cycle of fluorophores. After excitation to the S1 state, the fluorophores can either return to the ground state (S0) by emitting light of a longer wavelength or enter the nonfluorescent OFF state. Fluorophores in the OFF state are reactivated by UV light or spontaneously in the presence of reducing agents in the imaging buffer. (**b**) Experimental workflow for two-color dSTORM imaging and data analysis. (**c**) Representative images of two-color dSTORM recordings. GlyR α1 subunits were labeled with a single primary antibody, followed by a combination of two secondary antibodies conjugated with Alexa Fluor 647 and Cy3B dyes. From left to right: epifluorescence reference images, acquisition of image stacks with single-fluorophore blinking events, pointillist dSTORM images reconstructed from single-fluorophore localizations (after drift correction), and rendered super-resolution images in the far-red (Alexa 647) and in the red channel (Cy3B). Arrowheads point to beads in the region of interest (ROI) used for drift correction. Scale: 2 μm

imaging, a sparse subset of fluorophores is activated into the ON state by a low dose of UV light or spontaneously in the presence of reducing agents in the imaging buffer to achieve a temporal separation of the signals. These single detections are fitted mathematically during image processing, attaining a localization precision of ~10 nm. Super-resolution images can then be reconstructed by superimposing all the detections from the recording.

SMLM differs from conventional optical microscopy in several ways, and there are some definitions in SMLM that biologists should be aware of. The first one is the localization precision. It is the precision with which single detections are localized by fitting the signal to a Gaussian distribution or by using other algorithms. The localization precision can be approximated by $\Delta/\sqrt{N}$, where Δ is the full width at half-maximum (FWHM) of the point spread function (PSF) of the detections, and N is the number of collected photons [1, 2]. Note that the localization precision is also affected by background noise, in other words, the signal-to-noise ratio (SNR). The second parameter is the localization density, which is the number of localizations per spatial unit. It relies on the labeling density, as well as the blinking properties of the fluorophores. It has been proposed that the labeling density should be sufficiently high to meet the Nyquist-Shannon criterion [3, 4]. According to this concept, the sampling must be at least two times denser than the achieved spatial resolution. However, this is not always the case in SMLM, because fluorophore localizations are not classical samples as in conventional microscopy but stochastic events [5]. The detection of redundant information due to multiple switching cycles of the fluorophores means that the number of detections is much higher than the actual sampling of independent targets. Baddeley and Bewersdorf demonstrated that a significantly higher localization density, empirically set to exceed the Nyquist-Shannon criterion by fivefold, is therefore needed to attain the best spatial resolution [5]. The third definition is the spatial resolution, which is not as clear a concept any more as in conventional microscopy. Canonically, it refers to how well two objects can be separated, which is strongly dependent on the wavelength and the objective. In SMLM, the spatial resolution is primarily dependent on the localization precision of the detections. Due to the scarcity of the labeling and the stochasticity of the detections in SMLM, however, it is also strongly dependent on the labeling efficiency, i.e., the density of the probes.

SMLM techniques include stochastic optical reconstruction microscopy (STORM) [6] and photoactivated localization microscopy (PALM) [7, 8]. The main difference between STORM and PALM is the labeling technique: in STORM imaging, endogenous proteins are labeled with organic dyes, whereas for PALM the target proteins are tagged with photo-convertible fluorescent proteins. Both organic dyes and fluorescent proteins can be induced to blink under different conditions. In STORM, the blinking of fluorescent dyes requires particular imaging buffers and high illumination intensity. In PALM, the blinking of the photo-convertible fluorescent proteins is achieved by sequential activation with UV light in buffered salt solutions. PALM is thus preferable for live-cell imaging, since the STORM imaging buffers are detrimental to living cells. However, labeling with fluorescent proteins requires the expression of recombinant fusion proteins, which may potentially change the target protein's function. Moreover, organic dyes usually have more narrow excitation/emission spectra and higher photon emission compared to fluorescent proteins. Therefore, STORM is generally more suitable for multicolor imaging and to gain a better localization precision.

There are two versions of STORM imaging, N-STORM [6] and direct STORM (dSTORM) [9]. N-STORM adopts a pair of dyes, one with a short wavelength (green) and another with a longer wavelength (red), to act as activator and reporter, respectively. The pair of dyes are conjugated to the same antibody probe at close proximity. During imaging, a red laser is used to excite the red dyes and to switch them into the OFF state, while a green laser at low intensity excites the green dyes that in turn reactivate the red fluorophores. Note that no UV light is needed in N-STORM. For dual-color imaging, the two target proteins are labeled with the same reporter dye (e.g., Alexa Fluor 647), but different activator dyes are used to selectively visualize the two targets [10]. There are, however, a few drawbacks of using activator/reporter dye pairs in STORM. Firstly, it is difficult to control precisely the ratio of activator and reporter dyes on the same antibody. Secondly, by using two wavelengths for a single target protein, N-STORM leaves limited options of dyes for multicolor imaging. Thirdly and most importantly, the commonly used reporter dye Alexa Fluor 647 can blink spontaneously in the imaging buffer without laser activation, making N-STORM susceptible to channel cross talk. This is particularly problematic for dense structures such as synaptic protein assemblies, because there is no efficient way to identify and remove the contaminating detections. In dSTORM, only single fluorophores (reporter dyes) are used for imaging, and their activation is tuned by a low dose of UV light and reducing agents in the imaging buffer. The spectral separation of the fluorophores affords a larger choice of dyes, making dSTORM more suitable for dual-color super-resolution imaging of synaptic proteins.

1.2 SMLM Applications in Neurobiology

SMLM techniques have been successfully applied to many biological structures, providing new insights into their subcellular organization and functions [5, 11, 12]. Baddeley and Bewersdorf classified the analyzed structures into three categories of increasing difficulty: (1) well-isolated repetitive or stereotypic structures, (2) complex but isolated structures, and (3) complex and closely spaced structures [5]. This classification is a good way to illustrate that the labeling, imaging, and data analysis need to be optimized for a given molecular complex. Stereotypic structures allow combining the information from many complexes to reconstruct composite images, and hence are less demanding in terms of the labeling density and detection efficiency. Variable structures such as synapses need to be treated individually. Therefore, it is not surprising that most SMLM studies to date have been conducted on stereotypic structures such as filaments and nuclear pore complexes (NPC), and that only few detailed studies have been carried out on complex structures such as gene loci or synapses.

One of the most striking discoveries that have been made with SMLM imaging was the identification of periodic actin rings in neuronal dendrites and axons [13]. These actin rings are evenly distributed along axons and connected by spectrin tetramers, forming a periodic cytoskeletal structure. The same periodic organization was found in the axon initial segment (AIS), providing a robust structural scaffold [14]. The dendritic spine neck also contains periodic actin structures, as demonstrated by another super-resolution imaging technique, namely stimulated emission depletion (STED) microscopy [15]. The discovery of actin rings in neurons has thus led to substantial further research that has fundamentally changed our understanding of the morphogenesis of neurons. Another successful application of SMLM is the description of the eightfold symmetry of the NPC and the orientation of its Y-shaped component [16–18]. Due to their stereotypic structure and well-known features, NPCs and microtubule filaments have been commonly used to assess the performance of newly developed fluorophores, probes, algorithms, etc.

Synapses are particularly exciting subjects for SMLM due to their small size and high complexity, and important insights into their ultrastructure have thus been gained. Dani and colleagues have examined pre- and postsynaptic protein assemblies with SMLM, showing the layered molecular organization of different synaptic components [19]. This arrangement may explain the different kinetic properties of excitatory scaffold proteins, where proteins that are located at a greater distance from the synaptic membrane have faster exchange rates [20]. It has also been shown by SMLM that postsynaptic scaffold proteins are organized in subsynaptic domains (SSDs) that undergo dynamic changes and that are thought to regulate the strength of neuronal transmission at excitatory [21–23] and inhibitory synapses [24–27]. SSDs were also observed in vivo and their existence has been confirmed using different super-resolution imaging techniques [28–32] (discussed in [33]). It was further proposed that pre- and postsynaptic SSDs are aligned in so-called trans-synaptic nanocolumns [34]. Synaptic adhesion molecules are likely to coordinate these trans-synaptic complexes [23, 35–38]. Finally, SMLM has been used to determine the absolute numbers of scaffold proteins and neurotransmitter receptor complexes at synapses, providing new types of quantitative information about the regulation of synaptic structure and function [24, 39, 40].

1.3 Difficulties of Applying Two-Color dSTORM to Synaptic Proteins

SMLM has shown great promise in revealing the internal organization of synapses and its dynamic rearrangements during synaptic plasticity. However, SMLM is a complex technology, and several intrinsic limitations need to be taken into consideration for the experimental design (*see* Subheading 2) [41]. In addition, the unique character of synaptic protein clusters poses particular challenges to the acquisition of reliable SMLM data and to obtaining statistically meaningful results.

First of all, the size of synapses is of the same order as the diffraction limit, with only a few hundred nanometers in diameter, which makes it difficult to obtain isolated blinking events during imaging (reviewed in [33]). Even when the overall blinking in the field of view is sparse, there may still be overlapping blinking events at synaptic sites. The assignment of multiple signals to a single detection necessarily causes wrong representations in the super-resolution image. Second, synapses have large structural variability in terms of cluster shape and molecule numbers. As a consequence, individual synaptic clusters cannot be combined and population averaging is not applicable. Each synaptic cluster needs to be treated individually, which makes the data analysis more challenging. Thirdly, even though synapses are densely packed compartments, the copy numbers of many of their components are quite small, with receptors in the order of tens and scaffold proteins in the order of hundreds of molecules per synapse. Small numbers of target proteins make it difficult to attain sufficiently dense labeling in order to efficiently sample the synaptic structure and to reconstruct a representative super-resolution image. Fourth, synaptic proteins undergo dynamic changes and reorganization during synaptic plasticity, which adds to the difficulties of SMLM data interpretation. Given the variability and plasticity of synapses, large sample sizes (number of synapses) are needed to draw meaningful conclusions. Taken together, the characterization of the internal synaptic organization from dSTORM data is subject to the difficulties faced by closely spaced complex structures [5]. In addition to the aspects mentioned above, the detection of two synaptic components in parallel dramatically increases the complexity of the experiments, because it requires not only a good imaging quality in the two channels, but also the implementation of more complicated data processing and analysis tools.

Owing to the complexity of the technology and the unique properties of any research target, each application of SMLM has to be optimized and the imaging quality closely controlled (Subheading 4.3). We strongly recommend that a suitable biological question is formulated in the first place, and that factors such as the spatial resolution, sampling efficiency, and data output of the SMLM recordings are weighed against the reproducibility and expected magnitude of the observed effect. In our case, we hypothesized that different neurotransmitter receptors and scaffold proteins occupy specific subsynaptic domains (SSDs) at inhibitory synapses [33]. Our aim was to compare the distributions of these inhibitory synaptic proteins on a scale of tens of nanometers using dual-color dSTORM imaging.

In spinal cord neurons, fast inhibition is mainly mediated by two types of receptors, glycine receptors (GlyRs) and γ-aminobutyric acid (GABA) type A receptors ($GABA_ARs$). Their neurotransmitters glycine and GABA are co-released from the same presynaptic terminals [42, 43] and the two types of receptors coexist at the same postsynaptic densities (PSD) [44–48]. Both GlyRs and $GABA_ARs$ bind to the inhibitory synaptic scaffold protein gephyrin. They share a common binding site but have different binding affinities for gephyrin [49–52]. The accumulation of the two types of receptors at synapses can be differentially regulated; for example, phosphorylation of gephyrin at the amino acid residue S270 reduces $GABA_AR$-gephyrin binding, without affecting the diffusion properties of GlyRα1-containing receptors [53]. GlyR- and $GABA_AR$-mediated postsynaptic currents have different kinetic properties, with GlyR currents exhibiting fast decay times and $GABA_ARs$ decaying more slowly [43, 52, 54]. Together, the co-transmission and the independent regulation of GlyRs and $GABA_ARs$ at mixed inhibitory synapses may modulate the time course and amplitude of fast inhibition in spinal cord neurons. Recently, super-resolution fluorescence microscopy has revealed that gephyrin molecules form SSDs that are dynamically reorganized during inhibitory synaptic plasticity [24, 26, 27, 29]. It is tempting to speculate that these gephyrin SSDs mediate the subsynaptic distributions of the inhibitory neurotransmitter receptors. We therefore set out to investigate the spatial relationship between GlyRs, $GABA_ARs$, and gephyrin at mixed inhibitory synapses. Whether GlyRs and $GABA_ARs$ occupy the same or separate SSDs has profound consequences for the understanding of co-transmission at mixed inhibitory synapses.

The intrinsic challenges that we encountered when applying SMLM to inhibitory synaptic proteins led us to the method described herein (Fig. 1b, c). We found that practical guidelines for biologists to set up quality-controlled SMLM experiments were sorely missing. In this chapter, we describe the critical considerations for implementing two-color dSTORM of synaptic proteins and provide biology-oriented strategies to assess the imaging quality. We also present a workflow for data analysis with freely accessible and ready-to-use software. We hope that these guidelines can help biologists set up their own SMLM workflow to study synaptic structures or similarly complex cellular compartments.

2 Sample Preparation for Two-Color dSTORM

2.1 Labeling Strategies

With a localization precision of single molecules on the order of 10 nm, the size of the probes used for labeling can impact the quality of dSTORM data in several ways (Fig. 2). Commercial primary and secondary antibodies are easy to access and convenient for immunostaining; hence they are commonly used to label endogenous proteins in SMLM experiments. However, the average size of full-length IgG is about 10 nm, which is similar to the single-molecule localization precision in dSTORM. The combination of primary and secondary antibodies can add a distance of up to 20 nm between the epitope and the fluorophores. If the orientation of the antibodies bound to the target proteins is random, the detected synaptic clusters can be blurred and the resolution of the super-resolution images is lessened. Moreover, the bulky size of antibodies can limit the labeling density due to the masking of epitopes in the crowded synaptic environment, which reduces the sampling efficiency. And finally, the cross-linking effect of primary and secondary antibodies may potentially induce clustering artifacts that only become apparent on the nanometer scale, especially when using live staining of membrane proteins such as neurotransmitter receptors (Fig. 2a).

Different approaches can be adopted to overcome the drawbacks of conventional antibodies in dSTORM. For example, antigen-binding fragments (Fab) can be used instead of whole IgG as secondary labels. Primary antibodies can also be conjugated directly with fluorescent dyes, omitting secondary antibodies altogether [36]. Alternatively, single-domain antibody fragments derived from camelid or shark antibodies, so-called nanobodies,

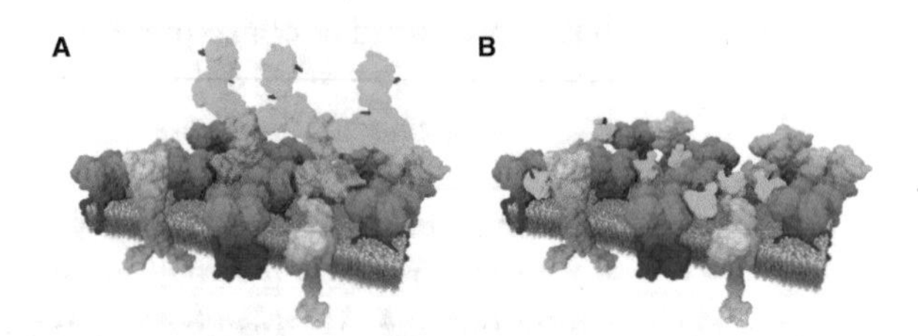

Fig. 2 The effect of the probe size. (**a**) The large size of primary and secondary antibodies limits their access to all the epitopes and keeps the fluorophores at a distance from the target proteins. Live labeling with antibodies may also cause artificial clustering of membrane proteins. (**b**) Small monovalent probes such as nanobodies circumvent these limitations, avoiding epitope masking, signal delocalization, and cross-linking. The epitope-binding interface is shown in magenta, organic dyes in red, labeling probes in green, and target proteins in blue. These illustrations were originally published by Maidorn et al. [55] (https:/doi.org/10.1042/BCJ20160366), and were reproduced with the kind permission of the authors and the publisher

have recently caught a lot of attention due to their favorable small size [55] (Fig. 2b). If combined with specific protein tags, dye-conjugated nanobodies can be used as a modular detection system that generates less epitope displacement resulting in a better spatial resolution [56–58]; however, this requires modification and recombinant expression of the target proteins. Unfortunately, none of these approaches can be easily implemented. Coupling primary antibodies with organic dyes is prone to reduce the binding efficacy and may even affect antibody specificity, whereas the production of specific nanobodies for a target protein can be time consuming. Nevertheless, the use of smaller probes is certainly preferable whenever they are available.

Due to the practical limitations mentioned above, we propose to start using conventional immunostaining with primary and secondary antibodies that have been previously used and validated by classical fluorescent microscopy, which is also the case in our experiments. Even though we obtain a slightly lower resolution with conventional antibody labeling of GlyRs and gephyrin clusters than with smaller probes, we expect that any significant changes in the synaptic distribution of the receptors and scaffold proteins in response to regulatory processes can still be detected. In addition to using commercial dye-conjugated secondary antibodies, we have also coupled unconjugated secondary antibodies with succinimidyl ester (NHS ester) derivatives of fluorescent dyes, in order to better control the dye/protein ratio (D/P ratio ranging from 1 to 4). By reducing the amount of redundant information arising from different dyes on the same antibody, this simple approach increases the sampling efficiency of independent target proteins.

2.2 Choice of Fluorescent Dyes and Imaging Buffers

Fluorescent dyes are the most critical factors for the imaging quality of multicolor dSTORM. As illustrated in Fig. 1a, fluorophores switch between ON and OFF state. The excited fluorophores (S1) can either emit red-shifted light and return to ground state (S0) or enter into the OFF state. Fluorophores in the OFF state can return to the ON state spontaneously (with low efficiency). This process can be accelerated by activation lasers and addition of reducing agents to the imaging buffer. Generally, all dyes can be activated by light of shorter wavelength and almost all dyes can be activated by UV light with different efficiency. The ON/OFF switching of fluorophores is stochastic, and the duration during which the fluorophores remain in the OFF state can last from sub-milliseconds to minutes or even hours. Since the ability to separate blinking events both spatially and temporally during imaging is key to obtaining accurate dSTORM data, the photophysical properties of the dyes have a decisive impact on the experimental outcomes. Artifacts that occur in the reconstructed images as a consequence of overlapping blinking events are difficult to be identified and removed during data processing, and can lead to false interpretations [59–61].

The blinking behavior of the dyes is dependent on the illumination intensity and the reducing agents in the imaging buffer. In dSTORM imaging, illumination with a high-intensity excitation laser is used to keep most fluorophores in the OFF state while inducing efficient blinking of the fluorophores [62, 63]. The presence of oxygen in the imaging buffer causes the irreversible bleaching of fluorophores and is therefore to be avoided. It has also been demonstrated that the blinking performance of many dyes is superior in low-oxygen buffers containing reducing agent such as thiols to assist the activation from the OFF state to the ON state [63–65]. Therefore, most dSTORM imaging buffers contain an enzymatic oxygen scavenging system as well as millimolar concentrations of the thiols β-mercaptoethylamine (MEA) or β-mercaptoethanol (*see* Appendix for buffer compositions). Most commonly, oxygen removal is achieved with a glucose oxidase-based enzymatic system (Gloxy buffer), which works optimally for carbocyanine dyes [66]. Rhodamine dyes usually do not blink well in the complete absence of oxygen; hence an oxyrase-based buffer (OxEA) with a residual oxygen level is preferred for multicolor dSTORM imaging with rhodamine dyes [67]. It should be noted that novel organic dyes are continuously being developed in an attempt to further improve the fluorophore performance [68–71], and there is quite a range of commercially available dyes for dSTORM. However, due to photochemical differences in addition to their spectral properties, identifying a pair of compatible dyes that behave equally well in the same imaging system is not trivial [66, 67].

A characterization of the blinking performance of the dyes is therefore indispensable and should precede the real experiments. Empirically, the performance of the dyes can be assessed according to the following principles: (1) at the beginning of the dSTORM recording, the activated fluorophores can be easily pushed into the OFF state in reducing and oxygen-depleted buffer conditions under strong illumination; (2) the fluorophores in the OFF state can be efficiently activated into the ON state by UV light or spontaneously; (3) the dyes can go through several ON/OFF switching cycles, which ensures that sufficient detections can be recorded; (4) the duration of the fluorescence emission should be short (not longer than a few image frames) in order to avoid repetitive detections of the same fluorophore during a single blinking event; (5) the dyes should be sufficiently bright to obtain a high localization precision of the single-molecule signals. Since all of the above parameters are dependent on the imaging conditions, the optimization of the laser intensities and the buffer composition can substantially improve the blinking performance of the fluorescent dyes.

As a starting point, we strongly recommend the carbocyanine dye Alexa Fluor 647 in the far-red channel. Alexa 647 is commonly used in dSTORM due to its excellent photophysical properties that satisfy the requirements mentioned above. Moreover, appropriate blinking of Alexa 647 can be achieved in various buffers, and there is generally very little background fluorescence in biological samples in the far-red channel. To determine the second color for two-color dSTORM, we have tested several rhodamine (Atto 488, Alexa Fluor 488, Alexa Fluor 568) and carbocyanine dyes (Cy3B) in different imaging buffers, to evaluate their compatibility with Alexa 647 in sequential two-color dSTORM. As expected, Gloxy and OxEA buffers performed better than MEA-containing buffers and PBS. In the green channel, Atto 488 performed better than Alexa 488 in Gloxy buffer. The concern of using green dyes is that biological samples often have a high level of fluorescence background in this part of the spectrum, and that the higher energy of the green light accelerates the bleaching of the fluorescent beads used as fiducial markers (Subheading 2.3). In the red channel, Alexa 568 blinks nicely in OxEA buffer but less well in Gloxy buffer. Cy3B on the other hand works better in Gloxy buffer. Also, Alexa 647 performs optimally in Gloxy buffer but does not exhibit extensive spontaneous blinking in OxEA buffer without UV activation. Therefore, we chose Alexa 647 and Cy3B in Gloxy buffer as the optimal imaging system. Since our two-color dSTORM recordings are done sequentially (Fig. 1b, c), this combination of dyes allows us to omit UV activation during the imaging of Alexa 647 due to its spontaneous blinking, thus avoiding any undesired bleaching of Cy3B or the beads. In the second stage, Cy3B is imaged in the red channel with an increasing 405 nm illumination to dynamically adjust the number of fluorophores in the ON state. It is recommended to prepare fresh imaging buffer before the dSTORM recordings, since aged buffers tend to induce less and slower blinking events. Alternatively, relatively stable glucose-oxidase-based imaging buffer is commercially available (e.g., Abbelight, smart kit).

2.3 Fluorescent Beads as Fiducial Markers

Drift correction and two-channel registration are essential steps for the reconstruction of super-resolution images and for co-localization analysis of synaptic components [72]. Fluorescent beads are generally used as fiducial markers in two-color dSTORM for both drift correction and image registration. Since the recordings can take up to 30 min, the microscope stage can drift dramatically in the x/y plane by up to tens of micrometers due to mechanical and thermal effects (Fig. 3). Although the autofocus system of our Nikon Ti microscope minimizes the drift in the axial direction, the lateral drift can lead to false structural representations. For densely labeled structures, the drift can be corrected using the structures themselves as landmarks. However, in the

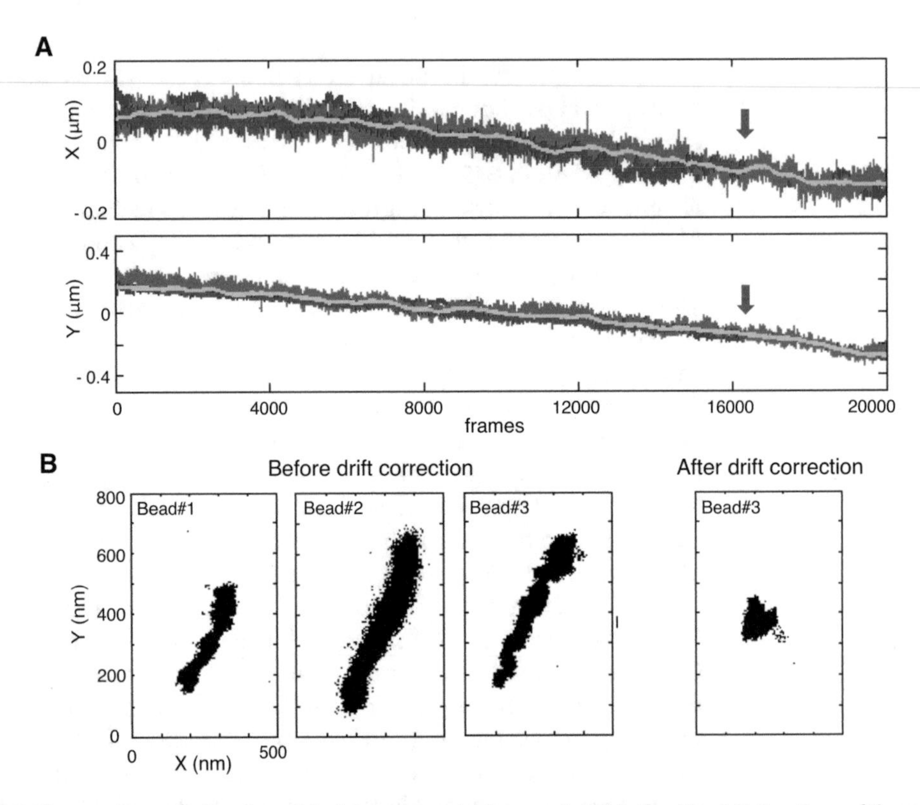

Fig. 3 Drift correction with beads as fiducial markers. (**a**) Apparent drift in the *X* and *Y* directions of three beads that were attached to the coverslip (red, green, and blue traces). The bright green trace represents the average drift of the three beads that is used for correction, calculated with a sliding window of 500 frames. (**b**) Projection of the detection coordinates of the three beads before drift correction (left panels) and bead #3 after drift correction (right). Note that the lateral drift of the beads can differ; bead #1 is lost after frame 16,000, and #2 is less bright and therefore has lower localization precision

case of sparse synaptic protein clusters, there are often too few detections for drift correction. Using beads for drift correction is therefore the best option. For sequential two-color dSTORM, two-channel registration is also mandatory, because both lateral drift and spectral differences result in a mismatch between the reconstructed images in the two channels. Using beads as independent fiducial markers induces less bias in the two-channel alignment than using the synaptic clusters themselves.

During sample preparation, we attach multicolor beads (TetraSpeck microspheres, 100 nm diameter, Thermo Fisher) nonspecifically to the coverslips. To this aim, the fixed and labeled samples are incubated for about 1 min with an excess of beads (~10^9/ml in PBS), rinsed, and mounted in imaging buffer. The number of attached beads strongly depends on the density of cellular material on the coverslip, and can be controlled by adjusting the dilution of the beads, the incubation duration, and the amount of rinsing. Too few beads will make it difficult to find suitable regions of interest

(ROIs), and too many beads can mask the signals of the fluorophores (Fig. 1c). Since the beads can bind nonspecifically to binding sites on the cell surface, some of the beads are not completely immobile and stable. This is seen as jittering of the beads during imaging, and the beads sometimes even detach and move out of the field of view (Fig. 3b, bead #1). Beads are generally very bright and produce precisely localized detections. However, when their signals are saturated, the positions of the beads are not fitted accurately, introducing errors in the drift correction. Consequently, strong laser illumination is needed at the beginning of the recording to partially bleach the beads. Note that the pre-bleaching of the beads can result in a loss of signals from the dye fluorophores. In short, saturated or moving beads should be avoided as markers, and using more than one bead in the ROI can improve the precision of drift correction (Fig. 3). In the ideal case, we aim to record about 2–3 stable and non-saturated beads per field of view.

2.4 Primary Spinal Cord Neuron Culture and Immunocytochemistry

All procedures using animals follow the regulations of the French Ministry of Agriculture and the Direction Départementale des Services Vétérinaires de Paris (Ecole Normale Supérieure, Animalerie des Rongeurs, license B 75-05-20). Primary spinal cord neuron cultures were prepared from embryonic Sprague-Dawley rats on day 14 of gestation (E14) as described previously [24], with some modifications. Cells were plated on 18 mm glass coverslips (thickness 0.16 mm, No. 1.5, VWR #6310153) that were pre-washed with ethanol and coated with 70 μg/ml poly-DL-ornithine. Dissociated neurons were seeded at a density of 4×10^4 cells/cm^2 in Neurobasal medium (Thermo Fisher) supplemented with B-27, 2 mM glutamine, 5 U/ml penicillin, and 5 μg/ml streptomycin. Neurons were cultured at 37 °C with 5% CO_2, and the medium was replenished twice a week by replacing half of the volume with BrainPhys neuronal medium (Stemcell Technologies) supplemented with SM1 and antibiotics. Spinal cord neurons developed mature neurites within 2 weeks in culture, at which point they were used for immunocytochemistry.

Cultured neurons were fixed with 100% methanol at −20 °C for 10 min or with pre-warmed (37 °C) 4% paraformaldehyde (PFA) in PBS for 15 min. Unspecific binding sites were blocked with 3% BSA in PBS (blocking buffer) for at least 30 min. The cells were then incubated with primary antibodies diluted in blocking buffer for 1 h at room temperature or overnight at 4 °C, followed by several washing steps in PBS and incubation with secondary antibodies in blocking buffer for 1 h at room temperature in the dark. In the experiments described here, the following primary antibodies were used: mouse monoclonal mAb7a (Synaptic Systems, #147011, at 1:500 dilution) and rabbit polyclonal (Synaptic Systems, #147002, 1:500) antibodies against gephyrin, as well as custom-made rabbit polyclonal antibody against the GlyR subunit α1 (A. Triller lab, #2353, 1:500 to 1:1000). The secondary antibodies were Alexa Fluor 647-conjugated donkey anti-mouse IgG (Jackson ImmunoResearch, #715-605-151, 1:500 to 1:1000) and Alexa Fluor 647-conjugated donkey anti-rabbit IgG (Jackson ImmunoResearch, #711-605-152, 1:500 to 1:1000). In addition, unconjugated donkey anti-mouse IgG (Jackson ImmunoResearch, #715-005-151) and donkey anti-rabbit IgG (Jackson ImmunoResearch, #711-005-152) secondary antibodies were coupled with Cy3B mono-reactive NHS ester (PA63101, GE Healthcare) according to the supplier's protocol, purified using size-exclusion columns (Illustra NAP-5 columns, #17085302, GE Healthcare) and used at a dilution of 1:50 to 1:100 for immunolabeling.

The choice of fixatives, concentrations of primary and secondary antibodies, as well as buffering solutions and incubation times can all affect dSTORM imaging [73–75]. Even though the primary neuron culture and the immunocytochemistry procedures are standard protocols, optimization of the sample preparation is important for dSTORM, since the requirements are not the same as for conventional fluorescence microscopy. For example, Cy3B-tagged secondary antibodies with low dye/protein ratio (D/P $\approx$ 1–4) give more discrete blinking profiles in our dSTORM recordings, even if the total fluorescence intensity of the labeled synapses is reduced. Fixatives such as PFA can contribute to background fluorescence. We have therefore quenched the fluorescence background with 50 mM NH_4Cl after fixation. Membrane molecules may retain a certain level of residual mobility after chemical fixation that can be detected in super-resolution images [76]. In these cases, it is recommended to increase fixation times, as long as this does not interfere with the access of the antibodies to the epitopes. Where necessary, membrane proteins such as neurotransmitter receptors can also be stained by live-cell immunocytochemistry to enhance epitope recognition. However, care should be taken to avoid the artificial clustering of membrane proteins due to the cross-linking effect of secondary antibodies [77]. Finally, samples should always be protected from light and kept at 4 °C for no longer than 2 days prior to dSTORM imaging.

3 Sequential Two-Color dSTORM Imaging

3.1 Microscope Setup

The STORM setup is built on an inverted Nikon Eclipse Ti microscope with a perfect focus system (PFS) to minimize the axial drift of the microscope stage, and equipped with a 100× oil immersion objective (HP APO TIRF 100× oil, NA 1.49, Nikon). The setup includes several continuous laser lines with emission wavelengths at 640 nm, 561 nm, 532 nm, 488 nm, and 405 nm (coherent), with nominal maximum power of 1 W, 1 W, 500 mW, 500 mW, and

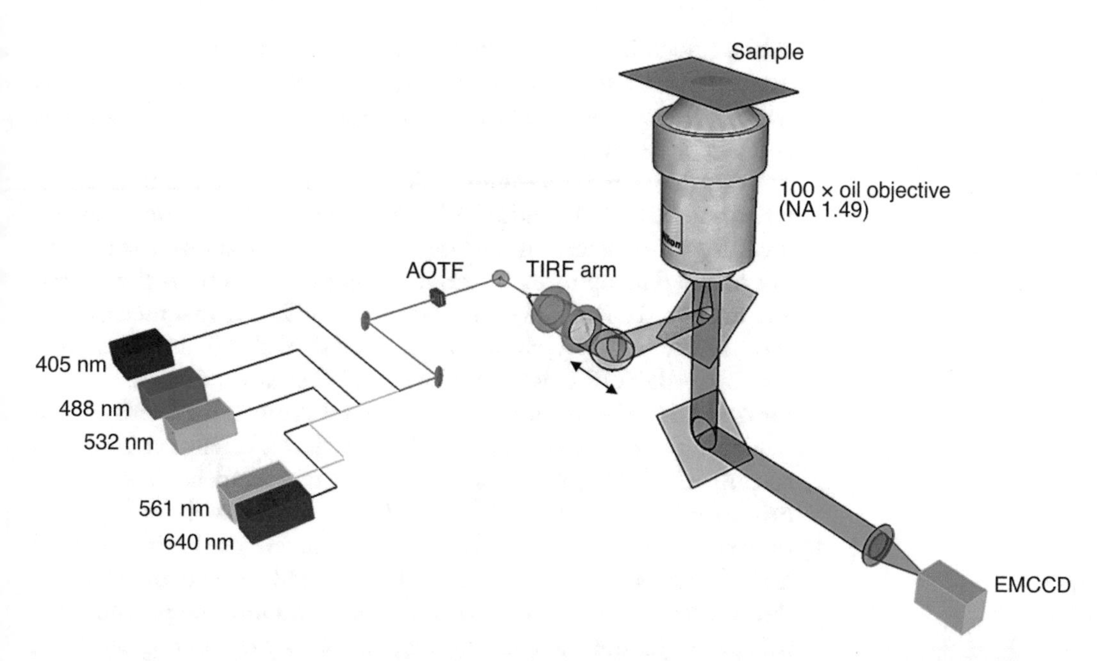

Fig. 4 dSTORM microscope setup with its main components. The lasers are collimated in an external platform and controlled with an acousto-optic tunable filter (AOTF) that sets the illumination intensity and temporal sequence. The laser beam is expanded with a total internal reflection fluorescence (TIRF) arm that focuses the light on the back focal plane of the objective and controls the inclination angle of the excitation light with a translatable mirror (indicated by an arrow). The light emitted by the fluorophores on the coverslip is captured with an EMCCD camera. This illustration was kindly provided by Ignacio Izeddin (ESPCI, Paris)

120 mW, respectively (Fig. 4). The lasers are combined in an external platform and controlled via an acousto-optic tunable filter (AOTF) to apply fast illumination pulses and to set the laser intensity. The laser beam is expanded with a total internal reflection fluorescence (TIRF) arm that controls the inclination angle of the excitation light, and is focused on the back focal plane of the objective. The emitted light is captured with an electron-multiplying charge-coupled device (EMCCD) camera (Andor iXon 897 Ultra, 512 × 512 pixels) using a multiple-wavelength dichroic mirror and appropriate band-pass filters (Semrock FF02-684/24 for Alexa 647, FF01-607/36 for Cy3B) in the emission light path. Conventional fluorescence images are taken with a mercury lamp (Intensilight C-HGFIE, Nikon) that emits a broad spectrum of wavelengths, using the corresponding band-pass filters in the excitation light path (FF01-650/13 for Alexa 647, FF01-560/25 for Cy3B). All the elements are controlled by NIS-Elements software (Nikon).

3.2 Imaging Procedure

The dSTORM imaging workflow is illustrated in Fig. 1b, c. On the day of imaging, labeled neuron cultures on glass coverslips are incubated with 100 nm beads, and mounted on glass slides with a cavity (diameter 15–18 mm, depth 0.6–0.8 mm) containing freshly prepared Gloxy imaging buffer. Coverslips are sealed with silicone rubber (Picodent twinsil speed 22). The sealing of the sample is important because it minimizes the access of oxygen to the sample, which prevents the acidification through the activity of the glucose oxidase in the buffer.

We then search for regions of interest (ROIs) containing several synaptic clusters and fiducial beads, under illumination with the mercury lamp instead of the lasers in order to minimize the bleaching of the fluorophores. Reference images are taken in the far-red and in the red channel with the lamp. The effective magnification of these images is 100×, resulting in a pixel size of 160 nm. Sequential two-color dSTORM is then carried out first in the far-red channel for Alexa 647 signals, followed by the red channel for Cy3B. In the beginning of each recording, the field of view is briefly pre-bleached to push the fluorophores into the OFF state and to dampen the brightness of the beads. At least 20,000 frames with an exposure time of 50 ms are recorded in each channel. We use highly inclined laser illumination rather than TIRF to reduce the background fluorescence while avoiding interferences and inhomogeneous illumination. In the far-red channel, the 640 nm laser is used for illumination at an intensity of approximately 1 kW/cm^2, resulting in the spontaneous blinking of Alexa 647 fluorophores in the absence of 405 nm laser activation in Gloxy buffer. In the red channel, Cy3B dyes are imaged with the 561 nm laser at an intensity of 2 kW/cm^2. The blinking of Cy3B is supported by continuous low-intensity activation with the 405 nm laser (~0.03–0.3 kW/cm^2). The total acquisition time needed for each ROI is about 1 h.

3.3 Determining the Length of Imaging

The primary goal of SMLM imaging is to record as many blinking events as possible from the fluorophores. However, it is practically impossible to detect all the signals in dSTORM because of the random ON/OFF switching cycles of the organic dyes. Some fluorophores can stay in the OFF state for long periods that exceed the recording time. Moreover, the multiple switching cycles can yield redundant information in which the same fluorophore is represented several times. The target proteins are usually labeled with several fluorophores through primary and secondary antibody staining, which produces further redundant information about the position of the protein. In addition, the number of detection artifacts (i.e., nonspecific fluorescence background) increases with the number of image frames, whereas the actual fluorophore blinking events decrease over time. Therefore, it is neither necessary nor advisable to extend the duration of the recordings and aim to detect all the blinking events from the fluorophores. It is only important to record sufficient independent localizations from which representative structures can be reconstructed. We have therefore adopted the following reasoning to determine the imaging length of our recordings: What is the minimal number of frames from which the reconstructed structures have visibly good agreement with the images reconstructed from longer recordings?

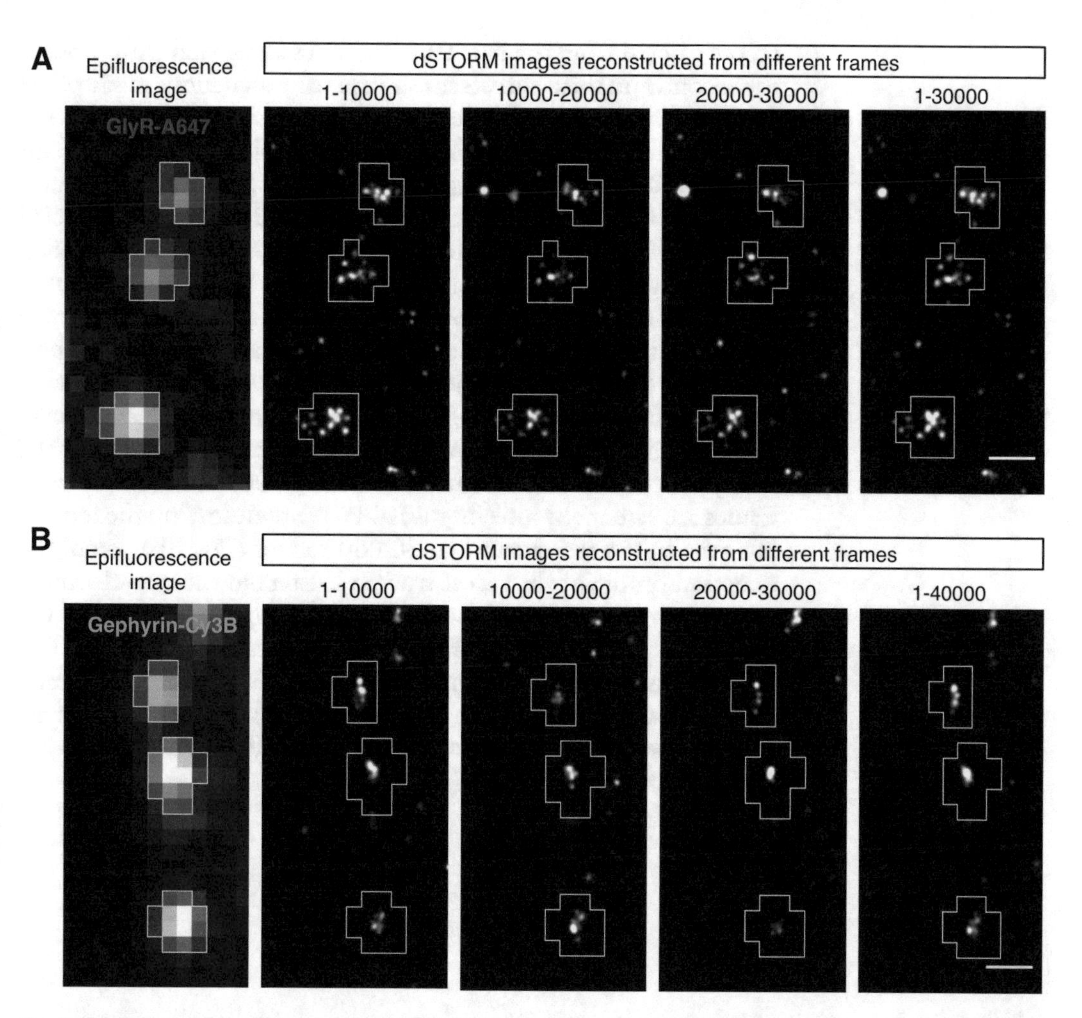

Fig. 5 Determining the optimal length of dSTORM recordings. (**a**) Long dSTORM recordings of 30,000 frames of GlyRs labeled with Alexa 647 (GlyR-A647) were acquired under optimized imaging conditions until there were few remaining blinking events. The reconstructed GlyR clusters from three independent substacks of 10,000 frames were compared to define the ideal length of recording. A good agreement between all the three substacks was observed. In our experiments, 20,000 frames were therefore considered to be sufficient to reconstruct synaptic clusters with Alexa 647 fluorophores. (**b**) For Cy3B-labeled gephyrin clusters, the first two substacks were similar and matched the reconstructed image of the entire stack (in this case 40,000 frames); however, the third substack showed incomplete sampling. To ensure that sufficient detections of Cy3B were collected, we therefore determined an optimal imaging length of 30,000 frames for this channel. Yellow boxes show the masks of individual synaptic clusters produced from the epifluorescence images. Scale: 500 nm

For two-color dSTORM imaging of GlyRs and gephyrin, labeled with Alexa 647 and Cy3B, respectively, we recorded 30,000 frames in the far-red and 40,000 frames in the red channel, until there remained few blinking events at the end of the recordings (Fig. 5). We then split the whole stack into consecutive sub-stacks of 10,000 frames from which we reconstructed separate images and compared their resemblance. GlyR clusters reconstructed from the first 10,000 frames agree well with those in the second and third sub-stacks. Image reconstructions made of the first 20,000 frames give very similar representations as 30,000 frames. This illustrates that Alexa 647 fluorophores undergo multiple ON/OFF switching cycles that give rise to relatively stable long-term blinking, and that the subsynaptic distribution of labeled GlyRs at synapses can be captured within 20,000 frames (Fig. 5a). In the red channel, the gephyrin clusters from the first two sub-stacks of 10,000 images are again quite similar; however, the reconstruction of the third sub-stack produces a less detailed representation. Clearly, Cy3B fluorophores have fewer blinking cycles, meaning that the number of independent detections declines over time, despite the use of the 405 nm activation laser. As a result, the gephyrin clusters reconstructed from 30,000 and from 40,000 frames are essentially indistinguishable. We therefore set the recording time in the red channel to 30,000 frames (Fig. 5b). From the temporal profile of the reconstructed clusters in the two channels, we can also gain some insights into the blinking performance of the dyes. If the clusters reconstructed from the sub-stacks are similar, this implies that the switching cycles of the fluorophores are relatively dispersed throughout the recording (also *see* Fig. 6b), confirming that the two dyes are suitable for dSTORM imaging of synaptic proteins.

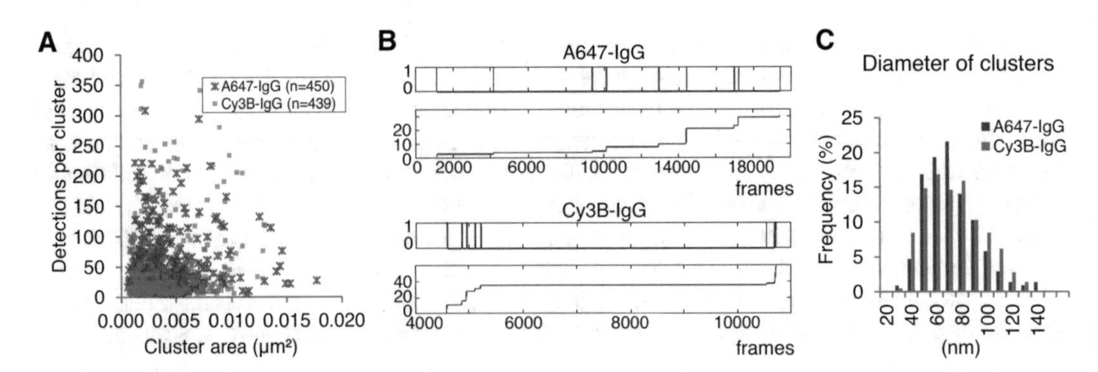

Fig. 6 Blinking profiles of the dyes conjugated to secondary antibodies. (**a**) Two-color dSTORM imaging was done on sparsely distributed secondary antibodies coupled with Alexa 647 (A647-IgG) or Cy3B (Cy3B-IgG). We observed a large variability in the number of detections per dye-coupled IgG, with clusters of up to hundreds of detections. (**b**) Representative time traces of single A647-IgG and Cy3B-IgG detections in dSTORM show that the blinking of Alexa 647 is more dispersed than that of Cy3B. The upper boxes show the number of detections per frame (zero or one), whereas the lower boxes show the cumulative detections throughout the recording. (**c**) Dye-coupled IgGs give rise to apparent nanoclusters with diameters between 40 and 120 nm, as a result of repetitive detections with limited localization precision

4 Data Analysis and Quality Assessment

4.1 Image Preprocessing

Critical processing steps of the raw imaging data include single-particle detection, drift correction, two-channel registration, and image reconstruction (Fig. 1b). Single-particle detection is realized by Gaussian fitting of the PSF of each fluorophore signal to determine the molecules' coordinates. To this aim, we use an adapted version of the multiple-target tracing (MTT) program [78] in Matlab 2012b (MathWorks), as described by Izeddin et al [3]. The lateral drift is corrected with the program PALMvis [79] in Matlab 2012b using beads as fiducial markers. A sliding window of 100–1000 frames is applied to calculate the average position of the beads throughout the recording, and the systematic x/y drift over time is estimated with a temporal resolution of 5–50 s based on the average displacement of the beads (Fig. 3). The calculated drift of the stage is then applied to the whole image to correct the positions of all the single-fluorophore localizations in the movie. At this stage, we obtain the final coordinates of the localizations, and the two-channel registration can now be done by aligning the beads that were chosen as fiducial markers. Nevertheless, we can go one step further and render the SMLM images as density-based representations of the fluorophore detections. The drift-corrected coordinates are converted into rendered images by representing each localization as a Gaussian distribution with the same intensity and with a standard deviation ($\sigma = 15$ nm) that is derived from the average localization precision of the detections. Super-resolution images are generally displayed with a pixel size of 10 nm to balance file size versus representation precision. In the rendered super-resolution images, the pixel intensities ultimately represent the density of localizations and the images have zero background. Finally, the two super-resolution images are aligned using the TurboReg plug-in [80] in Fiji [81].

Alternative tools are available for all of the processing steps described above [82]. For example, ThunderSTORM [83] and QuickPALM [84] are commonly used for signal detection and SMLM image reconstruction, and are freely accessible as plug-ins in Fiji. For more information about the software and programs designed for SMLM, please also see the website of the Biomedical Imaging Group of the EPFL (Lausanne, Switzerland; http://bigwww.epfl.ch/smlm/software).

4.2 Statistical Analysis

Both types of SMLM datasets, namely the coordinate-based pointillist data and the rendered super-resolution image data can be used for statistical analysis (Fig. 1c). Each type of data has its own strengths and drawbacks for analysis [85]. The coordinate-based pointillist data in principle make the best use of the single-molecule localizations, and the number of detections at a given cluster reflects the amount of target proteins. However, the localization precision of the detections is usually neglected in the analytical methods developed so far. Moreover, cluster segmentation, feature extraction, and spatial correlation analysis based on the pointillist data may demand qualified programming skills. On the other hand, the rendered super-resolution images give more intuitive information, and can be analyzed using common image-processing tools such as Fiji that are widely known to researchers. In the rendered images, pixel intensities represent the density of the localizations, taking into account the localization precision. Since the SMLM coordinates are not the actual positions of the fluorophores but only experimentally acquired coordinate measurements, they can also be understood as localization probabilities. The inclusion of the localization precision in image rendering thus produces more realistic representations. In addition, pixels that do not contain any localizations have the value of zero. In other words, there is no background in the rendered super-resolution images. Thus there is no need for background subtraction or thresholding, which is one of the major difficulties in fluorescent image processing. Furthermore, the pixel intensity of the clusters can be manually adjusted during the rendering step to modify the brightness and contrast of the resulting images. As a consequence, synaptic clusters rendered from small number of localizations in one image may have the same intensity as clusters rendered from a higher number of localizations in another image.

Different algorithms and software are available for analyzing coordinate-based pointillist data. For instance, density-based spatial clustering of applications with noise (DBSCAN) and Voronoï tessellation have been used for spatial clustering and segmentation of SMLM data [86]. The advantage of the two algorithms is that they do not make any assumptions about the shape of the structures, which is very suitable for segmenting synaptic clusters in SMLM data. To quantify the co-localization of two proteins, coordinate-based co-localization (CBC) [87, 88] enables single-molecule correlation analysis of dual-color pointillist data. This algorithm has been implemented in different open-source software for SMLM data analysis, such as Clus-Doc and LAMA [89–92]. However, using these tools can be problematic, because they are often not suited to the specific demands. In our case, we did not find any suitable analysis tool that would allow us to perform individual cluster-based analysis, which is essential for the comparison of synaptic GlyR and gephyrin clusters. In contrast, individual synaptic clusters in the rendered super-resolution images can be easily obtained with classical image processing. With the powerful image analysis platforms Fiji [81] and Icy [93], image-based analysis of individual synaptic clusters is relatively straightforward. Therefore, we suggest to begin with image-based SMLM analyses.

In the following, we describe an image-based analysis workflow using exclusively open-source tools that allow biologists to extract relevant information. With the help of simple macros in Fiji [81] to record commands and construct basic programs, a set of experimental data, namely two-color recordings of 10 cells with about 100 synapses, can be analyzed in 1 day using this workflow.

First, the conventional fluorescence reference images (taken with the lamp) are converted into binary masks with the Icy spot detector plug-in, choosing appropriate parameters to select well-focused and brightly labeled synaptic puncta (Fig. 5). The combined area of the synaptic masks in the two channels is obtained by calculating the sum of the binary images with Fiji, and serves to identify the synaptic clusters in the super-resolution images. Since nonspecific background fluorescence can also generate clusters of detections in the dSTORM images, using the mask to preselect synaptic clusters is necessary. However, it should be kept in mind that this process potentially introduces a certain bias in the analysis, since small and weakly labeled synaptic clusters may be lost. This is not an issue in our case, because our aim was to directly compare the subsynaptic distributions of different synaptic components within the same synapse. The dual-color dSTORM recordings are therefore considered as paired data. The selection of synaptic clusters based on the labeling intensity can thus be justified.

Since our super-resolution images are acquired with an inclined laser illumination and a different dichroic mirror, there may be a substantial mismatch between these images and the conventional reference images. Drift correction and chromatic differences add to the mismatch between the various images. Therefore, in the second step, the super-resolution images and the combined mask image are aligned with TurboReg plug-in in Fiji. Also, the pixel size of the low-resolution mask (160 nm) has to be artificially enhanced to match that of the rendered dSTORM images (10 nm) by splitting each pixel into 16 × 16 identical pixels. After image registration, the masks of the individual synapses are applied to the super-resolution images in Fiji in order to crop the individual synaptic clusters. At this point, we obtain individual synaptic cluster pairs of GlyRs and gephyrin, on which we can perform further cluster analysis.

To characterize the spatial correlation of GlyR and gephyrin clusters, we employ the intensity correlation analysis (ICA) plug-in in Fiji as described by Li et al. [94, 95]. Intensity correlation analysis compares the variations of the intensities for each pixel between two images in order to identify correlated changes that reflect the co-localization of the two proteins (Fig. 7a). Intensity correlation quotients (ICQ) are derived from the number of pixels that increase or decrease in synchrony in the two channels. ICQ values range between −0.5 and +0.5, where −0.5 indicates a mutually exclusive distribution, zero a random distribution, and +0.5 a

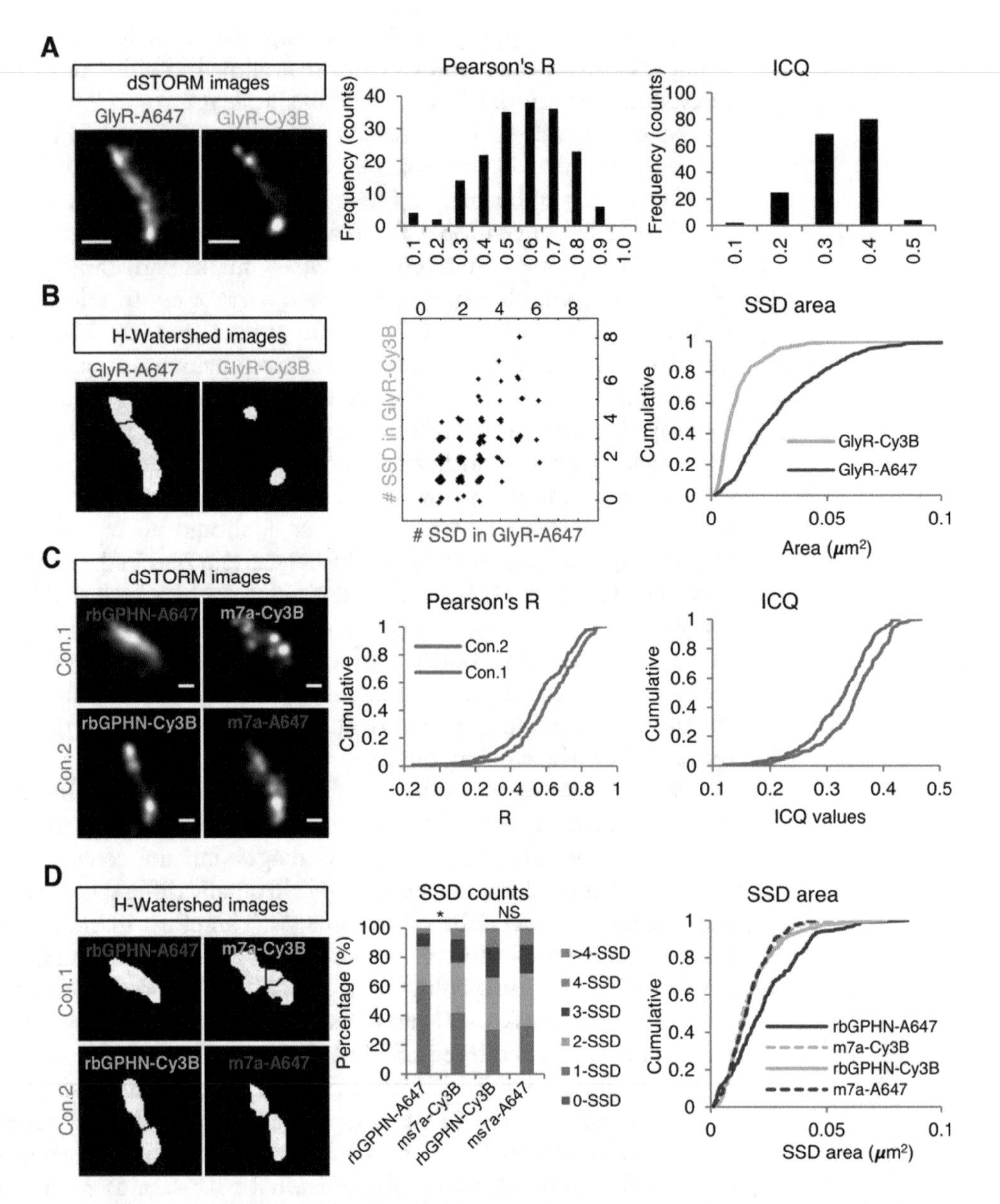

Fig. 7 Impact of dyes and antibodies on the visualization of SSDs with dSTORM. (**a**, **b**) Co-localization control: dSTORM imaging of GlyR clusters in spinal cord neurons (fixed with 4% PFA) labeled with the same primary antibody and with two different secondary antibodies (GlyR-A647 and GlyR-Cy3B). (**a**) GlyR clusters detected with Alexa 647 and Cy3B showed a high degree of spatial correlation, as judged by their Pearson's coefficient R (mean $\pm$ SD: 0.52 $\pm$ 0.18) and ICQ values (0.29 $\pm$ 0.07). (**b**) Subsynaptic domains (SSDs) of GlyRs were analyzed by H-watershed segmentation, using the same parameters for both channels. We observed a moderate positive correlation between the numbers of SSDs detected with Alexa 647 and Cy3B (middle panel, linear regression: $y = 0.63x + 0.75$, $R^2 = 0.30$). The SSD areas detected with Alexa 647 were significantly larger than those of Cy3B (Kolmogorov-Smirnov test, $p < 0.0001$, $n = 180$ synapses with 767 SSDs for GlyR-A647 and 874 SSDs for GlyR-Cy3B, from two independent experiments). Scale bar: 200 nm. (**c**, **d**) Dye reversal experiments: two-color dSTORM imaging of gephyrin clusters in spinal cord neurons (methanol fixation) labeled with two different primary antibodies against gephyrin, followed by two

perfect positive correlation between the two intensity distributions. In other words, this parameter measures the covariation of signals and is not very sensitive to the pixel intensity in the rendered images. In addition to ICQ values, the ICA plug-in also calculates the Pearson correlation coefficient (Pearson's R) that measures the linear correlation of the intensity of each pixel between the two images as additional information.

To characterize the internal, subsynaptic distribution of GlyRs and gephyrin, we carried out H-watershed analysis to segment the clusters using the interactive H-watershed plug-in developed by Benoit Lombardot (http://imagej.net/Interactive_Watershed) in Fiji. H-watershed finds local maxima with only one parameter, H, allowing the detection of subcluster peaks and the separation of the subclusters (Fig. 7b). After segmentation, the features of the binary subclusters are extracted in Fiji using extended particle analyzer (EPA) [96]. We thus obtained detailed information about SSDs of GlyRs and gephyrin at synapses under various experimental conditions. For example, changes in the number and area of the SSDs per synapse may reveal how the distribution of GlyRs and gephyrin within the PSD is regulated during inhibitory synaptic plasticity [33]. Even though H-watershed is an effective way to separate subsynaptic domains, it can overestimate the SSD counts, especially in images with relatively low numbers of localizations in which the pixel intensities were enhanced during image rendering.

4.3 Strategies for Imaging Quality Assessment

The evaluation of individual factors (dyes, buffers, probes, laser power, and so on) is important for understanding their influence on the quality of the dSTORM data. However, the final image quality results from the combined effect of these factors that are heavily interdependent. Therefore, an integrated assessment of the imaging data is necessary to estimate the capability of the protocol to answer the biological question or hypothesis. We propose to carry out the following types of control experiments to evaluate the performance of the chosen experimental approach (Figs. 6 and 7).

Fig. 7 (continued) combinations of secondary antibodies. (**c**) The antibody combination rbGPHN-Cy3B/m7a-A647 ($n = 174$ synapses) gave a significantly higher spatial correlation between the two channels than the combination with reversed dyes (rbGPHN-A647/m7a-Cy3B, $n = 148$), as shown by both the Pearson's coefficient (Mann-Whitney U test, $p < 0.01$) and ICQ ($p < 0.01$). (**d**) H-watershed segmentation was done with different parameters optimized for each channel (Alexa 647 and Cy3B). SSD counts per synapse were higher when clusters were detected with Cy3B than with Alexa 647 in condition 1 (paired t-test, $p < 0.001$), but not in condition 2 ($p = 0.37$). SSD areas detected by rbGPHN-A647 labeling were significantly larger than with m7a-Cy3B in condition 1 (Kolmogorov-Smirnov test, $p < 0.0001$, $n = 228$ SSDs for rbGPHN-A647 and 274 for m7a-Cy3B), while no difference was detected with the other antibody combination ($n = 388$ for rbGPHN-Cy3B and 369 for m7a-A647). Data were from three imaging experiments using two different spinal cord cultures. Scale bar: 100 nm

First, we recorded the *blinking behavior of the dye-coupled secondary antibodies* in order to estimate the degree of redundant detections and the appearance of fake nanoclusters. We carried out two-color dSTORM imaging of two secondary IgGs, conjugated with either Alexa 647 or Cy3B dye. Clean glass coverslips were incubated for 1 h with Alexa 647-tagged donkey anti-mouse IgG and Cy3B-tagged donkey anti-rabbit IgG that were diluted in 3% BSA and mounted with beads and imaging buffer. We then performed sequential two-color dSTORM imaging as described above and characterized the clusters produced by single secondary antibodies in the two channels (Fig. 6a). Both A647-IgG and Cy3B-IgG gave rise to nanometer-sized clusters with a wide variability in the number of detections per cluster as well as cluster size. The average number of detections per dye-coupled IgG is a measure that can be used to estimate the number of antibodies that are bound to a given structure. As such, it provides an approximate calibration to determine the absolute copy number of target proteins [22, 38]. From the temporal profiles of the blinking events, we can further infer the photophysical behavior of the dyes (Fig. 6b). For Alexa 647, the blinking events are brief and dispersed across the entire recording. Cy3B fluorophores on the other hand display relatively few blinking cycles, some of which have persistent ON times, resulting in large bursts of repetitive detections during a single blinking event. As a consequence, the number of Cy3B detections is not linearly dependent on the labeling efficiency (in particular for proteins with low copy numbers), meaning that Cy3B detections are less easily converted into molecule numbers as Alexa 647. The diameter of the clusters produced by single IgGs is around 40–120 nm (Fig. 6c), which is similar to the measured size of the SSDs of synaptic proteins [21–23, 33, 34]. Therefore, the number of detections and the size of the clusters produced by the individual probes used for labeling have to be taken into consideration during data analysis and interpretation when characterizing SSDs of synaptic components.

Next, we assessed the *reproducibility of two-color dSTORM imaging* of synaptic proteins by labeling GlyR clusters in spinal cord neurons with a single primary antibody against the GlyR α1 subunit and with two different secondary antibodies, namely Alexa 647-conjugated donkey anti-rabbit and Cy3B-conjugated donkey anti-rabbit IgG. We analyzed the spatial correlation of GlyR clusters detected in the two channels. As expected, both Pearson's coefficients and ICQ values indicated a strong positive correlation, demonstrating that Alexa 647 and Cy3B produce similarly good representations of the same GlyR cluster (Fig. 7a). We then looked into the internal organization of the GlyR clusters detected by the two dyes. The super-resolution clusters were segmented using H-watershed to identify GlyR SSDs (Fig. 7b). The number of SSDs detected with Alexa 647 and Cy3B only showed moderate correlation and Cy3B tended to produce more subclusters. Concomitantly, the sizes of the SSDs (in μm^2) detected by Alexa 647 were significantly larger than the ones detected by Cy3B. This shows that the different photophysical properties of Alexa 647 and Cy3B dyes have considerable impact on the representation of subsynaptic structures in dSTORM images.

Finally, we tested *the effect of dye reversal on the analytical results* (Fig. 7c, d). We used two different primary antibodies that recognize the same synaptic protein, gephyrin. Mouse monoclonal mAb7a (m7a) and rabbit polyclonal antibodies against gephyrin (rbGPHN) were labeled with two combinations of secondary antibodies: in one case with Cy3B-conjugated donkey anti-mouse IgG and Alexa 647-conjugated donkey anti-rabbit IgG (condition 1), and in the other case with Alexa 647-conjugated donkey anti-mouse IgG and Cy3B-conjugated donkey anti-rabbit IgG (condition 2). We then compared the spatial correlation and the SSDs of the synaptic gephyrin clusters in the two channels between the two conditions. We found that the second experimental condition yielded a significantly stronger correlation than the first condition, as judged from Pearson's R and ICQ values (Fig. 7c). Since we expect the images of the synaptic gephyrin clusters in the two channels to correspond closely, this result suggests that the blinking of the dyes has a stronger effect in condition 1, creating a pronounced imbalance between the two dSTORM representations.

As to the subsynaptic organization of the gephyrin scaffold, the SSD counts per synapse detected with Cy3B were higher than those of Alexa 647-labeled clusters in condition 1, but not in condition 2 (Fig. 7d). Also, the sizes of the SSDs probed with rbGPHN-A647 were much larger than with m7a-Cy3B (condition 1), an effect that was not seen in the second condition with reversed dyes. These data once again show the strong effect of the dyes, in combination with the antibodies, on the representation of subsynaptic structures. The fact that the SSD properties in condition 2 did not show any significant differences between the two dSTORM images is consistent with the higher spatial correlation obtained with this dye/antibody combination (Fig. 7c, d). It could thus be argued that this condition is less affected by the blinking properties and better suited for the analysis of gephyrin SSDs. Given that m7a and rbGPHN target different epitopes of gephyrin, however, it cannot be ruled out that the two antibodies are actually distributed differently at synapses, which could be exacerbated by the effect of the dyes, and may explain the mismatch in condition 1. It is part of our ongoing research to address the question whether the two epitopes of gephyrin indeed occupy specialized subsynaptic domains at inhibitory synapses. In any case, these data illustrate that ICA analysis and H-watershed segmentation are effective tools that characterize different, yet not fully independent properties of synapses. Our results also raise the necessity for control experiments with reversed dyes to scrutinize the experimental findings.

5 Conclusion

SMLM is a powerful technique that enables the characterization of biological samples with a resolution on the nanometer scale. Based on our own experiences we strongly recommend to consider the technical difficulties and limitations of the technique in the experimental design, in particular when SMLM is applied to complex biological structures. The methodological framework discussed in this chapter is applicable to synapses and similarly complex structures alike. We also propose that the quality of the reconstructed images and the relevance of the extracted parameters be critically assessed before drawing far-reaching conclusions and that, where possible, the results of these assessments and control experiments should be included in scientific publications.

Acknowledgment

We thank Manuel Maidorn and Felipe Opazo for the illustrations in Fig. 2 [55], and Ignacio Izeddin for Fig. 4. Our research is funded by grants (to Antoine Triller, IBENS, Paris) from the Agence Nationale de la Recherche (ANR-12-BSV4-0019-01, ANR-11-IDEX-0001-02, ANR-10-LABX-54) and the European Research Council (ERC, PlastInhib). X.Y. is supported by the China Scholarship Council (CSC).

Appendix: Imaging Buffers

Gloxy buffer: 0.5 mg/ml glucose oxidase, 40 μg/ml catalase, 0.5 M D-glucose, 50 mM β-mercaptoethylamine (MEA), in PBS pH 7.4, degassed with N_2 [3].

OxEA buffer: 3% (v/v) oxyrase (EC-Oxyrase, Sigma-Aldrich), 1 M sodium DL-lactate, 50 mM MEA, in PBS pH 8.0 [67].

References

1. Thompson RE, Larson DR, Webb WW (2002) Precise nanometer localization analysis for individual fluorescent probes. Biophys J 82:2775–2783
2. Huang B (2010) Super-resolution optical microscopy: multiple choices. Curr Opin Chem Biol 14:10–14
3. Izeddin I, Specht CG, Lelek M, Darzacq X, Triller A, Zimmer C, Dahan M (2011) Super-resolution dynamic imaging of dendritic spines using a low-affinity photoconvertible actin probe. PLoS One 6:e15611
4. van de Linde S, Löschberger A, Klein T, Heidbreder M, Wolter S, Heilemann M, Sauer M (2011) Direct stochastic optical reconstruction microscopy with standard fluorescent probes. Nat Protoc 6:991–1009
5. Baddeley D, Bewersdorf J (2018) Biological insight from super-resolution microscopy: what we can learn from localization-based images. Annu Rev Biochem 87:965–989
6. Rust MJ, Bates M, Zhuang X (2006) Stochastic optical reconstruction microscopy (STORM)

provides sub-diffraction-limit image resolution. Nat Methods 3:793–795

7. Betzig E, Patterson GH, Sougrat R, Lindwasser OW, Olenych S, Bonifacino JS, Davidson MW, Lippincott-Schwartz J, Hess HF (2006) Imaging intracellular fluorescent proteins at nanometer resolution. Science 313:1642–1646
8. Hess ST, Girirajan TPK, Mason MD (2006) Ultra-high resolution imaging by fluorescence photoactivation localization microscopy. Biophys J 91:4258–4272
9. Heilemann M, van de Linde S, Schuttpelz M, Kasper R, Seefeldt B, Mukherjee A, Tinnefeld P, Sauer M (2008) Subdiffraction-resolution fluorescence imaging with conventional fluorescent probes. Angew Chem Int Ed Engl 47:6172–6176
10. Bates MW, Huang B, Dempsey GT, Zhuang X (2007) Multicolor super-resolution imaging with photo-switchable fluorescent probes. Science 317:1749–1753
11. Zhong H (2016) Applying super-resolution localization-based microscopy to neurons. Synapse 69:283–294
12. Sydor AM, Czymmek KJ, Puchner EM, Mennella V (2015) Super-resolution microscopy: from single molecules to supramolecular assemblies. Trends Cell Biol 25:730–748
13. Xu K, Zhong G, Zhuang X (2013) Actin, spectrin, and associated proteins form a periodic cytoskeletal structure in axons. Science 339:1–11
14. Leterrier C, Potier J, Caillol G, Debarnot C, Noroni FR, Dargent B (2015) Nanoscale architecture of the axon initial segment reveals an organized and robust scaffold. Cell Rep 13:2781–2793
15. Bär J, Kobler O, Van Bommel B, Mikhaylova M (2016) Periodic F-actin structures shape the neck of dendritic spines. Sci Rep 6:1–9
16. Loschberger A, van de Linde S, Dabauvalle M-C, Rieger B, Heilemann M, Krohne G, Sauer M (2012) Super-resolution imaging visualizes the eightfold symmetry of gp210 proteins around the nuclear pore complex and resolves the central channel with nanometer resolution. J Cell Sci 125:570–575
17. Szymborska A, De Marco A, Daigle N, Cordes VC, Briggs JAG, Ellenberg J (2013) Nuclear pore scaffold structure analyzed by super-resolution microscopy and particle averaging. Science 341:655–658
18. Broeken J, Johnson H, Lidke DS, Liu S, Nieuwenhuizen RPJ, Stallinga S, Lidke KA, Rieger B (2015) Resolution improvement by 3D particle averaging in localization microscopy. Methods Appl Fluoresc 3:1–20
19. Dani A, Huang B, Bergan J, Dulac C, Zhuang X (2010) Super-resolution imaging of chemical synapses in the brain. Neuron 68:843–856
20. Kuriu T, Inoue A, Bito H, Sobue K, Okabe S (2006) Differential control of postsynaptic density scaffolds via actin-dependent and -independent mechanisms. J Neurosci 26:7693–7706
21. MacGillavry HD, Song Y, Raghavachari S, Blanpied TA (2013) Nanoscale scaffolding domains within the postsynaptic density concentrate synaptic AMPA receptors. Neuron 78:615–622
22. Nair D, Hosy E, Petersen JD, Constals A, Giannone G, Choquet D, Sibarita J-B (2013) Super-resolution imaging reveals that AMPA receptors inside synapses are dynamically organized in nanodomains regulated by PSD-95. J Neurosci 33:13204–13224
23. Haas KT, Compans B, Letellier M, Bartol TM, Grillo-Bosch D, Sejnowski TJ, Sainlos M, Choquet D, Thoumine O, Hosy E (2018) Pre-post synaptic alignment through neuroligin-1 tunes synaptic transmission efficiency. Elife 7:1–22
24. Specht CG, Izeddin I, Rodriguez PC, El Begeiry M, Rostaing P, Darzacq X, Dahan M, Triller A (2013) Quantitative nanoscopy of inhibitory synapses: counting gephyrin molecules and receptor binding sites. Neuron 79:308–321
25. Rodriguez PC, Almeida LG, Triller A (2017) Continuous rearrangement of the postsynaptic gephyrin scaffolding domain: a super-resolution quantified and energetic approach. BioRxiv
26. Pennacchietti F, Vascon S, Nieus T, Rosillo C, Das S, Tyagarajan SK, Diaspro A, Del Bue A, Petrini EM, Barberis A, Zanacchi FC (2017) Nanoscale molecular reorganization of the inhibitory postsynaptic density is a determinant of GABAergic synaptic potentiation. J Neurosci 37:1747–1756
27. Crosby KC, Gookin SE, Garcia JD et al (2019) Nanoscale subsynaptic domains underlie the organization of the inhibitory synapse. Cell Rep 26:3284–3297
28. Broadhead MJ, Horrocks MH, Zhu F et al (2016) PSD95 nanoclusters are postsynaptic building blocks in hippocampus circuits. Sci Rep 6:1–14
29. Dzyubenko E, Rozenberg A, Hermann DM, Faissner A (2016) Colocalization of synapse marker proteins evaluated by STED-

microscopy reveals patterns of neuronal synapse distribution in vitro. J Neurosci Methods 273:149–159
30. Hruska M, Henderson N, Le Marchand SJ et al (2018) Synaptic nanomodules underlie the organization and plasticity of spine synapses. Nat Neurosci 21:671–682
31. Wegner W, Mott AC, Grant SGN et al (2018) In vivo STED microscopy visualizes PSD95 sub-structures and morphological changes over several hours in the mouse visual cortex. Sci Rep 8:1–11
32. Masch J-M, Steffens H, Fischer J et al (2018) Robust nanoscopy of a synaptic protein in living mice by organic-fluorophore labeling. Proc Natl Acad Sci U S A 115:E8047–E8056
33. Yang X, Specht CG (2019) Subsynaptic domains in super-resolution microscopy: the treachery of images. Front Mol Neurosci 12:1–8
34. Tang AH, Chen H, Li TP et al (2016) A trans-synaptic nanocolumn aligns neurotransmitter release to receptors. Nature 536:210–214
35. Perez de Arce K, Schrod N, Metzbower SWRR et al (2015) Topographic mapping of the synaptic cleft into adhesive nanodomains. Neuron 88:1165–1172
36. Chamma I, Letellier M, Butler C et al (2016) Mapping the dynamics and nanoscale organization of synaptic adhesion proteins using monomeric streptavidin. Nat Commun 7:1–15
37. Chamma I, Levet F, Sibarita J-B et al (2016) Nanoscale organization of synaptic adhesion proteins revealed by single-molecule localization microscopy. Neurophotonics 3:041810
38. Trotter JH, Hao J, Maxeiner S, et al (2019) Synaptic neurexin-1 assembles into dynamically regulated active zone nanoclusters. J Cell Biol. pii: jcb.201812076
39. Fricke F, Beaudouin J, Eils R, Heilemann M (2015) One, two or three? Probing the stoichiometry of membrane proteins by single-molecule localization microscopy. Sci Rep 5:1–8
40. Patrizio A, Renner M, Pizzarelli R et al (2017) Alpha subunit-dependent glycine receptor clustering and regulation of synaptic receptor numbers. Sci Rep 7:1–11
41. Deschout H, Zanacchi FC, Mlodzianoski M et al (2014) Precisely and accurately localizing single emitters in fluorescence microscopy. Nat Methods 11:253–266
42. Jonas P, Bischofberger J, Sandkühler J (1998) Co-release of two fast neurotransmitters at a central synapse. Science 281:419–424
43. Aubrey KR (2016) Presynaptic control of inhibitory neurotransmitter content in VIAAT containing synaptic vesicles. Neurochem Int 98:94–102
44. Triller A, Cluzeaud F, Korn H (1987) Gamma-aminobutyric acid-containing terminals can be apposed to glycine receptors at central synapses. J Cell Biol 104:947–956
45. Bohlhalter S, Mohler H, Fritschy J (1994) Inhibitory neurotransmission in rat spinal cord: co-localization of glycine- and GABAA-receptors at GABAergic synaptic contacts demonstrated by triple immunofluorescence staining. Brain Res 642:59–69
46. Todd AJ, Watt C, Spike RC, Sieghart W (1996) Colocalization of GABA, glycine, and their receptors at synapses in the rat spinal cord. J Neurosci 16:974–982
47. Dumoulin A, Lévi S, Riveau B et al (2000) Formation of mixed glycine and GABAergic synapses in cultured spinal cord neurons. Eur J Neurosci 12:3883–3892
48. Shrivastava AN, Triller A, Sieghart W (2011) GABA(A) receptors: post-synaptic co-localization and cross-talk with other receptors. Front Cell Neurosci 5:1–12
49. Meyer G, Kirsch J, Betz H, Langosch D (1995) Identification of a gephyrin binding motif on the glycine receptor b subunit. Neuron 15:563–572
50. Maric H, Mukherjee J, Tretter V et al (2011) Gephyrin-mediated γ-aminobutyric acid type A and glycine receptor clustering relies on a common binding site. J Biol Chem 286:42105–42114
51. Tretter V, Mukherjee J, Maric H-M et al (2012) Gephyrin, the enigmatic organizer at GABAergic synapses. Front Cell Neurosci 6:1–16
52. Alvarez FJ (2017) Gephyrin and the regulation of synaptic strength and dynamics at glycinergic inhibitory synapses. Brain Res Bull 129:50–65
53. Niwa F, Patrizio A, Triller A, Specht CG (2019) cAMP-EPAC-dependent regulation of gephyrin phosphorylation and GABAAR trapping at inhibitory synapses. iScience 22:453–465
54. Russier M, Kopysova IL, Ankri N et al (2002) GABA and glycine co-release optimizes functional inhibition in rat brainstem motoneurons in vitro. J Physiol 541:123–137
55. Maidorn M, Rizzoli SO, Opazo F (2016) Tools and limitations to study the molecular composition of synapses by fluorescence microscopy. Biochem J 473:3385–3399
56. Ries J, Kaplan C, Platonova E et al (2012) A simple, versatile method for GFP-based super-

resolution microscopy via nanobodies. Nat Methods 9:582–584
57. Platonova E, Winterflood CM, Junemann A et al (2015) Single-molecule microscopy of molecules tagged with GFP or RFP derivatives in mammalian cells using nanobody binders. Methods 88:89–97
58. Pleiner T, Bates M, Trakhanov S et al (2015) Nanobodies: site-specific labeling for super-resolution imaging, rapid epitope-mapping and native protein complex isolation. Elife 4:1–21
59. Annibale P, Vanni S, Scarselli M et al (2011) Identification of clustering artifacts in photo-activated localization microscopy. Nat Methods 8:527–528
60. Annibale P, Vanni S, Scarselli M et al (2011) Quantitative photo-activated localization microscopy: unraveling the effects of photo-blinking. PLoS One 6:1–8
61. Burgert A, Letschert S, Doose S, Sauer M (2015) Artifacts in single-molecule localization microscopy. Histochem Cell Biol 144:123–131
62. van de Linde S, Wolter S, Heilemann M, Sauer M (2010) The effect of photoswitching kinetics and labeling densities on super-resolution fluorescence imaging. J Biotechnol 149:260–266
63. van de Linde S, Krstić I, Prisner T et al (2011) Photoinduced formation of reversible dye radicals and their impact on super-resolution imaging. Photochem Photobiol Sci 10:499–506
64. Vogelsang J, Kasper R, Steinhauer C et al (2008) A reducing and oxidizing system minimizes photobleaching and blinking of fluorescent dyes. Angew Chem Int Ed Engl 47:5465–5469
65. Ha T, Tinnefeld P (2012) Photophysics of fluorescence probes for single molecule biophysics and super-resolution imaging. Annu Rev Phys Chem 63:595–617
66. Dempsey GT, Vaughan JC, Chen KH et al (2011) Evaluation of fluorophores for optimal performance in localization-based super-resolution imaging. Nat Methods 8:1027–1036
67. Nahidiazar L, Agronskaia AV, Broertjes J et al (2016) Optimizing imaging conditions for demanding multi-color super resolution localization microscopy. PLoS One 11:1–18
68. Zheng Q, Juette MF, Jockusch S et al (2014) Ultra-stable organic fluorophores for single-molecule research. Chem Soc Rev 43:1044–1056
69. Zhegalova NG, He S, Zhou H, Kim DM, Berezin MY (2014) Minimization of self-quenching fluorescence on dyes conjugated to biomolecules with multiple labeling sites via asymmetrically charged NIR fluorophores. Contrast Media Mol Imaging 9:355–362
70. Grimm JB, English BP, Choi H et al (2016) Bright photoactivatable fluorophores for single-molecule imaging. Nat Methods 13:985–988
71. Lehmann M, Lichtner G, Klenz H, Schmoranzer J (2016) Novel organic dyes for multicolor localization-based super-resolution microscopy. J Biophotonics 9:161–170
72. Annibale P, Scarselli M, Greco M, Radenovic A (2012) Identification of the factors affecting co-localization precision for quantitative multicolor localization microscopy. Opt Nanoscopy 1:1–13
73. Richter KN, Revelo NH, Seitz KJ et al (2017) Glyoxal as an alternative fixative to formaldehyde in immunostaining and super-resolution microscopy. EMBO J 37(1):139–159
74. Stanly TA, Fritzsche M, Banerji S et al (2016) Critical importance of appropriate fixation conditions for faithful imaging of receptor microclusters. Biol Open 5:1343–1350
75. Whelan DR, Bell TDMM (2015) Image artifacts in single molecule localization microscopy: why optimization of sample preparation protocols matters. Sci Rep 5:1–10
76. Tanaka KAK, Suzuki KGN, Shirai YM et al (2010) Membrane molecules mobile even after chemical fixation. Nat Methods 7:865–866
77. Brünig I, Scotti E, Sidler C, Fritschy JM (2002) Intact sorting, targeting, and clustering of γ-aminobutyric acid A receptor subtypes in hippocampal neurons in vitro. J Comp Neurol 443:43–55
78. Sergé A, Bertaux N, Rigneault H, Marguet D (2008) Dynamic multi-target tracing to probe spatiotemporal cartography of cell membrane. Nat Methods 5:687–694
79. Lelek M, Di Nunzio F, Henriques R et al (2012) Superresolution imaging of HIV in infected cells with FlAsH-PALM. Proc Natl Acad Sci U S A 109:8564–8569
80. Thevenaz P, Ruttiman UE, Unser M (1998) A pyramid approach to sub-pixel registration based on intensity. IEEE Trans Image Process 7:27–41
81. Schindelin J, Arganda-Carreras I, Frise E et al (2012) Fiji: an open-source platform for biological-image analysis. Nat Methods 9:676–682
82. Sage D, Kirshner H, Pengo T et al (2015) Quantitative evaluation of software packages for single-molecule localization microscopy. Nat Methods 12:717–724

83. Ovesný M, Křížek P, Borkovec J et al (2014) ThunderSTORM: a comprehensive ImageJ plug-in for PALM and STORM data analysis and super-resolution imaging. Bioinformatics 30:2389–2390
84. Henriques R, Lelek M, Fornasiero EF et al (2010) QuickPALM: 3D real-time photoactivation nanoscopy image processing in ImageJ. Nat Methods 7:339–340
85. Coltharp C, Yang X, Xiao J (2014) Quantitative analysis of single-molecule superresolution images. Curr Opin Struct Biol 28:112–121
86. Nicovich PR, Owen DM, Gaus K (2017) Turning single-molecule localization microscopy into a quantitative bioanalytical tool. Nat Protoc 12:453–461
87. Malkusch S, Endesfelder U, Mondry J et al (2012) Coordinate-based colocalization analysis of single-molecule localization microscopy data. Histochem Cell Biol 137:1–10
88. Georgieva M, Cattoni DI, Fiche J et al (2016) Nanometer resolved single-molecule colocalization of nuclear factors by two-color super resolution microscopy imaging. Methods 105:44–55
89. Pageon SV, Nicovich PR, Mollazade M et al (2016) Clus-DoC: a combined cluster detection and colocalization analysis for single-molecule localization microscopy data. Mol Biol Cell 27:3627–3636
90. Malkusch S, Heilemann M (2016) Extracting quantitative information from single-molecule super- resolution imaging data with LAMA-LocAlization Microscopy Analyzer. Sci Rep 6:1–4
91. Levet F, Hosy E, Kechkar A et al (2015) SR-Tesseler: a method to segment and quantify localization-based super-resolution microscopy data. Nat Methods 12:1065–1071
92. Andronov L, Orlov I, Lutz Y et al (2016) ClusterViSu, a method for clustering of protein complexes by Voronoi tessellation in super-resolution microscopy. Sci Rep 6:1–9
93. De Chaumont F, Dallongeville S, Chenouard N et al (2012) Icy: an open bioimage informatics platform for extended reproducible research. Nat Methods 9:690–696
94. Li Q, Lau A, Morris TJ et al (2004) A syntaxin 1, Gαo, and N-type calcium channel complex at a presynaptic nerve terminal: analysis by quantitative immunocolocalization. J Neurosci 24:4070–4081
95. Khanna R, Li Q, Sun L et al (2006) N type Ca2 + channels and RIM scaffold protein covary at the presynaptic transmitter release face but are components of independent protein complexes. Neuroscience 140:1201–1208
96. Brocher J (2014) Qualitative and quantitative evaluation of two new histogram limiting binarization algorithms. Int J Image Process 8:30–48

Chapter 10

Single-Molecule Localization Microscopy Propelled by Small Organic Fluorophores with Blinking Properties

Akihiko Morozumi, Mako Kamiya, and Yasuteru Urano

Abstract

Super-resolution fluorescence imaging by single-molecule localization microscopy (SMLM) has paved the way for a better understanding of cellular structures and processes at the molecular scale. Since SMLM relies on the fluorescence switching and detection of single molecules of fluorescent probes, the probe performance is of crucial importance. Small organic fluorophores have contributed greatly to SMLM, owing to their brightness, photostability, and potential for chemical modification and functionalization. Here, we provide an overview of the techniques of SMLM, including a discussion of the characteristics of existing small organic fluorophores. In particular, we highlight a novel class of spontaneously blinking fluorophores that we have recently developed, describing some of their applications, together with details of the experimental methods.

Key words Single-molecule localization microscopy, Fluorescent probes, Small organic fluorophores, Fluorescence blinking, Spontaneously blinking fluorophores

1 Introduction

Super-resolution fluorescence imaging provides microscopic images with a resolution well below the diffraction limit, and is a powerful tool for investigating the details of cellular structures and processes in both live and fixed cells. Among various families of super-resolution techniques, single-molecule localization microscopy (SMLM) has become one of the most widely used. SMLM achieves super-resolution through determination of the precise positions (localizations) of single fluorescent probe molecules that label the observation target. To enable this, it is required that only a sparse subset of molecules is somehow allowed to fluoresce at a time, with the rest in a dark state. In this situation, the location of a fluorescent probe molecule can be pinpointed by fitting a Gaussian function to its image, allowing the centroid position to be determined with nanometric precision. These fluorophores are then bleached or driven into some temporary dark state. This process is

Nobuhiko Yamamoto and Yasushi Okada (eds.), *Single Molecule Microscopy in Neurobiology*, Neuromethods, vol. 154,
https://doi.org/10.1007/978-1-0716-0532-5_10,

repeated many times, so that a sufficient number of molecular localizations can be collected. Finally, a super-resolution image is reconstructed as a composite image of all the localizations.

SMLM includes several technical variants, such as photoactivated localization microscopy (PALM) [1], fluorescence photoactivation localization microscopy (FPALM) [2], stochastic optical reconstruction microscopy (STORM) [3], direct STORM (dSTORM) [4–6], ground-state depletion microscopy followed by individual molecule return (GSDIM) [7], blink microscopy [8, 9], point accumulation for imaging in nanoscale topography (PAINT) [10], and PAINT-related techniques [11–15]. These SMLM methods all share the basic principle described above, but differ mainly in what sorts of probes are preferably employed, or the means used to achieve spatiotemporally sparse and stochastic fluorescence. Fluorescent probes suitable for SMLM are discussed later.

SMLM has also been extended to three dimensions by introducing optical devices that allow for axial localization of single molecules. Such optical approaches include astigmatism induced by a cylindrical lens [16, 17], point-spread-function (PSF) engineering with double-helix shaping [18], multiplane detection [19], and an interferometric scheme based on a dual-objective configuration [20].

Multicolor SMLM has also been developed, providing a powerful means for colocalization and interaction analysis. The properties of probes, as well as the optical setups, are crucial for successful multicolor SMLM, and this aspect is discussed further in Subheading 3, accompanied with a description of our work in this area.

Some of the most fascinating developments proposed recently are based on the idea that each individual probe molecule can report not only its position, but also some additional information, possibly describing the nano-environment or a nano-event that the molecule experiences. For instance, a spectrally resolved SMLM method was used for super-resolution mapping of the hydrophobicity or polarity on membranes by detecting the single-molecule spectra of solvatochromic Nile Red fluorophores concurrently with their localizations [21, 22].

1.1 Small Organic Fluorophores for SMLM

A wide variety of fluorescent probes have been used for SMLM, and they fall largely into three categories: small organic fluorophores, fluorescent proteins, and quantum dots. Here, we focus on small organic fluorophores, the most widely used and noteworthy category, after briefly mentioning fluorescent proteins. Fluorescent proteins capable of photoactivation or photoconversion have been used for SMLM, typically in PALM and FPALM. Fluorescent proteins have the advantage of absolute specificity, but also have the disadvantages of poor brightness, poor photostability, and relatively large size. In some cases, they can be utilized in a hybrid fashion in combination with small fluorophores, offering a practical approach to multicolor SMLM.

In general, small organic fluorophores are superior to fluorescent proteins in terms of brightness, photostability, size, and flexibility of modification. Brightness is crucial, because the precision of single-molecule localization strongly depends on the number of photons detected per single switching event. Photostability, coupled with brightness, defines the photon budget, or the total number of photons emitted before bleaching, which directly determines the localization precision of an irreversibly activatable probe, or translates into how many switching cycles are allowed for a reversibly switchable probe. Thus, photostability is an important factor. Small organic fluorophores have a roughly ten times greater photon budget than fluorescent proteins.

Small size is a fundamental advantage of small organic fluorophores. For one thing, smaller size enables denser labeling of a target structure, leading to higher spatial resolution, as suggested by the Nyquist-Shannon sampling theorem [23, 24]. For another, a smaller probe is likely to cause less perturbation of the structure, localization, or function of the target. In practice, however, these two aspects have to be reconsidered in the context of the effects of the labeling methods.

Small organic fluorophores are also advantageous in that they can be flexibly and rationally designed and modified for fine-tuning or functionalization. At present, a colorful palette of small fluorophores is available, spanning the whole spectrum. Other properties and functions can also be modulated, enhanced, or newly introduced. For example, certain small structural modifications can enhance the brightness and photostability of conventional organic fluorophores in various classes, including rhodamines, oxazines [25, 26], and cyanines [27]. Incorporation of particular kinds of partial structures offers a variety of targeting and labeling strategies, as described in the following subsection. Furthermore, it is possible to achieve activatability in response to photoirradiation [28–32] or binding to targets [33–36], as well as the capability of spontaneous blinking [37, 38].

1.2 Labeling Strategies with Small Organic Fluorophores

In contrast to fluorescent proteins, small organic fluorophores are not genetically encodable, and therefore they usually require a means to label targets specifically, which is often a challenging aspect of SMLM with small fluorophores. Immunostaining is one of the methods of choice in some contexts. Immunostaining, in many cases, involves the use of a secondary antibody, which can be labeled using a small fluorophore functionalized with a reactive group such as an active ester group. Although antibodies offer high specificity, they also present some drawbacks. First, since antibodies cannot permeate the cell membrane, their targets are limited to fixed cells or extracellular proteins of live cells. The large size of antibodies also makes it difficult to label targets located in close proximity and at high density. To partially overcome these problems, single-domain antibodies have recently been exploited [39–41].

For live-cell labeling, the most versatile and reliable method utilizes self-labeling protein tags such as SNAP-tag [29, 33, 37, 42–45], CLIP-tag [46], HaloTag [25, 26, 37, 47, 48], and TMP-tag [49–51]. A tag protein is genetically encoded as a fusion with a target protein, and forms a covalent linkage to a small fluorophore equipped with a corresponding substrate or ligand. Although these conjugates can be as large as fluorescent proteins, such tag protein-fluorophore conjugates work as brighter alternatives to fluorescent proteins. Peptide tags of smaller size are also available. For example, tetracysteine motifs can be inserted in the protein of interest and coupled with high-affinity-binding fluorescent reagents containing two trivalent arsenics, such as FlAsH [52, 53] and ReAsH [54]. Alternatively, genetically encoded oligohistidine sequences (His_6-, His_{10}-tags) can be utilized in combination with multivalent *N*-nitrilotriacetic acid (NTA)-fluorophore conjugates [55–57]. Furthermore, even less bulky labeling can be performed by genetically based insertion of unnatural amino acids with reactive functional groups, followed by coupling with fluorescent reactive counterparts via click chemistry [58].

There are also other approaches. For example, some fluorophores, including MitoTracker, LysoTracker, and ER-Tracker, simply permeate through the plasma membrane, accumulate in specific organelles, and exhibit photoswitching behavior that enables SMLM imaging [59, 60]. Alternatively, incorporating particular structures can direct the fluorophores to specific targets in living cells. For example, a silicon-substituted rhodamine (SiR) coupled to a taxane, phallotoxin, or the core moiety of Hoechst dyes can site-specifically label microtubules, actins [34, 36], or DNA [35], respectively. Such a strategy can be combined with click chemistry to assemble two reagents in situ [61]. In a typical approach, one reagent is carried to a specific organelle such as endoplasmic reticulum or mitochondria by a ceramide or cationic rhodamine moiety, respectively, and this reagent also contains a reactive *trans*-cyclooctene moiety. The other reagent is a fluorophore equipped with a reactive tetrazine moiety, which reacts with the former reagent to accomplish site-specific labeling.

1.3 Fluorescence Switching of Small Organic Fluorophores

As mentioned earlier, SMLM techniques presuppose that only a small subset of fluorescent probe molecules is in the fluorescent state at any given time during data acquisition. The means used to ensure such a situation can be broadly classified into irreversible and reversible approaches. In accordance with that classification, we describe below the small organic fluorophores applicable to SMLM and their fluorescence switching behaviors.

1.3.1 Irreversible Photoactivation

Irreversibly photoactivatable small fluorophores, also known as caged fluorophores, have been used for SMLM since their potential utility was suggested in the original report on PALM [1]. The scheme of SMLM with irreversibly photoactivatable probes is as follows. All probe molecules start in the unactivated (caged) non-fluorescent state, and small fractions of the molecules are photo-activated (uncaged) so that they can be localized, and are then bleached sequentially. Photoactivation is usually done with UV or near-UV light.

Caged rhodamines comprise a major class of photoactivatable small fluorophores. Rhodamines can be caged, thereby taking a nonfluorescent spiro lactone form, by using photocleavable nitro-benzyl groups [28–30]. Upon irradiation with UV light, the caging group is cleaved, and a fluorescent xanthene moiety is regenerated. Recently, rhodamine NN dyes [31, 62–64] have been developed as a novel class of caged rhodamines, which have a small photolabile diazoketone caging group incorporated into a spirocyclic structure. Other photoactivatable fluorophores include an azide-based dicya-nomethylenedihydrofuran (DCDHF) fluorogen [32, 65]. That is, the push-pull fluorophore DCDHF is caged by replacing the electron-donating amine with an azide, which has poor electron-donating ability, but which can be converted back to an amine upon irradiation with violet light. Some conventional fluorophores can also be caged by the addition of chemical reagents, such as reductants and phosphines, prior to use. Rhodamines [66] and cyanines [67] are reduced by sodium borohydride ($NaBH_4$) to take nonfluorescent leuco forms, which, in turn, are activated with UV or violet light. Some cyanines such as Cy5 and Alexa 647 can also be caged by attachment of tris(2-carboxyethyl)phosphine (TCEP) [68], and in this case, the caging and uncaging can be reversible.

1.3.2 Photoinduced Reversible Blinking

In many cases of reversible fluorescence switching or "blinking," all of the initially fluorescent probe molecules are first turned off by irradiation with intense laser light. Subsequently, a small number of molecules are turned on back by photoactivation or thermally.

SMLM with small fluorophores originated from the fluorescence switching of cyanine dyes [69, 70]. The first reported small fluorophore-based SMLM method, termed STORM, used a cyanine dye pair consisting of Cy5 as a reporter and Cy3 as an activator [3]. Here, the blinking of the reporter Cy5 is achieved by iterating the following scheme. Red laser illumination allows Cy5 to fluoresce and can also drive the dye to a stable dark state. Subsequent exposure to green laser light converts Cy5 back to the fluorescent state, and its recovery rate depends on the proximity of the activator dye Cy3. Other activator-reporter pairs, with Cy2, Cy3, or Alexa405 as an activator and Cy5, Cy5.5, Cy7, or Alexa647 as a reporter, are also usable, offering the possibility of multicolor SMLM [71].

Later, an activator-free method named dSTORM was introduced [4] based on the fact that cyanines can exhibit reversible blinking without any activator dye [70]. The scheme of dSTORM was subsequently extended to encompass various conventional fluorophores, including rhodamines and oxazines [5]. In parallel with dSTORM, a similar method called GSDIM also utilized the fact that some rhodamines blink reversibly without activation light to enable SMLM imaging [7]. The photochemical behavior shown by conventional fluorophores in dSTORM and GSDIM can be summarized as follows. When irradiated intensely in the presence of reducing agents such as thiols, the fluorophores are forced into certain long-lived dark states via the first excited singlet state. The chemical species of the dark states depend on the class of fluorophores. Rhodamines, as a result of the reduction of their triplet state, form semi-reduced radical anions, whereas oxazines are further reduced to leuco forms. In contrast, cyanines form thiol adducts from their radical anions. These dark states have the common feature that the original conjugated π-electron system of the fluorophore is broken, resulting in blue-shifted absorption. The dark states can revert to the singlet ground state upon oxidation by molecular oxygen. This recovery of the original fluorescent state occurs thermally or can be facilitated by irradiation with UV or blue light corresponding to the absorption wavelength of the dark state. Oxidation and reduction by other chemicals such as methyl viologen and ascorbic acid, respectively, can also induce and control the blinking of fluorophores, as exploited in blink microscopy [9, 72].

Reversible fluorescence switching can also be offered by photochromic compounds, whose absorption spectra reversibly change under UV or visible light. The advantage here is that their blinking only relies on photoirradiation, eliminating the need for chemical additives as inducers of blinking. Among the photochromic compounds are rhodamine amides. The nonfluorescent closed isomer can be converted into the fluorescent open isomer by irradiation with UV light (for single-photon absorption) or red light (for two-photon absorption). The resultant fluorescent isomer can thermally revert to the initial isomer within a period of milliseconds to minutes, depending on the conditions, especially the solvent. Despite their potential usability as reversibly blinking probes, rhodamine amides have in many cases been used as irreversibly photoactivatable probes; that is, the activated fluorophores were bleached before thermal reversion. Recently, however, reversible and spontaneous blinking has been achieved with a new rhodamine amide, in which even the activation process occurs thermally, and this blinking was shown to be usable for SMLM in fixed cells [73]. Diarylethenes are another representative class of photochromic compounds. Their fluorescence ability is reversibly switched off and on, typically by visible and UV light, respectively. Although they could potentially be used for SMLM, their application to biological samples was restricted due to their poor water solubility, as with other photochromic compounds. However, water-soluble diarylethenes have recently been developed and successfully applied to SMLM in fixed cells [74]. Interestingly, blue light facilitates their interconversion in both directions: it not only switches off the fluorescent isomer, but also switches on the nonfluorescent isomer. This feature allows reversible blinking and SMLM imaging to be achieved using a single laser.

We should also briefly mention a different type of fluorescence switching, which is based on stochastic binding of probes to a target, and which has been exploited for PAINT and related techniques. This kind of "blinking" builds on the fact that a freely diffusing fluorescent molecule cannot be captured or imaged, whereas the molecule becomes detectable when it is immobilized on the surface of an object. Thus, the binding event translates into "switching on," and the "on" state is ended by a "switching off" event, that is, bleaching or detachment from the object. Since the first report of PAINT [10], various developments have been reported. The specificity and versatility of targeting have been enhanced by using fluorophores coupled with antibodies, small ligands for peptide tags [11], DNA [12, 14, 75], and protein fragments [15]. Control of the reversibility of binding events has also been investigated [14, 15, 75]. Various fluorogenic dyes have been employed in PAINT-related studies. Representatives are DNA-intercalating dyes such as YOYO-1 and PicoGreen, whose fluorescence ability is activated upon binding to double-stranded DNA [13, 76]. Malachite green and thiazole orange exhibit similar activation when coupled with certain single-chain variable-fragment antibodies termed fluorogen-activating proteins or peptides (FAPs) [77, 78]. This kind of fluorogenic scheme is also promising as a genetically encodable labeling system for SMLM.

1.3.3 Spontaneous Blinking

Fluorescence blinking of most conventional small organic fluorophores relies on photoinduced processes, as described above. Therefore, one has to expose the experimental sample to intense light irradiation, especially in order to force most of the probe molecules into a nonfluorescent state. In addition, chemical additives such as reductive thiols and/or enzymatic oxygen scavengers are often required to induce well-defined blinking and to provide a long photobleaching lifetime. These treatment and factors are essentially incompatible with live-cell experiments. Moreover, in the case of using two or more types of fluorophores for multicolor SMLM in either live or fixed cells, it can be troublesome to find optimal conditions for all the fluorophores. Therefore, small fluorophores that blink spontaneously have substantial advantages for SMLM.

Recently, we have developed a novel class of small fluorophores that spontaneously blink in the absence of intense light irradiation or any additive. We focused on the fact that certain rhodamine derivatives, in which a carboxy group is substituted with a more nucleophilic group, such as a hydroxy group, exist in a thermal equilibrium between a fluorescent open form and a nonfluorescent spirocyclized form. We carefully optimized the chemical structure to obtain blinking properties suitable for SMLM in terms of the percentage of the fluorescent form in the equilibrium and the duration of the temporary fluorescent state. Our first product of this kind was a silicon-substituted rhodamine (SiR) bearing a hydroxy group, named HMSiR [37]. HMSiR was confirmed to blink appropriately in the absence of any additive and without exposure to intense illumination, and it has been successfully applied to live-cell experiments in various contexts. Subsequently, we developed another fluorophore termed HEtetTFER [38], which can be regarded as a derivative of rhodamine B. Although it has not yet been used in live-cell applications, its utility for fixed-cell SMLM has been established. Furthermore, the combination of HMSiR and HEtetTFER enables simple dual-color SMLM without the need for any special buffer. In Subheading 3, we describe the methodology, and present typical results of some model experiments that take advantage of our spontaneously blinking fluorophores.

2 Materials and Equipment

2.1 Reagents

- Dulbecco's modified Eagle's medium (DMEM) containing glucose (at a high concentration), L-glutamine, and phenol red (Wako Pure Chemical).
- Dulbecco's modified Eagle's medium (DMEM) containing glucose (at a high concentration) and L-glutamine, but no phenol red (Gibco).
- Fetal bovine serum (FBS) (Gibco).
- Penicillin-streptomycin (PS) solution (Gibco).
- TrypLE Express not containing phenol red (Gibco).
- 8-Well Lab Tek II Chambered Coverglass (No. 1.5 borosilicate) (Nunc).
- Cellmatrix Type I-C (Nitta Gelatin).
- Glycine (Wako Pure Chemical).
- 0.01% Poly-L-lysine (Sigma).
- β-Tubulin-Halo plasmid.
- X-tremeGENE HP DNA Transfection Reagent (Roche).

- Dimethyl sulfoxide (DMSO, Luminasol) (Dojindo).
- Phosphate-buffered saline (PBS) (Gibco).
- Leibovitz's L-15 (Gibco).
- Sodium phosphate buffer prepared from sodium phosphates (Wako Pure Chemical).
- PD MiniTrapTM G-25 or PD-10 column (GE Healthcare).
- Secondary antibodies:
  - Goat anti-mouse IgG (whole molecule) (Sigma).
  - Affinipure donkey anti-rabbit IgG (H+L) (Jackson ImmunoResearch Laboratories).
- Primary antibodies:
  - Rabbit polyclonal anti-Tom20 (FL-145) antibody (IgG) (Santa Cruz Biotechnology).
  - Mouse monoclonal anti-β-tubulin antibody (IgG) (Sigma).
  - Anti-Green Fluorescent Protein, Monoclonal Antibody (mouse, IgG) (Wako Pure Chemical).
- 4% Paraformaldehyde phosphate buffer solution (Wako Pure Chemical).
- 25% Glutaraldehyde solution (Wako Pure Chemical).
- Sodium borohydride ($NaBH_4$) (Wako Pure Chemical).
- Bovine serum albumin (BSA) (Wako Pure Chemical).
- Triton X-100 (MP Biomedicals).
- TetraSpeck Microspheres, 0.1 μm, fluorescent blue/green/orange/dark red (Invitrogen).
- Methanol (Wako).
- Immersion oil for microscopy (Nikon or Olympus).

2.2 Fluorescent Probes

Spontaneously blinking probes can be synthesized according to the methods we reported previously [37, 38]. Some of them can also be purchased from Goryo Chemical. The spontaneously blinking core fluorophores were established through optimization of the molecular structures in terms of the proportion of the fluorescent form in the equilibrium and the duration of the temporary fluorescent state. The probes themselves should be prepared as suitable derivatives of the core fluorophores, depending on the target and the labeling method. The probes used for the model experiments described in Subheading 3 are as follows:

- Derivatives with an amine-reactive group for labeling antibodies:
  - HMSiR-NHS
  - HEtetTFER-sulfoNHS

- A derivative with a ligand to be recognized by HaloTag for live-cell experiments:
  - HMSiR-Halo.

2.3 Cells

- Vero cells.
- HeLa cells expressing POM121-GFP or Nup107-GFP fusion protein.

For long-term storage, cells are frozen in cryopreservation medium and kept in liquid nitrogen. Once the frozen cells are thawed, the cells can be cultured and passaged up to 10–20 times. The cells are cultured in Dulbecco's modified Eagle's medium (DMEM; high glucose) containing L-glutamine (Wako Pure Chemical), and supplemented with 10% fetal bovine serum (FBS) and 1% penicillin-streptomycin (PS) solution at 37 °C in humidified air containing 5% CO_2. To minimize the background fluorescence, phenol red-free medium (Gibco) is preferred. Cells are regularly passaged using TrypLE Express.

For imaging experiments, cells are seeded into 8-well Lab Tek II Chambered Coverglass. For some kinds of cells, it is recommended to coat the surface of the cover glass beforehand with some coating material, such as collagen, which is left in the chambers for 10–15 min and then removed. To avoid nonspecific adsorption of fluorophores, the coating process is preferably done by using a solution of 2 M glycine [6, 45] or 0.01% poly-L-lysine for 30–60 min, followed by washing with phosphate-buffered saline (PBS) several times. An alternative way to reduce the background signals is to detach the cells (after labeling) from the glass surface by trypsinization and transfer them into a new Lab-Tek II chamber. This step is taken 1–3 h before imaging, so that the cells have time to reattach to the glass surface.

2.4 Microscope

2.4.1 TIRF/Quasi-TIRF Illumination System

SMLM imaging is carried out using an N-STORM system (Nikon) equipped with the following:

- Inverted fluorescence microscope (Eclipse Ti-E; Nikon).
- Oil-immersion objective (CFI Apo TIRF 100× Oil, NA 1.49; Nikon).
- Irradiation lasers:
  - 647 nm (Visible Fiber Laser, 2RU-VFL-P-200-647; MPB Communications).
  - 488 nm (Sapphire 488 LP; Coherent).
- Cooled electron-multiplying charge-coupled device (EMCCD) camera (iXon3 DU897; Andor).
- Software for system control and image analysis (NIS-Elements Advanced Research; Nikon).

The pixel size of the EMCCD camera and the magnification of the objective lens translate into the final pixel size of 160 nm on the sample plane. During data acquisition, a Perfect Focus System is used to maintain a constant focal plane. For dual-color SMLM, chromatic aberrations are corrected using a function offered by the software, which processes an experimental data set according to a reference data set pre-obtained by imaging multicolored fluorescent particles (TetraSpeck Microspheres).

2.4.2 Spinning-Disk Confocal System

SMLM with a spinning-disk confocal microscope is performed on a customized system composed of the following:

- Inverted fluorescence microscope (IX81; Olympus) equipped with a confocal scanner unit (CSU-X1; Yokogawa Electric).
- Oil-immersion objective (APON 60× OTIRF, NA 1.49; Olympus).
- Irradiation lasers (640 nm; Coherent).
- Cooled EMCCD camera (iXon3 DU-897; Andor).
- Software for system control (MetaMorph; Molecular Devices).

A 2× magnification lens is inserted between the microscope and the spinning disk to match the pinhole size to the magnification of the objective lens, and the final value of the pixel size corresponds to 100 nm on the sample plane. The system is controlled with MetaMorph software, and the raw images obtained are analyzed to provide super-resolution reconstructed images using open-source software, rapidSTORM.

3 Methods

3.1 Time-Lapse SMLM in Live Cells

Spontaneously blinking fluorophores are fundamentally advantageous in that they do not require severe imaging conditions, such as intense illumination, addition of thiols, or use of an oxygen depletion system. Since cytotoxicity and photobleaching are minimized, this approach is favorable for live-cell imaging, particularly in a repetitive time-lapse manner, as illustrated by the following example [37].

In this model experiment, microtubules in live mammalian cells are labeled with HMSiR by means of a self-labeling protein tag and are imaged under near-physiological conditions. β-Tubulin fused with HaloTag is expressed in Vero cells, and is labeled by incubating the cells in culture medium containing HMSiR-HaloTag ligand conjugate. Importantly, SMLM is performed in ordinary culture medium without an oxygen depletion system or thiols, under much lower intensity illumination than generally used for dSTORM. HMSiR indeed exhibits spontaneous blinking, and affords a reconstructed image of microtubules with a resolution beyond the diffraction limit (Fig. 1a, b, d). Time-lapse images can be acquired at 5- or 10-min intervals, and the motion of the microtubules can be tracked for over 1 h (Fig. 1c). Higher temporal resolution can also be achieved by sliding-window image processing [37, 79]. Typically, tens of thousands of consecutive frames are recorded at 30 ms/frame, and super-resolution images are reconstructed from 1000 frames (corresponding to 30-s acquisition time) with 500-frame overlaps (Fig. 1e). The resulting video-like sequence, consisting of a stack of reconstructed super-resolution images, visualizes the dynamics of the polymerization and depolymerization of tubulins at a temporal resolution of 15 s/image (Fig. 1f).

3.1.1 Procedure for Labeling β-Tubulin-Halo in Live Vero Cells

1. At 1–2 days before imaging, transfect Vero cells transiently with β-tubulin-Halo plasmid using a transfection reagent (X-tremeGENE HP DNA Transfection Reagent) according to the manufacturer's protocol (**Note 1a**). Incubate the cells at 37 °C under 5% CO_2 overnight.
2. Dissolve the HaloTag-compatible derivative of HMSiR (HMSiR-Halo) in dimethyl sulfoxide (DMSO) to obtain a stock solution at a convenient concentration (i.e., 100 μM or 1 mM).
3. After washing the cells once with culture medium or PBS, incubate the cells in culture medium to which a stock solution of the probe has been added to give a final concentration of 100 nM (**Note 1a, b**).
4. To reduce the background signals from the fluorophores attached nonspecifically to the glass surface, detach the cells and transfer them to a new Lab-Tek II chamber.
5. Leave the sample in the incubator for 3 h to allow the cells to adhere to the cover glass.

3.1.2 Procedure for Live-Cell SMLM Imaging of β-Tubulin-Halo in Vero Cells

1. Replace the medium of the sample with an imaging medium: Leibovitz's L-15 supplemented with 10% FBS.
2. Carry out data acquisition for SMLM with the N-STORM system at room temperature (24 °C) (**Note 2**).
   - For prolonged repetitive time-lapse SMLM, record typically 1000 consecutive frames of images at 15–30 ms/frame, applying continuous illumination at 647 nm with a low (40–100 W/cm^2) intensity. Repeat this several times at 5- or 10-min intervals.
   - For fast time-lapse SMLM, record typically tens of thousands of consecutive frames of images at 15–30 ms/frame, applying continuous illumination at 647 nm with a low (40–100 W/cm^2) intensity.

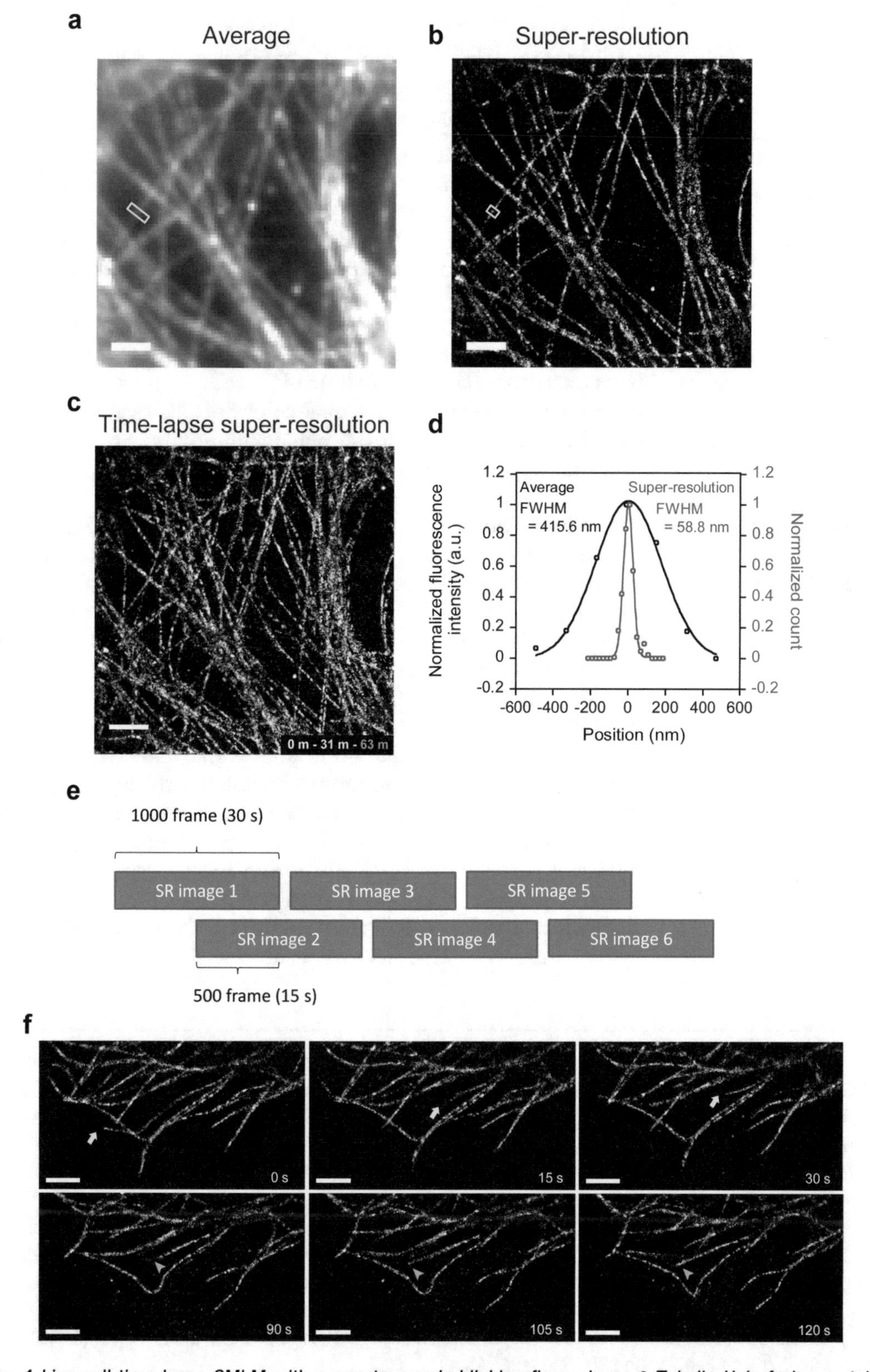

Fig. 1 Live-cell time-lapse SMLM with a spontaneously blinking fluorophore. β-Tubulin-Halo fusion protein expressed in Vero cells was labeled with HMSiR-Halo. (**a–d**) Long-lasting repetitive time-lapse SMLM of

3. Analyze the raw images to reconstruct super-resolution images with NIS-Elements Advanced Research software.
   - For repetitive time-lapse SMLM, reconstruct super-resolution images from each set of 1000 frames.
   - For fast time-lapse SMLM, reconstruct super-resolution images from sets of consecutive 1000 frames with 500-frame overlaps (frames 1–1000; 501–1500; 1001–2000; 1501–2500; and so on), and concatenate them to form a sequence of super-resolution images.

3.2 Simple Dual-Color SMLM Without Special Optical Devices or Buffer Optimization

We have so far reported two spontaneously blinking fluorophores, HMSiR [37] and HEtetTFER [38], whose colors are different from each other. The combination of these fluorophores has proven to be an attractive option for dual-color SMLM, clearly displaying the advantage of spontaneously blinking fluorophores over conventional fluorophores [38].

3.2.1 Multicolor SMLM with Conventional Small Organic Fluorophores

We begin by outlining multicolor SMLM with conventional small organic fluorophores. The first reported multicolor SMLM employed activator-reporter dye pairs [71] that have different activator dyes, and so can be selectively activated at different wavelengths. This approach can avoid chromatic aberrations by using the same reporter dye, but tends to suffer from color cross talk due to false and nonspecific activation. More simply, multicolor dSTORM with activator-free fluorophores is possible. However, it is difficult to find multiple fluorophores that blink appropriately together under the same experimental conditions, especially in the same medium or buffer, and chromatic aberrations are also an issue. It can be helpful to use a spectral demixing technique [43, 80, 81], by which partially overlapping emission spectra can be separated. For this purpose, a fluorescence signal is split into two spectral channels and assigned to a fluorophore on the basis of the ratiometric intensity. This approach allows the use of multiple

Fig. 1 (continued) microtubules. (**a**, **b**) Averaged image (**a**) and super-resolution image (**b**) obtained with HMSiR. (**c**) Sequential acquisition of super-resolution images at 0 min (white), 31 min (yellow), and 63 min (green). Each super-resolution image was reconstructed from 1000 frames recorded at 30 ms/frame, corresponding to the acquisition time of 30 s. (**d**) Transverse profiles of fluorescence intensity and localizations corresponding to the regions boxed in yellow in (**a**) and (**b**), respectively. (**e**, **f**) Fast time-lapse SMLM of microtubules. (**f**) Sequential super-resolution images. Ten thousand frames were recorded at 30 ms/frame. Each super-resolution image was reconstructed from 1000 frames, corresponding to the acquisition time of 30 s. The video-like sequence was composed of the reconstructed super-resolution images with 500-frame overlaps, as shown in (**e**). The super-resolution images at the indicated points of time were extracted for display. Yellow arrows: Retracting tubules. Green arrowheads: Extending tubules. Excited at 647 nm (40 W/cm^2). Measured in Leibovitz's L-15 medium supplemented with 10% FBS. Scale bars, 2 μm (adapted from Uno et al. [37] with permission from Springer Nature)

chemically similar fluorophores that may work well together in the same buffer, but whose spectra partially overlap each other. In addition, it can also minimize chromatic aberrations. Similar benefits can be gained by recently developed methods that incorporate single-molecule spectroscopic technology, referred to as spectrally resolved SMLM (SR-SMLM) techniques [82, 83]. They enable the concurrent acquisition of the localization and the emission spectrum of a single molecule by inserting a dispersive device into the detection path. However, the last two approaches require special optical setups and relatively complicated analysis.

3.2.2 Dual-Color SMLM with Spontaneously Blinking Fluorophores

One of the major advantages of employing spontaneously blinking fluorophores for multicolor SMLM is that they simplify the experimental protocols and conditions, removing the need for optimization of the imaging medium. As discussed above, when a conventional fluorophore is used for dSTORM/GSDIM-type SMLM methods, the composition of the imaging medium or buffer must be optimized carefully to induce appropriate blinking. In the case of dual-color imaging, especially, it is particularly troublesome to ensure that two fluorophores with well-separated spectral properties can blink appropriately in the same medium or buffer. Spontaneously blinking fluorophores can circumvent this difficulty, since their optimized blinking occurs as their intrinsic behavior.

An example is provided by the following model experiment on fixed cells, performed with the combination of the two spontaneously blinking fluorophores HMSiR and HEtetTFER [38]. Since their absorption/fluorescence spectra are completely separable from each other (Fig. 2a), the dual-color imaging can be performed using a commercially available setup without any special optical device to deal with spectral overlapping. Here, mitochondria and microtubules are visualized by immunostaining Tom20 and β-tubulin proteins, using antibodies labeled with HMSiR and HEtetTFER, respectively. Notably, SMLM imaging of the two targets is achieved in a simple buffer without any agent to induce blinking (Fig. 2b).

3.2.3 Procedure for Labeling Antibodies with Reactive Derivatives of Spontaneously Blinking Fluorophores

For indirect immunostaining, we first label secondary antibodies with spontaneously blinking fluorophores, and then use the resulting labeled antibodies to stain the corresponding primary antibodies.

1. Dissolve the succinimidyl esters of spontaneously blinking fluorophores (HMSiR-NHS, HEtetTFER-sulfoNHS) individually in DMSO to obtain stock solutions at convenient concentrations (i.e., 1 or 10 mM).
2. Incubate a secondary antibody with each fluorophore (in an excess amount, or typically at 20 equivalents with respect to the

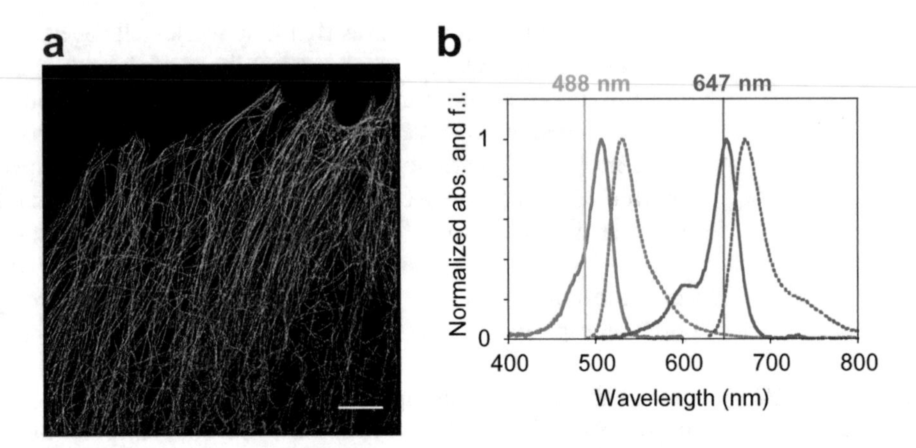

Fig. 2 Dual-color SMLM with two different-colored spontaneously blinking fluorophores. (**a**) Normalized absorption and fluorescence spectra of HMSiR (red) and HEtetTFER (green). Solid and dashed lines represent absorption and fluorescence spectra, respectively. (**b**) Dual-color SMLM of mitochondria (red) and microtubules (green) in a fixed Vero cell. Tom20 and β-tubulin were immunostained with HMSiR and HEtetTFER, respectively. 20,000 frames were recorded at 15 ms/frame for HMSiR and at 60 ms/frame for HEtetTFER. Excited at 647 nm (1000 W/cm^2) and 488 nm (240 W/cm^2) for HMSiR and HEtetTFER, respectively. Measured in PBS without any additive. Scale bar, 5 μm (adapted from Uno et al. [38] with permission from the Royal Society of Chemistry)

antibody) in 10–200 mM sodium phosphate buffer (pH 7.5–8.5) at room temperature or 37 °C for 30 min. Prepare the following three conjugates:

- Donkey anti-rabbit IgG labeled with HMSiR-NHS
- Goat anti-mouse IgG labeled with HMSiR-NHS
- Goat anti-mouse IgG labeled with HEtetTFER-sulfoNHS

3. Purify the labeled antibodies by using a PD MiniTrapTM G-25 or PD-10 column with PBS as the eluent.
4. Determine the degree of labeling (DOL) ratio (fluorophore [mol]/antibody [mol]) by measuring the absorbance of the labeled IgG at a pH where the fluorophores exist in colored open form and by assuming that the antibody is recovered without loss at the purification step (**Note 3**). For the model experiments demonstrated below, the DOLs are as follows:
   - HMSiR-labeled anti-rabbit IgG: 1.5
   - HMSiR-labeled anti-rabbit IgG: 1.9
   - HEtetTFER-labeled anti-mouse IgG: 5.1

3.2.4 Procedure for Immunostaining of Tom20 and β-Tubulin in Vero Cells

1. The day before imaging, seed Vero cells into 8-well Lab-Tek II chamber slides at 37 °C under 5% CO_2.
2. After washing the cells with pre-warmed PBS, fix them with 3% paraformaldehyde and 0.1% glutaraldehyde in PBS at room temperature for 10 min.

3. Replace the fixative solution with a solution of 0.1% sodium borohydride ($NaBH_4$) in PBS, prepared immediately before use, and leave the sample at room temperature for 7 min.
4. After washing the cells with PBS three times, permeabilize them, and block them in blocking buffer (3% bovine serum albumin (BSA) and 0.5% (v/v) Triton X-100 in PBS) at room temperature for 20 min.
5. Replace the blocking buffer with a solution of primary antibodies against the target antigens, and incubate the sample at room temperature for 30 min. Specifically, the solution of primary antibodies contains the following:
   - Rabbit polyclonal anti-Tom20 IgG, 1:50 dilution in blocking buffer
   - Mouse monoclonal anti-β-tubulin IgG, 1:100 dilution in blocking buffer
6. Wash the cells by replacing the former solution with washing buffer (0.2% BSA and 0.1% (v/v) Triton X-100 in PBS) and incubating at room temperature for 10 min. Repeat this washing step another two times.
7. Replace the washing buffer with a solution of fluorophore-labeled secondary antibodies against the primary antibodies, and incubate the sample at room temperature for 30 min. Specifically, the solution of secondary antibodies contains the following:
   - HMSiR-labeled anti-rabbit IgG (DOL: 1.5, diluted with blocking buffer to a final concentration of 10 μg/mL)
   - HEtetTFER-labeled anti-mouse IgG (DOL: 5.1, diluted with blocking buffer to a final concentration of 10 μg/mL)
8. Wash the cells with washing buffer three times for 10 min each, followed by PBS once quickly.
9. Post-fix the cells by replacing the PBS with a solution of 3% paraformaldehyde and 0.1% glutaraldehyde in PBS and incubating at room temperature for 10 min. Then wash the cells with PBS three times.

3.2.5 Procedure for Dual-Color SMLM Imaging

Carry out data acquisition for SMLM in PBS at room temperature (24 °C) using the N-STORM system. This model experiment is performed in a sequential manner. That is, HMSiR-labeled Tom20 is imaged first, and then HEtetTFER-labeled β-tubulin is imaged.

1. Record 20,000 frames of images of Tom20 at 15 ms/frame, applying continuous illumination at 647 nm with an intensity of 1000 W/cm^2 (**Note 2**).

2. Record 20,000 frames of images of β-tubulin at 60 ms/frame, applying continuous illumination at 488 nm with an intensity of 240 W/cm^2 (**Note 2**).
3. Analyze the raw images to reconstruct super-resolution images with NIS-Elements Advanced Research software. After ordinary localization analysis, process the dual-color super-resolution localization sets further to correct the effects of chromatic aberration. Then superimpose the two reconstructed images to obtain a dual-color super-resolution image.

3.3 SMLM Deep Inside Cells with a Spinning-Disk Confocal Microscope

The spontaneously blinking feature can also be utilized to deepen the target region covered by SMLM, expanding the compatibility with illumination schemes. The total internal reflection fluorescence (TIRF) or quasi-TIRF microscopy mode is usually employed for SMLM in order to ensure a high signal-to-background ratio, which is a prerequisite for precise localization of single fluorophores. However, these illuminating modes limit the imaging region to a thin layer above the surface of the cover glass. For the imaging of structures far above the cover glass at high speed, spinning-disk confocal microscopy has been widely used. However, this method is conventionally incompatible with dSTORM/GSDIM-type methods, because the illumination beam is divided into arrays of microbeams, whose intensity is too weak to drive the fluorophores into the dark state. In contrast, the continuous blinking of our spontaneously blinking fluorophores occurs without intense illumination. Therefore, it is possible to observe targets located far above the cover glass using a spinning-disk confocal microscope.

This is exemplified by the observation of nuclear pore complexes (NPCs) on the apical side of a nuclear membrane [37]. NPCs on the apical side, unlike those on the basal side, are too distant from the glass surface to be imaged with a high signal-to-background ratio in the TIRF or quasi-TIRF illumination scheme. The specific targets are POM121 and Nup107, two components of an NPC, which are individually immunostained with HMSiR-labeled antibody. The spontaneous blinking of HMSiR is indeed observed even from the apical side of the nucleus, and the reconstructed images exhibit ring structures with an average radius of 48.5 ± 1.4 nm for POM121 and 38.6 ± 1.2 nm for Nup107 (Fig. 3a–c). The difference in the radii is consistent with the accepted structural model of the NPC, in which POM121 forms the outer rim of the pore, and Nup107 forms an inner ring [84].

3.3.1 Procedure for Immunostaining of Nuclear Pore Proteins in HeLa Cells

1. Fix cells (HeLa cells expressing POM121-GFP or Nup107-GFP fusion protein) by replacing the medium with methanol at −20 °C and leaving the samples in a freezer at −20 °C for 30 min.

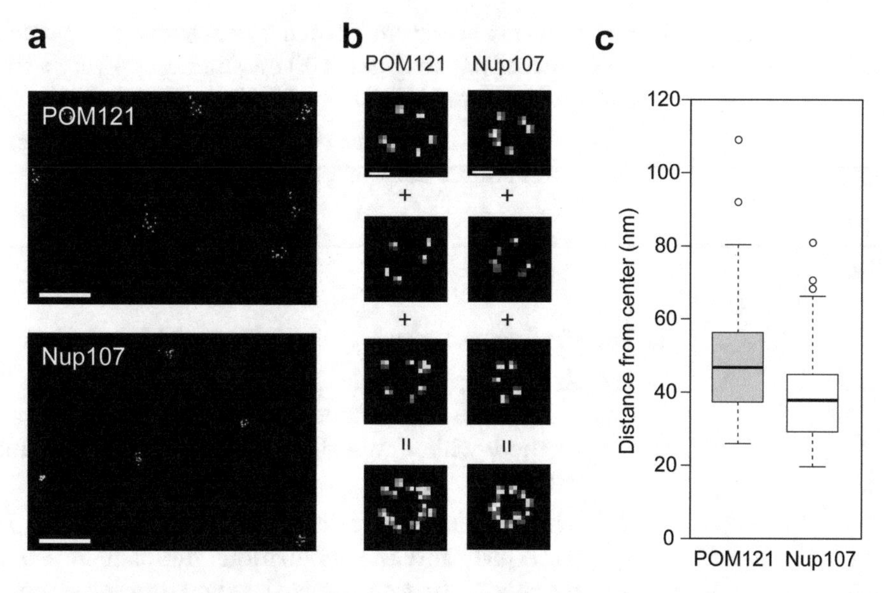

Fig. 3 SMLM deep inside cells with a spinning-disk confocal microscope. HeLa cells expressing POM121-GFP or Nup107-GFP fusion protein were fixed and immunostained using HMSiR-labeled antibody. (**a**) Nuclear pore proteins on the basal side of a nucleus. (**b**) Representative ring structures formed by nuclear pore proteins on the apical side of the nuclei, namely, on a single axial plane 4 μm above the cover glass. Each super-resolution image was reconstructed from 10 frames recorded at 100 ms/frame. Excited at 640 nm. Measured in PBS without any additive. (**c**) Box-and-whisker plot of the distance from the center to the nuclear pore proteins: 48.5 ± 1.4 nm for POM121 and 38.6 ± 1.2 nm for Nup107 (mean ± s.e., $N = 100$). Scale bars, 500 nm (**a**) and 50 nm (**b**). The localization frequency was color-coded according to the Red Hot look-up table (LUT) (adapted from Uno et al. [37] with permission from Springer Nature)

2. After washing the cells with PBS, block them in blocking buffer (1% BSA in PBS) at room temperature for 30 min.
3. Replace the blocking buffer with a solution of a primary antibody against the antigen (anti-GFP IgG, diluted with blocking buffer to a final concentration of 10 μg/mL), and incubate the sample at room temperature for 1 h.
4. After washing the cells with PBS three times, block them in blocking buffer at room temperature for 30 min.
5. Replace the blocking buffer with a solution of HMSiR-labeled secondary antibody against the primary antibody (HMSiR-labeled anti-mouse IgG, DOL: 1.9, diluted with blocking buffer to a final concentration of 10 μg/mL), and incubate the sample at room temperature for 40 min.
6. Wash the cells with PBS three times.

3.3.2 Procedure for SMLM Imaging of Nuclear Pore Proteins with a Spinning-Disk Confocal Microscope

Carry out data acquisition for SMLM in PBS using a customized microscope system composed of an inverted optical microscope (IX81; Olympus) and a spinning-disk confocal microscope system (CSU-X1; Yokogawa Electric).

1. For each ring structure formed by a nuclear pore protein in an NPC, record 10 images at 100 ms/frame, applying a spinning-disk multipoint illumination at 640 nm (**Note 2**).
2. Analyze the raw images to reconstruct super-resolution images with open-source software, rapidSTORM.

4 Notes

1. Transfection and labeling (**steps 1** and **3** in Procedure 3.1.1):
   (a) **Steps 1** and **3** can be performed simultaneously. That is, the cells can be incubated overnight in medium supplemented with a transfection reagent and a fluorescent probe.
   (b) The concentration of a fluorescent probe can be 100 nM to 1 μM, and the incubation time can be 30 min to overnight. In some cases, labeling at a lower concentration for a longer time (e.g., 100 nM, overnight) [37] may reduce nonspecific adsorption of the fluorophores on the cover glass and/or cellular components, leading to a higher signal-to-background ratio.
2. Acquisition/analysis conditions:

   The rationale for setting the illumination intensity, the frame rate, and the number of frames is as follows:
   (a) Illumination intensity:

   Since the blinking behavior of our spontaneously blinking fluorophores does not rely on photoinduced processes, the illumination intensity can be set by users depending on their experimental requirements or restrictions concerning spatial resolution, acquisition time, photobleaching of the fluorophores, and phototoxicity. For live-cell experiments, we usually begin with the intensity set as low as possible, typically 100 W/cm^2 or less.
   (b) Frame rate:

   In order to detect a single-molecule blinking event efficiently, the frame rate should roughly match the blinking rate of the fluorophore. Specifically, it would generally be appropriate that the average duration of the fluorescent state of the fluorophore corresponds to several frames on the camera. For HMSiR, whose average duration of the fluorescent state is 100–200 ms, an exposure time of 15–100 ms is preferred [37]. For HEtetTFER, the duration of the fluorescent state has not been fully evaluated but is likely longer than that of HMSiR. Therefore, we often employ a relatively long exposure time such as 60 ms [38]. Another point to consider is the mobility of the observation targets. For live-cell SMLM, especially of a fast-moving target, the frame rate should be relatively fast in order to follow the motion. It is important to note that if the target is moving too fast (typically, if the displacement within a single frame is well beyond the localization precision), SMLM becomes inapplicable.

(c) Number of frames to be recorded or analyzed:

Once raw images are obtained, one can conveniently process or analyze the data in any way, at any time, and any number of times. Thus, it is usually best to record the maximum number of frames that is convenient, and then to find the best means of analysis or visualization by varying the processing parameters, such as the number of frames to be analyzed. Two points to consider as to the number of frames are whether adequate molecular density is achieved and whether any significant motional artifact is excluded; there is generally a trade-off between the two. Similar considerations apply to time-lapse experiments, including the case of constructing fast time-lapse super-resolution image sequences for mobile targets. For more continuous observation and smoother visualization with a higher apparent temporal resolution, we often adopt sliding-window image processing [37, 79]. This analytical approach temporally "slides" an "analysis window" of a defined size (number of frames to be analyzed) with a constant step, and affords a quasi-continuous sequence of super-resolution images with certain overlaps. For example, super-resolution images are reconstructed from sets of 1000 consecutive frames with 500-frame overlaps (frames 1–1000; 501–1500; 1001–2000; 1501–2500; and so on), and then concatenated in chronological order to form a sequence of super-resolution images.

3. Determination of DOL (**step 4** in Procedure 3.2.3):

For the preparation step demonstrated here, the determination of DOL was based on the assumption that the antibody is recovered without loss during purification. Alternatively, the antibody itself can be quantified on the basis of its absorbance at 280 nm, which is calculated by subtracting the contribution of the labeling fluorophore from the absorbance at 280 nm of the whole sample solution.

5 Conclusion

SMLM with small organic fluorophores has proven to be useful for extremely high-resolution imaging of cellular structures and processes. The recent development of spontaneously blinking fluorophores represents a major advance in SMLM. As described here, spontaneously blinking fluorophores offer a simple, versatile, and live-cell-compatible tool for biological studies in various contexts. Further evolution of fluorescent probes based on spontaneously blinking fluorophores, in conjunction with advances in labeling technology, optical systems, data analysis, and experimental methodology, is expected to make SMLM a more powerful means for visualizing and understanding a wide range of biological structures and phenomena at an unprecedented level.

Acknowledgments

Live-cell SMLM and SMLM with spinning-disk confocal microscopy with our spontaneously blinking fluorophores introduced in this review were performed in collaboration with Prof. Yasushi Okada at The University of Tokyo. This research was supported in part by AMED/CREST (to Y.U.); by JST/PRESTO (JPMJPR14F8 to M.K.); by MEXT/JSPS KAKENHI, JP16H02606, and JP26111012 (to Y.U.); by JP15H05951 "Resonance Bio" (to M.K.); by Brain/MINDS (to Y.U.); by JSPS Core-to-Core Program, A. Advanced Research Networks; and by a JSPS stipend (to A.M.).

References

1. Betzig E, Patterson GH, Sougrat R et al (2006) Imaging intracellular fluorescent proteins at nanometer resolution. Science 313:1642–1645
2. Hess ST, Girirajan TP, Mason MD (2006) Ultra-high resolution imaging by fluorescence photoactivation localization microscopy. Biophys J 91:4258–4272
3. Rust MJ, Bates M, Zhuang X (2006) Sub-diffraction-limit imaging by stochastic optical reconstruction microscopy (STORM). Nat Methods 3:793–795
4. Heilemann M, van de Linde S, Schuttpelz M et al (2008) Subdiffraction-resolution fluorescence imaging with conventional fluorescent probes. Angew Chem Int Ed Engl 47:6172–6176
5. Heilemann M, van de Linde S, Mukherjee A et al (2009) Super-resolution imaging with small organic fluorophores. Angew Chem Int Ed Engl 48:6903–6908
6. van de Linde S, Loschberger A, Klein T et al (2011) Direct stochastic optical reconstruction microscopy with standard fluorescent probes. Nat Protoc 6:991–1009
7. Folling J, Bossi M, Bock H et al (2008) Fluorescence nanoscopy by ground-state depletion and single-molecule return. Nat Methods 5:943–945
8. Steinhauer C, Forthmann C, Vogelsang J et al (2008) Superresolution microscopy on the basis of engineered dark states. J Am Chem Soc 130:16840–16841
9. Vogelsang J, Cordes T, Forthmann C et al (2009) Controlling the fluorescence of ordinary oxazine dyes for single-molecule switching and superresolution microscopy. Proc Natl Acad Sci U S A 106:8107–8112
10. Sharonov A, Hochstrasser RM (2006) Wide-field subdiffraction imaging by accumulated binding of diffusing probes. Proc Natl Acad Sci U S A 103:18911–18916
11. Giannone G, Hosy E, Levet F et al (2010) Dynamic superresolution imaging of endogenous proteins on living cells at ultra-high density. Biophys J 99:1303–1310
12. Jungmann R, Steinhauer C, Scheible M et al (2010) Single-molecule kinetics and super-resolution microscopy by fluorescence imaging of transient binding on DNA origami. Nano Lett 10:4756–4761

13. Schoen I, Ries J, Klotzsch E et al (2011) Binding-activated localization microscopy of DNA structures. Nano Lett 11:4008–4011
14. Jungmann R, Avendano MS, Woehrstein JB et al (2014) Multiplexed 3D cellular super-resolution imaging with DNA-PAINT and Exchange-PAINT. Nat Methods 11:313–318
15. Kiuchi T, Higuchi M, Takamura A et al (2015) Multitarget super-resolution microscopy with high-density labeling by exchangeable probes. Nat Methods 12:743–746
16. Huang B, Wang W, Bates M et al (2008) Three-dimensional super-resolution imaging by stochastic optical reconstruction microscopy. Science 319:810–813
17. Xu K, Babcock HP, Zhuang X (2012) Dual-objective STORM reveals three-dimensional filament organization in the actin cytoskeleton. Nat Methods 9:185–188
18. Pavani SRP, Thompson MA, Biteen JS et al (2009) Three-dimensional, single-molecule fluorescence imaging beyond the diffraction limit by using a double-helix point spread function. Proc Natl Acad Sci U S A 106:2995–2999
19. Juette MF, Gould TJ, Lessard MD et al (2008) Three-dimensional sub-100 nm resolution fluorescence microscopy of thick samples. Nat Methods 5:527
20. Shtengel G, Galbraith JA, Galbraith CG et al (2009) Interferometric fluorescent super-resolution microscopy resolves 3D cellular ultrastructure. Proc Natl Acad Sci U S A 106:3125–3130
21. Bongiovanni MN, Godet J, Horrocks MH et al (2016) Multi-dimensional super-resolution imaging enables surface hydrophobicity mapping. Nat Commun 7:13544
22. Moon S, Yan R, Kenny SJ et al (2017) Spectrally resolved, functional super-resolution microscopy reveals nanoscale compositional heterogeneity in live-cell membranes. J Am Chem Soc 139(32):10944–10947
23. Shroff H, Galbraith CG, Galbraith JA et al (2008) Live-cell photoactivated localization microscopy of nanoscale adhesion dynamics. Nat Methods 5:417–423
24. van de Linde S, Heilemann M, Sauer M (2012) Live-cell super-resolution imaging with synthetic fluorophores. Annu Rev Phys Chem 63:519–540
25. Grimm JB, English BP, Chen J et al (2015) A general method to improve fluorophores for live-cell and single-molecule microscopy. Nat Methods 12:244–250, 243p following 250
26. Grimm JB, Muthusamy AK, Liang Y et al (2017) A general method to fine-tune fluorophores for live-cell and in vivo imaging. Nat Methods 14:987
27. Michie MS, Götz R, Franke C et al (2017) Cyanine conformational restraint in the far-red range. J Am Chem Soc 139:12406–12409
28. Grimm JB, Klein T, Kopek BG et al (2016) Synthesis of a far-red photoactivatable silicon-containing rhodamine for super-resolution microscopy. Angew Chem Int Ed 55:1723–1727
29. Banala S, Maurel D, Manley S et al (2012) A caged, localizable rhodamine derivative for superresolution microscopy. ACS Chem Biol 7:289–293
30. Grimm JB, Sung AJ, Legant WR et al (2013) Carbofluoresceins and carborhodamines as scaffolds for high-contrast fluorogenic probes. ACS Chem Biol 8:1303–1310
31. Belov VN, Wurm CA, Boyarskiy VP et al (2010) Rhodamines NN: a novel class of caged fluorescent dyes. Angew Chem Int Ed Engl 49:3520–3523
32. Lord SJ, Conley NR, Lee HL et al (2008) A photoactivatable push–pull fluorophore for single-molecule imaging in live cells. J Am Chem Soc 130:9204–9205
33. Lukinavicius G, Umezawa K, Olivier N et al (2013) A near-infrared fluorophore for live-cell super-resolution microscopy of cellular proteins. Nat Chem 5:132–139
34. Lukinavicius G, Reymond L, D'Este E et al (2014) Fluorogenic probes for live-cell imaging of the cytoskeleton. Nat Methods 11:731–733
35. Lukinavičius G, Blaukopf C, Pershagen E et al (2015) SiR–Hoechst is a far-red DNA stain for live-cell nanoscopy. Nat Commun 6:8497
36. Lukinavičius G, Reymond L, Umezawa K et al (2016) Fluorogenic probes for multicolor imaging in living cells. J Am Chem Soc 138:9365–9368
37. Uno S, Kamiya M, Yoshihara T et al (2014) A spontaneously blinking fluorophore based on intramolecular spirocyclization for live-cell super-resolution imaging. Nat Chem 6:681–689
38. Uno S, Kamiya M, Morozumi A et al (2018) A green-light-emitting, spontaneously blinking fluorophore based on intramolecular spirocyclization for dual-colour super-resolution imaging. Chem Commun 54:102–105
39. Ries J, Kaplan C, Platonova E et al (2012) A simple, versatile method for GFP-based super-resolution microscopy via nanobodies. Nat Methods 9:582

40. Platonova E, Winterflood CM, Ewers H (2015) A simple method for GFP- and RFP-based dual color single-molecule localization microscopy. ACS Chem Biol 10:1411–1416
41. Virant D, Traenkle B, Maier J et al (2018) A peptide tag-specific nanobody enables high-quality labeling for dSTORM imaging. Nat Commun 9:930
42. Keppler A, Gendreizig S, Gronemeyer T et al (2003) A general method for the covalent labeling of fusion proteins with small molecules in vivo. Nat Biotechnol 21:86–89
43. Testa I, Wurm CA, Medda R et al (2010) Multicolor fluorescence nanoscopy in fixed and living cells by exciting conventional fluorophores with a single wavelength. Biophys J 99:2686–2694
44. Jones SA, Shim S-H, He J et al (2011) Fast, three-dimensional super-resolution imaging of live cells. Nat Methods 8:499–505
45. Klein T, Loschberger A, Proppert S et al (2011) Live-cell dSTORM with SNAP-tag fusion proteins. Nat Methods 8:7–9
46. Gautier A, Juillerat A, Heinis C et al (2008) An engineered protein tag for multiprotein labeling in living cells. Chem Biol 15:128–136
47. Los GV, Encell LP, McDougall MG et al (2008) HaloTag: a novel protein labeling technology for cell imaging and protein analysis. ACS Chem Biol 3:373–382
48. Wilmes S, Staufenbiel M, Liße D et al (2012) Triple-color super-resolution imaging of live cells: resolving submicroscopic receptor organization in the plasma membrane. Angew Chem Int Ed 51:4868–4871
49. Miller LW, Cai Y, Sheetz MP et al (2005) In vivo protein labeling with trimethoprim conjugates: a flexible chemical tag. Nat Methods 2:255–257
50. Chen Z, Jing C, Gallagher SS et al (2012) Second-generation covalent TMP-tag for live cell imaging. J Am Chem Soc 134:13692–13699
51. Wombacher R, Heidbreder M, van de Linde S et al (2010) Live-cell super-resolution imaging with trimethoprim conjugates. Nat Methods 7:717–719
52. Lelek M, Di Nunzio F, Henriques R et al (2012) Superresolution imaging of HIV in infected cells with FlAsH-PALM. Proc Natl Acad Sci U S A 109:8564–8569
53. Griffin BA, Adams SR, Tsien RY (1998) Specific covalent labeling of recombinant protein molecules inside live cells. Science 281:269–272
54. Adams SR, Campbell RE, Gross LA et al (2002) New biarsenical ligands and tetracysteine motifs for protein labeling in vitro and in vivo: synthesis and biological applications. J Am Chem Soc 124:6063–6076
55. Wieneke R, Raulf A, Kollmannsperger A et al (2015) SLAP: small labeling pair for single-molecule super-resolution imaging. Angew Chem Int Ed 54:10216–10219
56. Kollmannsperger A, Sharei A, Raulf A et al (2016) Live-cell protein labelling with nanometre precision by cell squeezing. Nat Commun 7:10372
57. Karl G, Joest EF, Dietz MS et al (2018) Superchelators for advanced protein labeling in living cells. Angew Chem Int Ed 57:5620–5625
58. Lang K, Chin JW (2014) Cellular incorporation of unnatural amino acids and bioorthogonal labeling of proteins. Chem Rev 114:4764–4806
59. Shim S-H, Xia C, Zhong G et al (2012) Super-resolution fluorescence imaging of organelles in live cells with photoswitchable membrane probes. Proc Natl Acad Sci U S A 109:13978–13983
60. Carlini L, Manley S (2013) Live intracellular super-resolution imaging using site-specific stains. ACS Chem Biol 8:2643–2648
61. Takakura H, Zhang Y, Erdmann RS et al (2017) Long time-lapse nanoscopy with spontaneously blinking membrane probes. Nat Biotechnol 35:773
62. Kolmakov K, Wurm C, Sednev MV et al (2012) Masked red-emitting carbopyronine dyes with photosensitive 2-diazo-1-indanone caging group. Photochem Photobiol Sci 11:522–532
63. Belov VN, Mitronova GY, Bossi ML et al (2014) Masked rhodamine dyes of five principal colors revealed by photolysis of a 2-diazo-1-indanone caging group: synthesis, photophysics, and light microscopy applications. Chem Eur J 20:13162–13173
64. Roubinet B, Bischoff M, Nizamov S et al (2018) Photoactivatable rhodamine spiroamides and diazoketones decorated with "Universal Hydrophilizer" or hydroxyl groups. J Org Chem 83:6466–6476
65. Lee H-lD, Lord SJ, Iwanaga S et al (2010) Superresolution imaging of targeted proteins in fixed and living cells using photoactivatable organic fluorophores. J Am Chem Soc 132:15099–15101
66. Carlini L, Benke A, Reymond L et al (2014) Reduced dyes enhance single-molecule localization density for live superresolution imaging. Chemphyschem 15:750–755

67. Vaughan JC, Jia S, Zhuang X (2012) Ultrabright photoactivatable fluorophores created by reductive caging. Nat Methods 9:1181–1184
68. Vaughan JC, Dempsey GT, Sun E et al (2013) Phosphine quenching of cyanine dyes as a versatile tool for fluorescence microscopy. J Am Chem Soc 135:1197–1200
69. Bates M, Blosser T, Zhuang X (2005) Short-range spectroscopic ruler based on a single-molecule optical switch. Phys Rev Lett 94:108101
70. Heilemann M, Margeat E, Kasper R et al (2005) Carbocyanine dyes as efficient reversible single-molecule optical switch. J Am Chem Soc 127:3801–3806
71. Bates M, Huang B, Dempsey GT et al (2007) Multicolor super-resolution imaging with photo-switchable fluorescent probes. Science 317:1749–1753
72. van de Linde S, Sauer M (2014) How to switch a fluorophore: from undesired blinking to controlled photoswitching. Chem Soc Rev 43:1076–1087
73. Macdonald PJ, Gayda S, Haack RA et al (2018) Rhodamine-derived fluorescent dye with inherent blinking behavior for super-resolution imaging. Anal Chem 90:9165
74. Roubinet B, Weber M, Shojaei H et al (2017) Fluorescent photoswitchable diarylethenes for biolabeling and single-molecule localization microscopies with optical superresolution. J Am Chem Soc 139:6611–6620
75. Raab M, Schmied JJ, Jusuk I et al (2014) Fluorescence microscopy with 6 nm resolution on DNA origami. Chemphyschem 15:2431–2435
76. Flors C, Ravarani CNJ, Dryden DTF (2009) Super-resolution imaging of DNA labelled with intercalating dyes. Chemphyschem 10:2201–2204
77. Yan Q, Schwartz SL, Maji S et al (2014) Localization microscopy using noncovalent fluorogen activation by genetically encoded fluorogen-activating proteins. Chemphyschem 15:687–695
78. Saurabh S, Perez AM, Comerci CJ et al (2016) Super-resolution imaging of live bacteria cells using a genetically directed, highly photostable fluoromodule. J Am Chem Soc 138:10398–10401
79. Endesfelder U, van de Linde S, Wolter S et al (2010) Subdiffraction-resolution fluorescence microscopy of myosin–actin motility. Chemphyschem 11:836–840
80. Lampe A, Haucke V, Sigrist SJ et al (2012) Multi-colour direct STORM with red emitting carbocyanines. Biol Cell 104:229–237
81. Lehmann M, Gottschalk B, Puchkov D et al (2015) Multicolor caged dSTORM resolves the ultrastructure of synaptic vesicles in the brain. Angew Chem Int Ed 54:13230–13235
82. Zhang Z, Kenny SJ, Hauser M et al (2015) Ultrahigh-throughput single-molecule spectroscopy and spectrally resolved super-resolution microscopy. Nat Methods 12:935–938
83. Mlodzianoski MJ, Curthoys NM, Gunewardene MS et al (2016) Super-resolution imaging of molecular emission spectra and single molecule spectral fluctuations. PLoS One 11: e0147506
84. Hoelz A, Debler EW, Blobel G (2011) The structure of the nuclear pore complex. Annu Rev Biochem 80:613–643

Chapter 11

Highly Biocompatible Super-resolution Imaging: SPoD-OnSPAN

Tetsuichi Wazawa, Takashi Washio, and Takeharu Nagai

Abstract

Super-resolution microscopy facilitates observation with an optical microscope at a higher spatial resolution than the diffraction limit of light; however, super-resolution observation with high biocompatibility remains challenging. Leading super-resolution techniques such as reversible saturable/switchable optical fluorescence transition (RESOLFT) and single-molecule localization microscopy (SMLM) need to illuminate a sample at an appreciably high power density of illumination, i.e., from 0.1 kW/cm^2 to 1 GW/cm^2. Unfortunately, that high power density gives rise to phototoxicity in live cells, and this may prevent widespread use of super-resolution imaging in the life sciences. In this study we show a technique of super-resolution imaging that can be performed at a very low power density of illumination, SPoD-OnSPAN (super-resolution polarization demodulation/on-state polarization angle narrowing). This achieves super-resolution observations at a power density as low as 1 W/cm^2, and thereby high biocompatibility. The present technique is likely to be very useful for situations such as time-lapse super-resolution observations of live cells and tissues.

Key words Photoswitchable fluorescence protein, Biocompatibility, Polarization, Modulation, Regularized maximum likelihood

1 Introduction

Super-resolution imaging is a technique to perform observations with an optical microscope at a higher spatial resolution than the diffraction limit of light. Included in this technique are reversible saturable/switchable optical fluorescence transitions (RESOLFT) such as stimulated emission depletion (STED), single-molecule localization microscopy (SMLM) techniques such as photoactivated localization microscopy (PALM) and stochastic optical reconstruction microscopy (STORM), and structured illumination microscopy (SIM) [1–3]. Although RESOLFT and SMLM provide very high spatial resolution, they need a considerably high power density of illumination, i.e., 10^2–10^9 W/cm^2. Unfortunately, illumination at that high power density brings about phototoxicity in

Nobuhiko Yamamoto and Yasushi Okada (eds.), *Single Molecule Microscopy in Neurobiology*, Neuromethods, vol. 154, https://doi.org/10.1007/978-1-0716-0532-5_11,

live cells, such as anomalies in cellular morphology and apoptosis [4]. Therefore, this often hampers the time-lapse super-resolution observation of live cells. In contrast, SIM involves an even lower power density of illumination, but the spatial resolution is theoretically limited to about 100 nm [3]. In this study, we show an imaging technique of super-resolution polarization demodulation/on-state polarization angle narrowing (SPoD-OnSPAN) that is expected to resolve the problem so that we may carry out observations with very low phototoxicity and a spatial resolution better than 100 nm [5]. In this research work, we cover the technique's principles and usage, and finally, we demonstrate an application of the technique in the observation of live cells.

2 Brief Outline of SPoD-OnSPAN

SPoD-OnSPAN is a collection of multiple techniques composed of a fluorescence microscope with a unique polarized illumination, a specific image reconstruction calculation, and a photoswitchable fluorescent protein for labeling (Fig. 1). The microscope uses linearly polarized lights for illumination and their polarization planes are simultaneously rotated. Due to the directivity of a fluorescent protein in the absorption of polarized light, the fluorescence intensity periodically oscillates as the polarization planes are rotated, which is referred to as fluorescence modulation (*see* below). Because the phase of fluorescence modulation is dependent on the orientation of the fluorophore, fluorescent proteins with different orientations on a specimen plane emit fluorescence at different timings (Fig. 1). A series of images taken during the oscillation, which contains the information of the fluorescence modulation, is used to estimate the spatial distribution of fluorescent proteins on a specimen plane, i.e., a super-resolved image, by the image reconstruction calculation.

SPoD-OnSPAN was derived from techniques of SPoD (super-resolution polarization demodulation) and SPoD-ExPAN (SPoD/excitation polarization angle narrowing) developed by Walla and colleagues [6]. SPoD was devised to use only an excitation light for illumination in the configuration of Fig. 1 to perform super-resolution imaging, although the improvement of the spatial resolution was only modest. They also devised SPoD-ExPAN that used SPoD in combination with the STED process of fluorescent probes for higher spatial resolution than SPoD. Although SPoD-ExPAN was successful, a power density of illumination that was at times greater than 1 MW/cm^2 was necessary, and therefore, the problem of phototoxicity remained unresolved. Therefore, we worked on the development of highly biocompatible super-resolution imaging [5] to modify the principle of SPoD-ExPAN and combine SPoD-

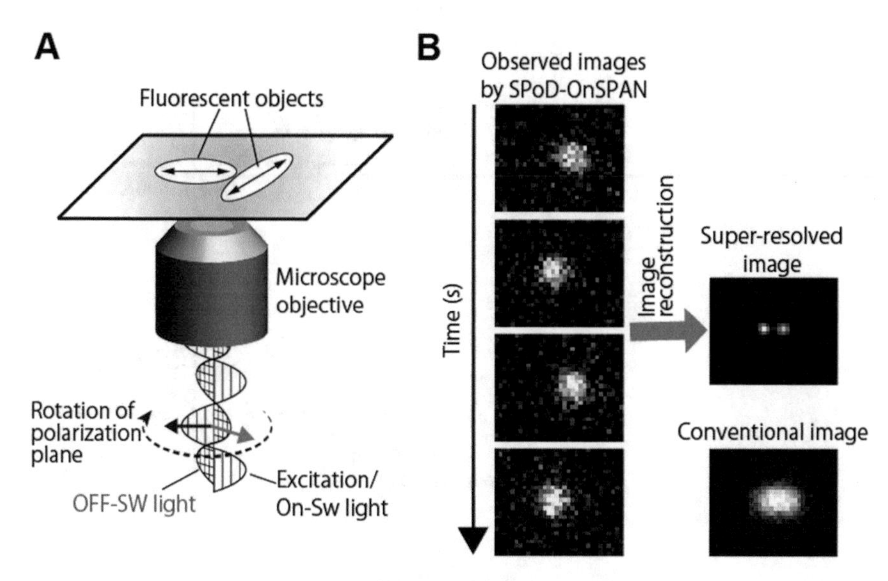

Fig. 1 Brief outline of SPoD-OnSPAN imaging. (**a**) Schematic diagram of illumination of a sample by SPoD-OnSPAN. A fluorescent sample is illuminated with linearly polarized lights, the polarization planes of which are rotated. (**b**) Schematic diagram of the output from the SPoD-OnSPAN observation

ExPAN with the use of a fast photoswitchable fluorescent protein known as Kohinoor [7]. We denote by SPoD-OnSPAN the present technique that is derived from SPoD-ExPAN and photoswitchable fluorescent protein.

3 Principles

3.1 Photoselection in Light Absorption

Suppose that a fluorescent molecule such as a fluorescent protein is irradiated with a linearly polarized excitation light (Fig. 2a). Let θ be an angle between the polarization plane of excitation light and the absorption transition dipole moment of the fluorophore that corresponds to the direction at which a fluorophore absorbs light at the highest probability. The probability of light absorption is then proportional to $\cos^2\theta$, and this is called the photoselection [8]. Because this light absorption process brings about the transition of the fluorophore to the excited state followed by radiative or radiationless deactivation of the excited state, the fluorescence intensity is also proportional to $\cos^2\theta$.

3.2 Positively Reversibly Photoswitchable Fluorescent Protein with Irradiation of Polarized Light

Reversibly photoswitchable fluorescent proteins (RPFPs) are a class of fluorescent proteins having the property that their fluorescent and nonfluorescent states are reversibly switchable by the irradiation of lights of different wavelengths [9]. If the irradiation of light of a specific color leads to conversion from the on state (fluorescent) to the off state (nonfluorescent), this process is called the off-switching, and this color of light is called the off-switching

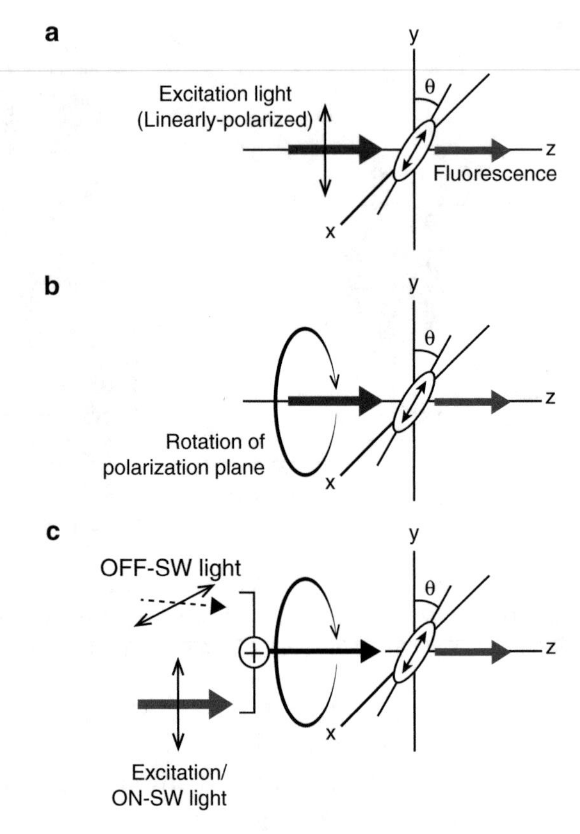

Fig. 2 Schematic drawing of illumination of a fluorescent dye with polarized lights. (**a**) A fluorescent dye illuminated with a linearly polarized excitation light. (**b**) A fluorescent dye illuminated with a linearly polarized excitation light, the polarization plane of which is rotated. (**b**) A fluorescent dye illuminated with a linearly polarized on-switching/excitation and off-switching lights, the polarization planes of which are rotated

light. In contrast, if the irradiation of another color of light leads to conversion from the off state to the on state, this is called the on-switching, and the light is called the on-switching light. Among RPFPs, positively RPFPs (pRPFPs), which undergo on-switching at the same wavelength as excitation but off-switching at a different wavelength, are used for SPoD-OnSPAN. For example, we used a pRPFP of Kohinoor [7] for SPoD-OnSPAN observation [5] with an illumination light at 405 nm for off-switching and another at 488 nm for on-switching and fluorescence excitation.

Because the photoswitching process follows the absorption of a switching light, the photoswitching rate is dependent on the probability of light absorption. Therefore, in an irradiation condition where the population of the ground state of fluorescent proteins is dominant, the photoselection applies to the photoswitching rate. Thus, the photoswitching rate is dependent not only on the power density of switching light but also on the angle between the polarization plane of switching light and the absorption transition dipole moment of the chromophore. Thereby, we have

$$k_S \propto P_S \cos^2\theta, \qquad (1)$$

where k_S is a rate of photoswitching, and P_S is the power density of switching light. This polarization-dependent photoswitching rate is used for SPoD-OnSPAN.

3.3 Fluorescence Response of FPs in SPoD and SPoD-OnSPAN

SPoD uses linearly polarized excitation light, the polarization plane of which is rotated with time (Fig. 2b). Let the rotation speed of the polarization plane be f (revolutions/s), and we express the time change of the fluorescence intensity of the fluorophore by

$$F(t) \propto \cos^2(2\pi f t - \theta) = \frac{1 + \cos(4\pi f t - 2\theta)}{2}, \qquad (2)$$

where t is a time (s). This means that the fluorescence emitted from the fluorophore is modulated at a frequency of $2f$ and a phase shift of -2θ. Because θ is dependent on the orientation of a fluorophore, every fluorophore on a specimen plane will show fluorescence modulation at a unique value of phase shift. Thus, by taking the time course of the fluorescence image of the fluorophores, we are able to spot fluorescent objects with respect to not only the positions in the image but also the phase shifts in the time course (Fig. 1). This is the basic principle of SPoD imaging. However, the time response of fluorescence to the rotation in Eq. 2 is quite diffuse, and therefore, a further effort has been devised in SPoD-OnSPAN to improve the resolution of this phase-sensitive detection of fluorescent objects.

SPoD-OnSPAN uses the photoswitching of pPRFPs to improve the resolution of the phase-sensitive detection over the SPoD method. In the illumination configuration for pRPFP, a linearly polarized excitation light, which also acts as on-switching light, is used in the same way as SPoD (Fig. 2b), and an off-switching light is added in the polarization angle perpendicular to that of the excitation/on-switching light (Fig. 2c). Thus, taking Eq. 1 into consideration, the rates of on- and off-switching, k_{on} and k_{off}, respectively, depend on time and illumination power in a manner such that

$$k_{on} \propto P_{on/ex} \cos^2(2\pi f t - \theta), \qquad (3)$$

$$k_{off} \propto P_{off} \cos^2\left(2\pi f t - \theta - \frac{\pi}{2}\right), \qquad (4)$$

where $P_{on/ex}$ and P_{off} are power densities (W/cm^2) of the excitation/on-switching and off-switching lights, respectively. The absorption transition dipole moments for the excitation and off-switching are likely to be parallel. Although, to be precise, the

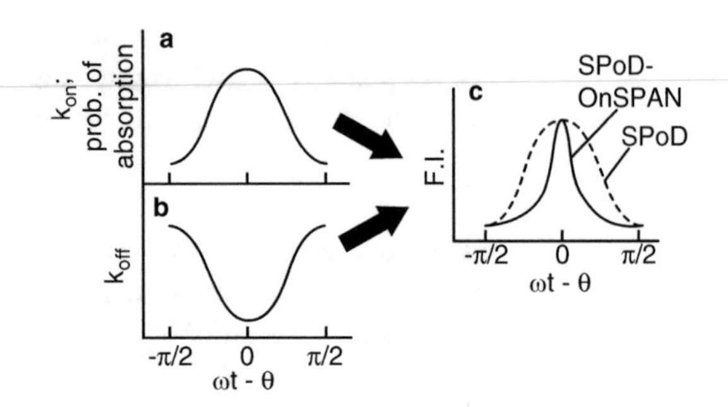

Fig. 3 Schematic diagram of the response of pRPFP fluorescence to the illumination of SPoD-OnSPAN in the configuration shown in Fig. 2c. (**a**) Plots of the rate of on-switching and the probability of light absorption as a function of the angle between the absorption transition dipole moment and the polarization plane of excitation/on-switching light ($\omega t - \theta$). (**b**) A plot of the rate of off-switching as a function of ($\omega t - \theta$). (**c**) A plot of fluorescence intensity as a function of ($\omega t - \theta$) on which both excitation/on-switching and off-switching lights have effects

absorption transition dipole moment of the on-switching may be to some extent tilted from that of off-switching, we assume in this work, by approximation, that they are also parallel.

The combination of the excitation/on-switching light and the off-switching light brings about the angular narrowing of the on state of pRPFP. Figure 3 shows the relationship of the optical response of a pRPFP with the angles between its absorption transition dipole moment and the polarization planes of illumination light. The probability of light absorption and the rate of on-switching have maxima when the absorption transition dipole moment and the polarization plane of the excitation/on-switching light are parallel ($\omega t - \theta = 0$, where $\omega = 2\pi f$) (Fig. 3a). However, because of the illumination configuration (Fig. 2c), the rate of off-switching is minimum when $(\omega t - \theta) = 0$, whereas the rate of off-switching becomes maximum when $(\omega t - \theta) = \pi/2$ (Fig. 3b). Because SPoD-OnSPAN uses both an excitation/on-switching light and an off-switching light simultaneously, the peak width of the fluorescence response narrows owing to the combination of these two effects (Fig. 3c). This narrowing of the polarization angle improves the sensitivity of the phase-sensitive detection of fluorescent objects. Although the peak width (Fig. 3c) of the fluorescence response narrows as the off-switching rate increases, the fluorescence intensity from the pRPFPs decreases as well, which likely leads to a lower quality of image. Thus, the illumination conditions should be optimized for high resolution and high image quality.

3.4 Image Reconstruction

In SPoD-OnSPAN, an image reconstruction calculation is carried out to derive a super-resolved image. This calculation uses a series of images taken by SPoD-OnSPAN observations and an image of a point spread function (PSF), i.e., an image of a fluorescent object sufficiently smaller than the diffraction limit of light. The basic approach in this image reconstruction is the solution of the linear inverse problem that solves a spatial distribution of fluorophores, i.e., a super-resolved image, such that its convolution image with the PSF is as close to the observed images as possible. For the calculation, we use a loss function M given by

$$M[g, b, I] = \int \int_0^T l(\mu(\mathbf{r}, t); I(\mathbf{r}, t)) dt d\mathbf{r} + \int \lambda_{g,\mathbf{r}} \int_0^T |g(\mathbf{r}, t)|_p dt d\mathbf{r} + \lambda_b \int \left|\tilde{b}(\mathbf{r})\right|_1 d\mathbf{r}, \quad (5)$$

where b is a background image, I is a function describing the modulation of illumination, μ is the series of observed images, l is a log-likelihood for a Poisson distribution, and $|*|_p$ and $|*|_1$ are L_p and L_1 norms, respectively [5]. The image g is the super-resolved image, and is solved by a regularized maximum likelihood estimation calculation. Although Hafi et al. [6] provided an image reconstruction program with L_1 regularization which used the L_1 norm for the second term in Eq. 5, the calculation often tends to result in overcorrection so that artifact patterns appear in the result of super-resolved image. To mitigate this problem, we introduced the L_p regularization into our image reconstruction program so that the L_p norm was used for the second term of Eq. 5, and thereby, we achieved image reconstruction of more accurate super-resolved images [5]. The factors λ_g and λ_b and the power p are experimentally adjusted parameters, and in observation of HeLa cells we typically use $p = 1.0$–2.0, $\lambda_g = 0.2$, and $\lambda_b = 5$.

4 Materials and Microscopy

4.1 Microscopy Setup

For SPoD-OnSPAN observation, a specially built microscope is used. In this work, we describe the details of our SPoD-OnSPAN microscope [5] designed for the observation of a fast photoswitchable fluorescent protein Kohinoor [7], as a typical example (Fig. 4). An inverted microscope body (IX83, Olympus, Hachioji, Japan; bodies of other manufacturers may also be used) is fixed on an optical bench (h-TDI, Herz, Yokohama, Japan), and combined with illumination optics and a camera device. A linearly polarized light from a 488 nm laser (SL1, Fig. 4b) (e.g., OBIS, Coherent, Santa Clara, CA, USA) is passed through a half-wave plate (HW1, Fig. 4b) (e.g., FRH-102, Sigma Koki, Saitama, Japan) to adjust the polarization angle. The beam is subsequently introduced into a combination of two plano-convex lenses (e.g., SLB-30-35PM and SLB-30-220PM, Sigma Koki) or a beam expander (e.g., LBE-10,

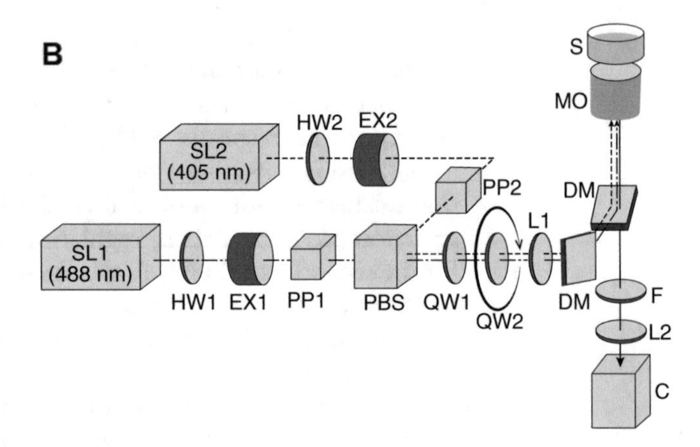

Fig. 4 SPoD-OnSPAN microscope. (**a**) Photograph of a SPoD-OnSPAN microscope for the observation of Kohinoor. (**b**) Block diagram of the microscope. *SL1, SL2* solid-state lasers; *HW1, HW2* half-wave plates; *EX1, EX2* beam expanders; *PP1, PP2* Glan-Thompson prisms; *PBS* polarizing beam splitter; *QW1, QW2* achromatic quarter-wave plates; *L1, L2* lenses; *DM* dichroic mirrors; *MO* microscope objective; *S* sample; *F* band-pass filter; *C* EMCCD camera

Sigma Koki) to expand the beam diameter (EX1, Fig. 4b) and then a polarizing prism (PP1, Fig. 4b) (such as a Glan-Thompson prism; e.g., GTPC-15-45.5SN, Sigma Koki) to improve its polarization ratio. Besides, a beam from a 405 nm laser (SL2, Fig. 4b) (OBIS, Coherent) is also modified in the same manner as above, and these two beams at 488 and 405 nm are combined with a polarizing beam splitter (PBS, Fig. 4b) (PBSW-20-3/7, Sigma Koki; wavelength range 370–750 nm). Because of combination by the polarizing beam splitter, the polarization angles of the lights of the two wavelengths are orthogonal to each other. The combined beam is directed into an achromatic quarter-wave plate (QW1, Fig. 4b) (AQWP05M-600, Thorlabs, Newton, NJ, USA; wavelength range 400–800 nm) so that the beam is circularly polarized, and then into another achromatic quarter-wave plate (QW2, Fig. 4b) (AQWP05M-600, Thorlabs) embedded in a hollow-shaft rotary motor (HM2853-E18H, Technohands, Yokohama, Japan). Thus, the light from the last quarter-wave plate is linearly polarized and the polarization plane is rotated when the motor is driven. This polarization-modulated light is reflected with a dichroic mirror (DM, Fig. 4b) to be directed into the inverted microscope and used as the illumination light. The dichroic mirror used here should be the same product as the one in a filter cube of the microscope (*see* below). The purpose of this pair of the same two dichroic mirrors is to improve the polarization ratio of illumination light [10].

Although lasers are used in the present example of setup, light-emitting diodes (LEDs) are also usable. In our previous study [5], we used UHP-LED-470 (center wavelength, 470 nm; Prizmatix, Holon, Israel) and UHP-LED-405 (center wavelength, 405 nm; Prizmatix) with liquid light guides. The beams from them were collimated by using plano-convex lenses (SLB-30-35PM, Sigma Koki), and were passed through band-pass filters (FF01-465/30-25 (Semrock) and D410/30 (Chroma, Bellows Falls, VT, USA), respectively). The beams were combined with the polarizing beam splitter and directed toward the inverted microscope as described above.

On the microscope body, lenses and filters are installed similar to a conventional fluorescence microscope. We use a dichroic mirror (DM, Fig. 4b) of FF518-Di01-25×36 (Semrock, Lake Forest, IL, USA) and an emission filter (F, Fig. 4b) of FF02-525/40-25 (Semrock) in the filter cube for the observation of Kohinoor. The guideline for the choice of emission filters and dichroic mirrors in SPoD-OnSPAN is quite similar to that in fluorescence microscopy. The emission filter is required to pass through the fluorescence but block the illumination light, Raman scattering of water, and noise such as autofluorescence and any other unwanted signals; the dichroic mirror needs to reflect the illumination light (e.g., 405 and 488 nm in the present case) and transmit the fluorescence at an incidence angle of 45°. For lenses, we usually use a high-magnification microscope objective (MO, Fig. 4b) such as UPLSAPO 100XO (NA, 1.40; Olympus) or UPLSAPO 60XO (NA, 1.35; Olympus) and a PE tube lens (L2, Fig. 4b) (magnification, 2–5×; Olympus) to achieve a total magnification suitable for a camera in use. Typically, we adjust the total magnification so that the pixel resolution (pixel pitch of a camera (nm/pixel)/total magnification) is 30–50 nm/pixel.

To capture images, we use a scientific-grade CMOS (sCMOS) or an electron-multiplying EMCCD camera such as the ORCA Flash 4.0 (Hamamatsu Photonics, Hamamatsu, Japan) or the iXON Ultra (Andor, Belfast, UK), but we can also use other cameras of equivalent specifications. A camera connected to the microscope body is operated in a time-lapse mode and an external trigger pulse triggers the acquisition of each frame. The acquisition of an image frame is triggered every time the quarter-wave plate is rotated by a predefined angle. A rotary encoder signal from the hollow-shaft motor is fed into a DAQ module (NI PCIe-6321, National Instruments, Austin, TX, USA) that performs counter measurement and digital output, to modify the encoder signal and generate the trigger pulse toward the camera. Most rotary encoders provide A, B, and Z signals: we coded a Vi in Labview platform (National Instruments) so as to use the Z signal to start the counter in the DAQ and the A or B signal to measure the rotation angle of the motor. The pulse signal generation is preferably performed by a rigorous mechanism of counter/digital IO circuits rather than software interrupts, because operation by software interrupts is susceptible to uncertainty in timing as they are affected by factors such as the background processes running on the Labview platform and the operating system.

4.2 Preparation of Transfected HeLa Cells

HeLa cells are cultured on a collagen-coated 35 mm glass-bottomed dish in Dulbecco's modified Eagle medium (DMEM) supplemented with 10% fetal bovine serum (FBS) at 37 °C overnight. The next day, the cells are transfected with 1.0 μg DNA vector using the calcium phosphate method [7]. The medium is refreshed after 8 h of incubation, and the cells are grown for an additional 24–48 h in a CO_2 incubator at 37 °C with 5% CO_2. Subsequently, the cells are washed twice with phenol red-free DMEM and observed in phenol red-free DMEM. Several types of expression vectors for Kohinoor-tagged proteins and polypeptides are available from a plasmid repository on Addgene (Plasmid, #67772–67781).

5 Protocol of Observation

5.1 Setup for Microscopy Observation

- **Checking magnification**: Use an objective micrometer (e.g., OB-M#, 1/100, Olympus) on a microscope stage to capture the image of the micrometer and confirm if the magnification or the pixel resolution (nm/pixel) on the sensor of the camera (C, Fig. 4b) is appropriate. If not, change the microscope objective or tube lens to adjust the magnification. Typically, we use a pixel resolution of 30–50 nm/pixel. The pixel resolution is dependent on not only the optical magnification at the sensor plane in the camera, but also the pitch of pixels on the camera sensor, and is expressed as $r = d/m$, where r, d, and m are the pixel resolution (nm/pixel), the pitch of pixels (nm/pixel), and the optical magnification, respectively. It is advisable to obtain an image of the micrometer, whenever a microscope objective, a tube lens, or a camera is changed. This image is useful as a record of total magnification.

- **Power density of illumination**: Measure the power of the beams (for the excitation/on-switching light and the off-switching light) emitted from the microscope objective with a laser power meter (e.g., S120VC and PM100D, Thorlabs). Next, measure the size of the illuminated area on the specimen plane to calculate the power densities (W/cm^2). The illuminated area can be visualized by observing a coverslip coated with fluorescent beads on the microscope (e.g., F8787, Thermo Fisher Scientific, Waltham, MA, USA) after irradiating it with each type of light. If a power density is not appropriate, we change the output power of the lasers to achieve optimal power densities.
- **Conditions of camera and motor**: Set an exposure time, the rotation speed of the hollow-shaft motor, the number of frames taken for one revolution of the quarter-wave plate, and the number of cycles of SPoD-OnSPAN observation. Also decide the camera parameters such as exposure time, readout time, total number of frames to be taken, gain of the analogue-to-digital converter, and EM gain (for EMCCD). The exposure time and readout time should be set so that frames are never dropped. To achieve this, parameters are adjusted such that

$$t_{\mathrm{ex}} + t_{\mathrm{r}} < \frac{60}{v \cdot N},$$

  where t_{ex} and t_{r} are the exposure time (s) and the readout time (s) for a frame, respectively; v is the rotation speed (rpm) of the quarter-wave plate; and N is the total number of frames to be captured during one revolution of the plate (*see* below). For example values of t_{ex}, t_{r}, N, and EM gain, *see* Table 1. It should be noted that the readout time t_{r} is dependent on the readout rate of the camera. Additionally, an external trigger mode is chosen in the camera setting.
- **Temperature control**: If a stage-top incubator (e.g., STX, Tokai Hit, Fujinomiya, Japan) is used, the incubator and the microscope objective should be preincubated in advance.

5.2 Procedure of SPoD-OnSPAN Observation

After placing a sample on a microscope stage, an experimenter may scan it to focus on an object to observe, and next commences the SPoD-OnSPAN observation. The acquisition of SPoD-OnSPAN images is performed according to the operation sequence shown in Fig. 5. When the observation is started, the microscope system waits for a moment while the QW2 in Fig. 4b settles on a start angle after which time it sends the camera a pulse signal to take an image. Subsequently, the pulse signal is sent to capture an image every time the polarization plane is rotated by a pre-defined angle until the total number of images reaches a limit. In our laboratory, 9–18 images are captured for one revolution of the quarter-wave plate to derive a reconstructed super-resolved image.

Table 1
Items and conditions used in the observation of Fig. 6

Items	Description
Microscope body	IX-71, Olympus
Microscope objective	UPlanSApo 100×/1.40, Olympus
Tube lens	PE 4×, Olympus
Filter, dichroic mirror	FF02-525/40-25, FF518-Di01-25×36, Semrock
Camera	EMCCD iXON3, Andor
Exposure time (t_{ex}), readout time (t_r), readout rate, EM gain	$t_{ex} = 0.1$ s, $t_r = 0.074$ s, readout rate = 20 MHz, EM gain = 300
Light sources (the observation was performed with LED light sources) [5]	UHP-LED-470 (center wavelength, 470 nm), UHP-LED-405 (center wavelength, 405 nm), Prizmatix
Power density of illumination at the specimen plane	1.3 W/cm^2 at 470 and 405 nm
Rotation speed of a quarter-wave plate	15 rpm
Number of frames to reconstruct a super-resolved image	18 frames
Parameter values for image reconstruction calculation	$L = 1$, $\lambda_g = 0.2$, $\lambda_b = 5$, $p = 1.05$, the number of loop iteration = 500

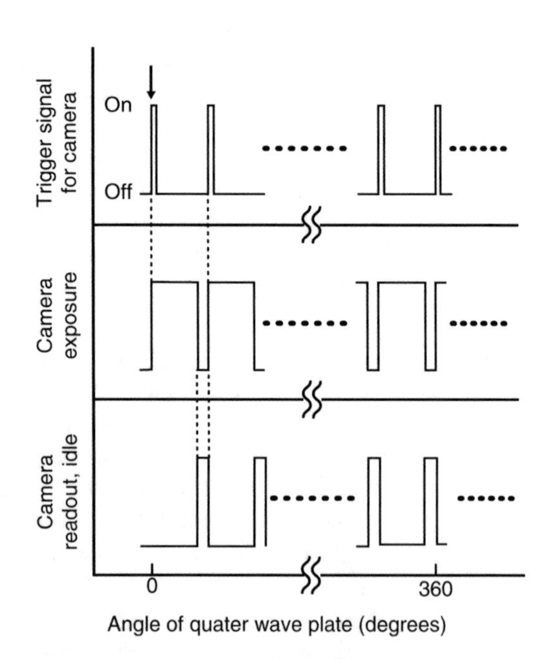

Fig. 5 Diagram of the timings for (top) a trigger signal, (middle) the exposure of camera, and (bottom) the readout and idle of camera

In addition to the SPoD-OnSPAN images described above, an image of fluorescent objects sufficiently smaller than the diffraction limit of light is captured. As fluorescent objects, fluorescent carboxylated beads of a diameter less than 50 nm from Thermo Fisher Scientific (Waltham, MA, USA) or Bangs Laboratories (Fishers, IN, USA) may be used. The fluorescent beads are diluted by 10^3–10^7 in an aqueous solution containing ~0.1 M KCl or NaCl and an appropriate buffer of neutral pH. The suspension is placed between a glass slide (e.g., S011110, Matsunami Glass, Kishiwada, Japan) and a coverslip (e.g., C018181, Matsunami Glass). The image is recorded at the same magnification as the sample observation and a subimage containing a single bead is cropped (typical image size, 32 × 32 or 64 × 64 pixels). Applying background subtraction and noise reduction with for instance a low-pass filter, the subimage is used as a PSF for image reconstruction.

5.3 Image Reconstruction

Image reconstruction software is used to derive a super-resolved image. Currently, software packages are available from the supplementary materials of our paper [5] and Hafi et al. [6]. These software products require a series of observed images, a PSF image, and a parameter set for calculation. These software applications work on a common desktop computer using Python 2.7 and the numpy, scipy, and matplotlib software libraries. The images should be grayscale; the format for the PSF image is raw in 32-bit floating point; the format for the series of images by SPoD-OnSPAN observation should be a 16-bit unsigned integer; the byte order of these image files should be little endian. In our program, the series of images in separate raw format files or a single multiple TIFF can be handled [5]. For the software developed by Hafi et al. [6], the series of images should be separate files in raw format. The format of images can be converted by software such as Image J, or a software package complementary to a camera in use. Some of the parameters in the parameter set are dependent on the camera in use and on observation conditions; for further details, the users should refer to the document in the supplementary material of the papers [5, 6]. The other parameters are for regularization: λ_g, λ_b, and p for the software in ref. 5; λ_1 and λ_2 for the software in ref. 6. If the experimenter is new to the present technique, it is advisable to begin with trying parameter values used in the supplementary materials to test image reconstruction, and then to try and adjust the parameter values for optimal results. As a typical example, when the reconstruction calculation of a super-resolved image with a size of 512 × 512 pixels is performed on a common desktop Windows PC, it takes 30 min to 1 h with the L_1-regularized method [6] and it takes severalfold longer with the L_p-regularized method [5].

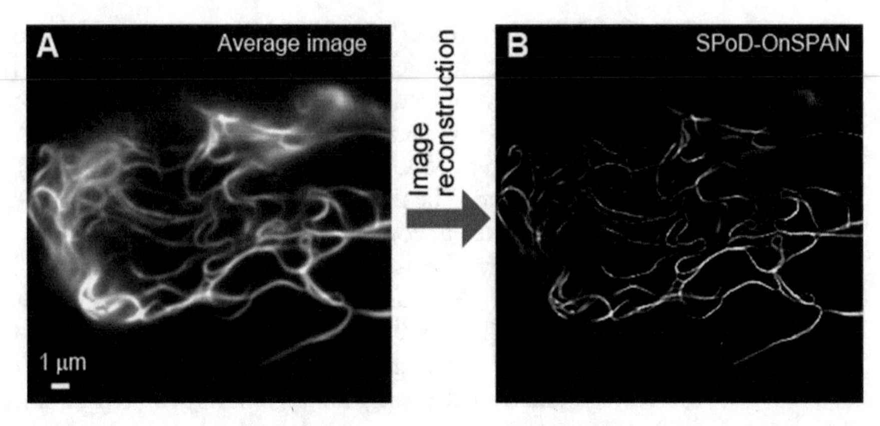

Fig. 6 An example of SPoD-OnSPAN observation Kohinoor-vimentin in a HeLa cell. (**a**) The average of 18 raw images. (**b**) A super-resolved image reconstructed from the series of 18 images. For experimental conditions, *see* Table 1

5.4 An Example of SPoD-OnSPAN Observation

In this section we show an example of SPoD-OnSPAN observation of vimentin intermediate filaments in a HeLa cell. We transfected HeLa cells with a vector to express a fusion protein of vimentin and Kohinoor (Kohinoor-vimentin) and incubated them. Thereafter, we performed their SPoD-OnSPAN observation (for experimental condition, *see* Table 1). Figure 6a shows the average of 18 raw images of Kohinoor-vimentin intermediate filaments in a live HeLa cell taken during the SPoD-OnSPAN observation, and it was equivalent to a conventional fluorescence image. Although the actual diameter of vimentin intermediate filaments is about 30 nm as studied by electron microscopy [11], the observed width of the single filaments in that image is about 250–300 nm in FWHM (full width of half maximum) because of the diffraction limit of light. Subsequently, the image dataset was subjected to our image reconstruction calculation [5] to yield a super-resolved image (Fig. 6b). As seen in the figure, the spatial resolution was improved and the width of single filaments in the super-resolved image was around 80 nm in FWHM. It should be noted that the super-resolution observation was performed at an illumination power density as low as 1.3 W/cm^2. This is at least two orders of magnitude lower than that for SMLM. Thus, even lower phototoxicity is anticipated in SPoD-OnSPAN observation than in the other super-resolution observations, and this is one advantage of SPoD-OnSPAN.

6 Conclusions and Perspectives

SPoD-OnSPAN imaging is the technique to achieve super-resolution observation with very low phototoxicity to live samples. This uses the fluorescence microscope with the polarized illumination, pRPFP, and an image reconstruction program. For observation, a sample labeled with pRPFP, which is expressed in cells, is used in the observation mechanism of SPoD-OnSPAN. A series of fluorescence images are captured of the sample that contains the fluorescence modulation information. This image data set is used to calculate a super-resolved image by the image reconstruction program. At present, typical spatial resolution for SPoD-OnSPAN is ~80 nm. The present technique would be useful for situations that require long-term observation of cells into the future.

It is desirable to develop a variety of pRPFPs for super-resolution imaging. Currently, pRPFPs which have been reported are Padron [12], Kohinoor [7], and rsCherry [13]. However, rsCherry would not suit SPoD-OnSPAN because its photoswitching rates and on/off contrast do not seem sufficient. Furthermore, the fluorescence emission bands of Padron and Kohinoor are both around 520 nm, corresponding to green. Thus, color variants of pRPFPs have yet to be developed for multiplexed super-resolution imaging. Additionally, we demonstrate in this chapter that Kohinoor can be used for SPoD-OnSPAN, and we presume that Padron would be suitable too. However, this does not mean that the photo-properties of Padron and Kohinoor are fully sufficient for SPoD-OnSPAN observations. In the future, a variety of excellent pRPFPs for super-resolution imaging need to be developed so that important properties such as photostability, switching fatigue resistance, maturation rate, and fluorescence brightness are further improved for higher spatial and temporal resolution, long-time observation, and higher biocompatibility.

The computation time for image reconstruction is also a vital issue, when the present technique is practiced. In fact, with currently available software packages [5, 6], image reconstruction calculations take an appreciably long time to derive a super-resolved image. Thus, we expect that innovative algorithms will soon be developed to allow calculations to be performed in a time short enough to make image reconstruction practical without sacrificing high accuracy.

Acknowledgments

This work was in part supported by a grant from CREST JST (Grant#, JPMJCR15N3) to T.N., and grants-in-aid from the Ministry of Education, Culture, Sports, Science and Technology, Japan (grant numbers, 23115003 to T.N., 16K07322 to T.Waz.).

References

1. Huang B, Bates M, Zhuang X (2009) Super resolution fluorescence microscopy. Annu Rev Biochem 78:993–1016
2. Eggeling C, Willig KI, Sahl SJ, Hell SW (2015) Lens-based fluorescence nanoscopy. Q Rev Biophys 48:178–243
3. Gustafsson MGL (2000) Surpassing the lateral resolution limit by a factor of two using structured illumination microscopy. J Microsc 198:82–87
4. Magidson V, Khodjakov A (2013) Circumventing photodamage in live-cell microscopy. Methods Cell Biol 114:545–560
5. Wazawa T, Arai Y, Kawahara Y, Takauchi H, Washio T, Nagai T (2018) Highly biocompatible super-resolution fluorescence imaging using the fast photoswitching fluorescent protein Kohinoor and SPoD-ExPAN with Lp-regularized image reconstruction. Microscopy 67:89–98
6. Hafi N, Grunwald M, van den Heuvel LS, Aspelmeier T, Chen J-H, Zagrebelsky M, Schütte OM, Steinem C, Korte M, Munk A, Walla PJ (2014) Fluorescence nanoscopy by polarization modulation and polarization angle narrowing. Nat Methods 11:579–584
7. Tiwari DK, Arai Y, Yamanaka M, Matsuda T, Agetsuma M, Nakano M, Fujita K, Nagai T (2015) A fast and positively photoswitchable fluorescent protein for ultralow-laser-power RESOLFT nanoscopy. Nat Methods 12:515–518
8. Lakowicz JR (2006) Principles of fluorescence spectroscopy. Springer, New York
9. Shcherbakova DM, Sengupta P, Lippincott-Schwartz J, Verkhusha VV (2014) Photocontrollable fluorescent proteins for superresolution imaging. Annu Rev Biophys 43:303–329
10. Galvez EJ (2001) Achromatic polarization-preserving beam displacer. Opt Lett 26:971–973
11. Wickert U, Mücke N, Wedig T, Müller SA, Aebi U, Herrmann H (2005) Characterization of the in vitro co-assembly process of the intermediate filament proteins vimentin and desmin: mixed polymers at all stages of assembly. Eur J Cell Biol 84:379–391
12. Andresen M, Stiel AC, Fölling J, Wenzel D, Schönle A, Egner A, Eggeling C, Hell SW, Jakobs S (2008) Photoswitchable fluorescent proteins enable monochromatic multilabel imaging and dual color fluorescence nanoscopy. Nat Biotechnol 26:1035–1040
13. Stiel AC, Andresen M, Bock H, Hilbert M, Schilde J, Schönle A, Eggeling C, Egner A, Hell SW, Jakobs S (2008) Generation of monomeric reversibly switchable red fluorescent proteins for far-field fluorescence nanoscopy. Biophys J 95:2989–2997

Chapter 12

Nanoscale Molecular Imaging of Presynaptic Active Zone Proteins in Cultured Hippocampal Neurons

Hirokazu Sakamoto, Shigeyuki Namiki, and Kenzo Hirose

Abstract

Neurotransmitter release occurs at a specialized region of the nerve terminal known as the presynaptic active zone, where a large number of synaptic proteins reside. The composition and nanoscale spatial organization of the active zone proteins are crucial determinants of presynaptic functions, and variety in the composition and organization may account for wide diversity of central synapses. Here, we describe stochastic optical reconstruction microscopy (STORM) imaging, a single-molecule localization-based super-resolution imaging, for analyzing nanoscale distributions of active zone proteins in cultured hippocampal neurons. STORM enables multicolor and three-dimensional analysis with remarkable spatial resolution, and thus resolves detailed nanostructures formed by the active zone proteins, such as Munc13-1 supramolecular nanoassemblies which correspond to the quantal synaptic vesicle release sites. In this chapter, we provide the reader with guidelines on how to stain presynaptic active zone proteins, to prepare specimens suitable for STORM imaging, and to perform STORM imaging in a quantitative manner.

Key words Synapses, Presynaptic active zone, Hippocampal neurons, Super-resolution imaging, STORM

1 Introduction

Synaptic transmission, which involves neurotransmitter release from presynaptic terminals and neurotransmitter binding to postsynaptic receptors, is fundamental to information processing between neuronal cells [1]. Typically, neurotransmitter release takes place at a specialized area of the presynaptic terminal known as the active zone which contains electron-dense materials precisely opposing to postsynaptic structures [2–4]. In central synapses, active zones are sub-micrometer planar structures which consist of a large number of synaptic proteins [5, 6]. There are several evolutionarily conserved proteins, including Munc13, RIM, RIMBP, α-Liprin, and CAST/ELKS, and two vertebrate-specific proteins, Bassoon and Aczonin/Piccolo in the active zone. Also, active zones contain membrane proteins essential for regulating synaptic vesicle

Nobuhiko Yamamoto and Yasushi Okada (eds.), *Single Molecule Microscopy in Neurobiology*, Neuromethods, vol. 154,
https://doi.org/10.1007/978-1-0716-0532-5_12,

exocytosis, including syntaxin and calcium channels. These active zone proteins are differentially expressed among many types of synapses, and the heterogeneity in composition and organization of these proteins might underlie a wide diversity of synaptic functions across synapses in brain [7, 8]. Therefore, experimental techniques for in-depth analysis of their spatial organization in synapses with high spatial resolution and high specificity are important for understanding synapse functions.

Immunostaining methods are widely used to study localization of protein molecules in biological specimens. However, conventional optical microscopy is not suitable for analyzing supramolecular organization of protein assemblies at the active zone because it has the limited resolution of ~300 nm due to the diffraction of light. Although electron microscopy with immuno-gold labeling might afford nanometer-scale localization precision, inefficiency in immuno-gold labeling as well as difficulty in three-dimensional (3D) analysis compromises its use. The recent development of super-resolution microscopy enables optical imaging of immunofluorescent specimens with sub-diffraction-limit resolution [9, 10], providing a convenient alternative to electron microscopy imaging with immuno-gold labeling. Among many super-resolution imaging methods, stochastic optical reconstruction microscopy (STORM), a single-molecule localization imaging technique [11–13], provides the highest spatial resolution. Furthermore, STORM enables multicolor and 3D imaging [14, 15] and has been successfully applied to visualize nanoscale organization of synaptic proteins in synapses [16–20]. In this chapter, we illustrate a detailed method for visualizing nanoscale supramolecular architectures of the active zone proteins in synapses of hippocampal cultures.

2 Materials

2.1 Primary Culture of Hippocampal Neurons

- Coated coverslips for culturing: Coverslips (25 × 25 mm No. 1, Matsunami glass) are cleaned by concentrated nitric acid with shaking for more than 18 h and then extensively washed with water. The cleaned coverslips are autoclaved in water for sterilization and then dried in a laminar flow hood. Sterilized cell culture stainless cylinders of 15 mm diameter are then attached on the center of the coverslips using sterilized high-vacuum grease (Dow Corning Toray). The enclosed region of the coverslip is coated with 10 μg/ml laminin (Thermo Fisher Scientific) and 25 μg/ml poly-L-lysine (Nacalai Tesque) for 1 day at 37 °C, then washed with phosphate-buffered saline (PBS), and stored at 4 °C until use.
- Isoflurane (Wako).
- Trypsin solution (Thermo Fisher Scientific).

- DNase I from bovine pancreas Type IV (Sigma-Aldrich).
- Trypsin inhibitor from Glycine max (Sigma-Aldrich).
- Cell strainer (70 μm, FALCON).
- Neuronal culture medium: Neurobasal A medium (Thermo Fisher Scientific) supplemented with B-27 supplement (Thermo Fisher Scientific), 0.5 mM Glutamax (Thermo Fisher Scientific), 1 mM sodium pyruvate (Wako), and 1% penicillin/streptomycin (Thermo Fisher Scientific).
- Cytosine β-D-arabinofuranoside (Sigma-Aldrich).
- Glial culture medium: DMEM (Wako) supplemented with 10% fetal bovine serum (FBS, Sigma-Aldrich), 4 mM L-glutamine (Wako), 1 mM sodium pyruvate (Wako), and 1% penicillin/streptomycin (Thermo Fisher Scientific).

2.2 Immunocytochemistry of Presynaptic Active Zone Proteins

- Paraformaldehyde (PFA, Wako): PFA is dissolved in PBS by heating to 65 °C. Although 10% PFA solution can be stored at −20 °C, it must be depolymerized by heating to 65 °C before use.
- Saponin (Sigma-Aldrich): 10% Stock solution in PBS can be stored at −20 °C until use.
- Albumin, bovine serum, F-V, pH 5.2 (BSA, Nacalai Tesque): BSA is dissolved in PBS with gentle shaking, and the solution should be centrifuged at 15,000 × *g* for 10 min. The collected supernatant can be stored at −20 °C.
- Primary antibodies: Anti-Munc13-1 (clone 11B-10G IgG1, ref. 19), anti-Bassoon (clone SAP7F407, IgG2a, Enzo Life Sciences), anti-RIM1 (140013, Synaptic Systems), and anti-syntaxin-1 (clone 78.2, IgG1, Synaptic Systems).
- Secondary antibodies: AffiniPure Goat Anti-Mouse IgG, Fcγ Subclass 1 Specific ML (Jackson ImmunoResearch), AffiniPure Goat Anti-Mouse IgG, Fcγ Subclass 2a Specific ML (Jackson ImmunoResearch), and AffiniPure Goat Anti-Rabbit IgG (H+L) ML (Jackson ImmunoResearch). For single-color STORM imaging [12, 13], the secondary antibodies are labeled with Alexa Fluor (Alexa-) 647 NHS Ester (Thermo Fisher Scientific) (*see* **Note 1**). The dye labeling efficiency is set to 2–3 Alexa-647 dyes per antibody. For two-color STORM imaging with activator dyes [14], the secondary antibodies are labeled with Alexa-647 NHS Ester and Alexa-405 NHS Ester (Thermo Fisher Scientific) or Alexa-647 NHS Ester and Alexa-488 NHS Ester (Thermo Fisher Scientific). The dye labeling efficiencies of the secondary antibodies are set to approximately 1 for Alexa-647, 2 for Alexa-405, and 3 for Alexa-488 (*see* **Note 2**).

2.3 Sample Preparation for STORM Imaging

- FluoSpheres Carboxylate-Modified Microspheres, 0.2 μm, red fluorescent (F-8810, Thermo Fisher Scientific).
- Triton X-100 (Sigma-Aldrich).
- STORM imaging buffer: 50 mM HEPES (pH 8.0, Nacalai Tesque), 10 mM NaCl (Nacalai Tesque), 60% sucrose (Wako), 10% glucose (Nacalai Tesque), 1% β-mercaptoethanol (Sigma-Aldrich), 0.5 mg/ml glucose oxidase (Nacalai Tesque), and 0.04 mg/ml catalase (Sigma-Aldrich). The enzymes (glucose oxidase and catalase) and β-mercaptoethanol should be added to the buffer immediately before imaging. For preparing enzyme stock solutions, glucose oxidase and catalase are dissolved in PBS, and centrifuged at 15,000 × *g*. The collected supernatant can be stored at −80 °C until use.
- Coverslips (15 mm Round No. 1, Matsunami glass).

2.4 STORM Imaging Data Acquisition

Imaging experiments are performed on a custom-built microscope with a ×100 oil immersion objective (1.40 NA, Olympus), an EMCCD camera (iXon3 860 operated by Solis software, Andor), a piezo-positioning stage (P-733.3 XYZ Piezo Nanopositioning Stage, PI), and an objective holder firmly fixed on rigid props that minimized mechanical sample drift during experiments (Fig. 1a). A simple schema of the optics for our STORM imaging system is shown in Fig. 1b. A 640 nm laser beam (100 mW, OBIS, Coherent) passing through a Cy5 excitation filter (FF02-628/40-25, Semrock) is used to excite Alexa-647 molecules. Fluorescence detection of Alexa-647 is achieved using a penta-band dichroic mirror (FF408/504/581/667/762-Di01-25x36, Semrock) and a band-pass emission filter (710QM80, Omega). A 405 nm laser beam (100 mW, OBIS, Coherent) is used to directly activate Alexa-647 molecules from the dark state to the emitting state, or to indirectly activate Alexa-647 molecules via Alexa-405. A 488 nm laser beam (50 mW, OBIS, Coherent) is used to activate Alexa-647 molecules via Alexa-488. Fluorescence images, formed by a collecting lens (AC254-250-A-ML, Thorlabs), are relayed onto the camera with a pair of lenses (AC254-100-A-ML, Thorlabs), achieving a pixel size of 192 nm. For 3D astigmatism-based calibration, a cylindrical lens (LJ1558RM-A, $f = 300$ mm, Thorlabs) is inserted into the emission detection path at ~30 mm before the EMCCD camera, providing astigmatism with approximately 0.59 μm difference in vertical and horizontal focal lengths on the image planes.

2.5 STORM Imaging Data Analysis

ImageJ/Fiji (https://imagej.nih.gov/ij/, https://fiji.sc/) is used for image processing including single-molecule localization and reconstruction. We use a customized plug-in based on a 2D Gaussian model with the least-square fitting method for determining x–y coordinates. For determining z coordinates, we use an astigmatism-

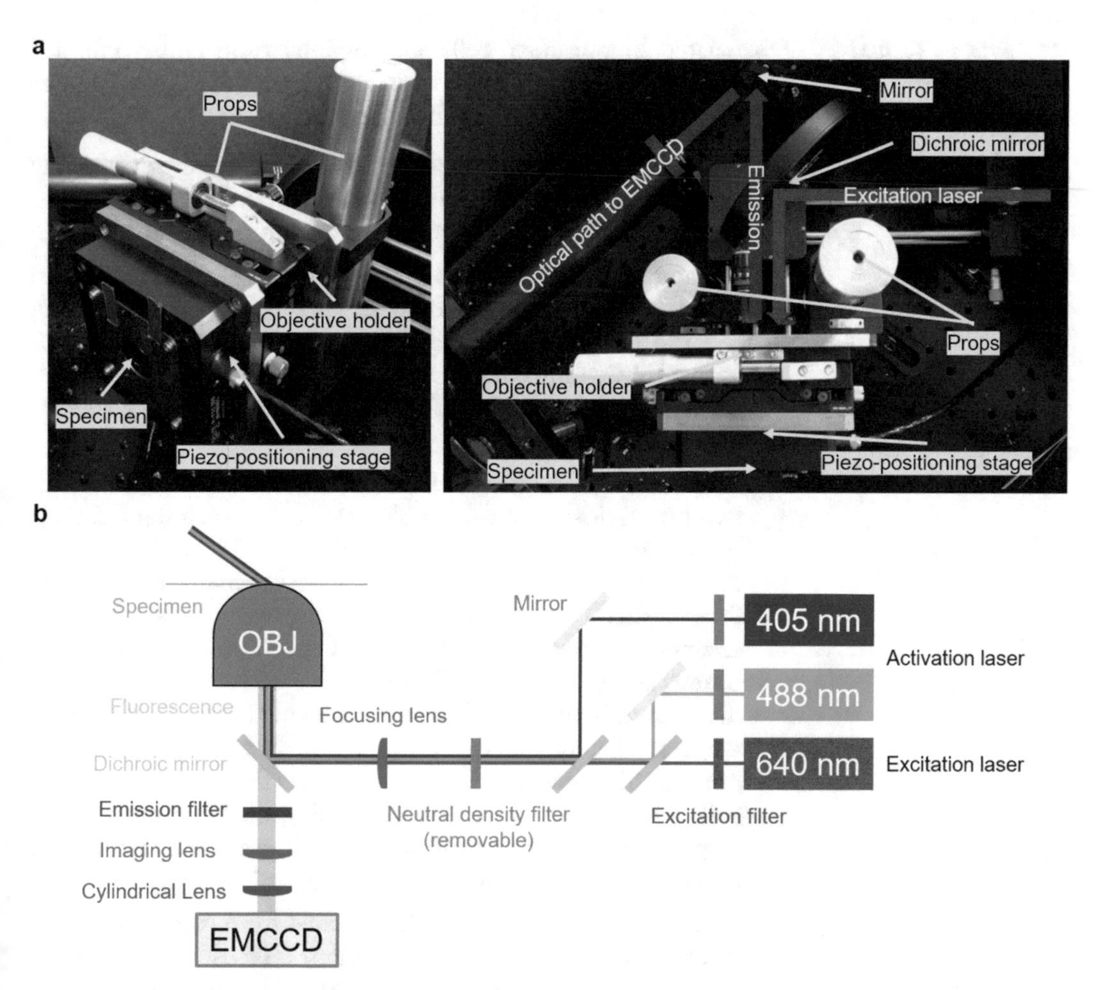

Fig. 1 A custom-built microscope for STORM imaging. (**a**) A piezo-positioning stage and the objective holder are firmly fixed on rigid props, minimizing mechanical sample drift during experiments. Specimens are magnetically attached on the stage using neodymium magnet sheets. (**b**) A schema of the optics for our 3D-STORM imaging system

based calibration. Although many types of open-source software for single-molecule localization may be available, we recommend that the reader uses a software based on a 2D Gaussian model.

3 Methods

All animals must be handled according to the protocols that have been approved by the institutional animal ethics committee. All protocols used in this study were performed in accordance with the policies of the Animal Ethics Committee of the University of Tokyo.

3.1 Primary Culture of Hippocampal Neurons

Dissociated hippocampal cells are obtained from embryonic day 20–21 (E20–21) Sprague-Dawley rats (Japan SLC), and maintained on precultured glial cells [21].

1. Deeply anesthetize a pregnant rat by isoflurane, and cut the carotid artery with a sharp scissor to remove blood, leading to euthanasia. Transfer embryos onto a clean petri dish.
2. Decapitate the embryos and place in a dish containing ice-cold PBS. Remove and transfer brains to another dish containing ice-cold PBS, and dissect the hippocampi under a stereomicroscope. Finely mince the hippocampi, and then treat them with 1% trypsin and 1 mg/ml DNase I for 5 min at RT in a laminar flow hood.
3. After blocking trypsin with DMEM containing 10% FBS for two times, dissociate cells in ice-cold PBS containing 0.3 mg/ml trypsin inhibitor and 0.5 mg/ml DNase I by pipetting with Pasteur pipettes (once by a regular Pasteur pipette and twice by fire-polished Pasteur pipettes). Filter dissociated cells with a cell strainer to remove aggregated cells, centrifuge at ~1300 × *g* for 6 min, and then resuspend in neuronal culture medium. (The dissociated hippocampal cells are used in two ways. First, they are immediately plated onto precultured glial cells to obtain primary hippocampal neurons. Second, they are seeded on a collagen-coated culture dish containing glial culture medium to prepare glial cell precultures for later use. Glial cultures are passaged once after a week and used after 2 weeks.)
4. Suspend precultured glial cells detached from culture dishes by trypsinization in glial culture medium. Transfer the glial cell suspension containing about 60,000 cells into the culture cylinder of 15 mm diameter which is attached on the laminin- and poly-L-lysine-coated coverslip. After 1-h incubation at 37 °C under 5% CO_2, exchange the medium once with PBS. Then after removing PBS, place 300 μl of dissociated hippocampal cells suspended in neuronal culture medium of 1.2×10^4 ml^{-1} (~3600 cells) onto the glial cell layer in the culture cylinder (*see* **Note 3**).
5. Maintain cultures at 37 °C under 5% CO_2. After 2-day culture in vitro, add 300 μl of neuronal culture medium containing 5 μM cytosine β-D-arabinofuranoside to neuronal cultures (a final concentration of 2.5 μM and 600 μl of total culture medium) to suppress a proliferation of glial cells. One week after plating, and every week thereafter (for 3–4 weeks), maintain neuronal cultures by taking away 150–300 μl of cultured medium and adding 300 μl of fresh medium.

3.2 Immunocytochemistry of Presynaptic Active Zone Proteins

Immunocytochemical procedures are performed in cell culture cylinders of 15 mm diameter which are rigidly attached on the center of coverslips. In this configuration, the requisite amount of solution at each step is ~150 μl per specimen.

1. Fix neuronal cells with 1% PFA in PBS for 15 min at room temperature (RT) or on ice (*see* **Note 4**) and then permeabilize with 0.1% saponin or 0.1% Triton X-100 in PBS for 10 min on ice (*see* **Note 5**).
2. After blocking with 0.3% BSA in PBS, incubate the specimens with primary antibodies in 0.3% BSA in PBS for 1 h at RT (*see* **Note 6**).
3. After washing with 150 μl of 0.3% BSA in PBS several times, incubate the specimens with fluorescent dye-labeled secondary antibodies in 0.3% BSA in PBS for 30 min at RT.
4. Finally, after extensive washing, postfix specimens with 4% PFA for 30 min at RT (*see* **Note 7**). Fixed specimens are stored at 4 °C in PBS until imaging.

3.3 Sample Preparation for STORM Imaging

Specimens are mounted in the STORM imaging buffer which contains β-mercaptoethanol and oxygen-scavenging agents. These additives are essential for stable blinking of fluorescent dyes, a requisite condition for STORM imaging. Microspheres (red fluorescent 200 nm beads) nonspecifically bound to immunostained specimens are used as fiducial markers.

1. Treat a postfixed specimen with 0.1% Triton X-100-containing PBS for 5 min at RT.
2. Suspend the microspheres in water at 1:100 and vortex intently. Then, dilute the suspended beads in PBS containing 50 mM $MgCl_2$ at 1:100 (final concentration, 1:10,000), quickly apply to the specimen, and incubate for 30 min at RT.
3. After washing with PBS, remove the cell culture cylinder from the coverslip while grease remains on the coverslip. Then wash the specimen with water to remove any contamination.
4. After removing excess water, apply the STORM imaging buffer onto the specimen (*see* **Note 8**). Then sandwich the specimen with a 15 mm round coverslip (secured by the remaining grease) and seal with nail polish (*see* **Note 9**).

3.4 STORM Imaging Data Acquisition

Resolution of STORM imaging largely relies on good photoswitching kinetics of Alexa-647 under strong laser excitation in the STORM imaging buffer. A typical imaging experiment comprises many cycles of imaging, in which only a sparse subset of fluorophores (e.g., Alexa-647 molecules) are activated. Low density of the activated fluorophores is required to separately detect fluorescent spots emitted from individual fluorophores.

1. Initially, search the imaging field of interest with a weak 640 nm excitation intensity of approximately 5 W/cm^2 by inserting appropriate neutral-density filters in the beam path. Then, increase the excitation intensity to approximately 5 kW/cm^2 to turn off most of Alexa-647 molecules, enabling single-molecule imaging (*see* **Note 10**). Take images of 128 × 128 pixels with setting the EM gain of EMCCD camera to 100–200, and store the raw image data as 16-bit tiff images. In-focus fluorescent beads, which appear as constant fluorescent spot signals, must be included in the imaging field for correction of sample drift.
2. For single-color imaging, use a 405 nm laser beam to directly activate Alexa-647 molecules from the dark state to the emitting state [12, 13]. Increase the excitation intensity of the 405 nm laser gradually from 0.1 to 100 W/cm^2 during imaging, so that fluorescent signals from activated Alexa-647 molecules are not spatially overlapped and the number of their localizations is kept constant (Fig. 2a). Take a series of 50,000–100,000 imaging frames at ~100 Hz (~10 ms exposure) for a single-image reconstruction.
3. For two-color STORM imaging with activator dyes [14], use secondary antibodies which are labeled with Alexa-647 and either of activator dyes, Alexa-405 or Alexa-488 (*see* **Note 11**). Activate Alexa-647 on different antibodies selectively by alternating between 405 and 488 nm lasers for excitation of Alexa-405 and Alexa-488, respectively. Increase the excitation intensities of the 488 and 405 nm lasers gradually from 0.25 to 32 W/cm^2 (Fig. 2b). Take sets of 5000 consecutive frames at 100 Hz alternately for two channels with a total of 40,000–50,000 imaging frames for each channel (*see* **Note 12**).

3.5 STORM Imaging Data Analysis

Reconstruction of a sub-diffraction-limit image for target molecules from raw image data is done on a personal computer with an appropriate software. Localizations (x, y, and z coordinates) of individual Alexa-647 molecules are determined based on single-molecule localization.

1. Before determining localizations of fluorophores, process the raw image data with a Gaussian spatial filter with a radius of one pixel (192 nm for our imaging setup).
2. Obtain x and y coordinates of individual Alexa-647 fluorophores by estimating center positions with least-square fitting of the 2D Gaussian model. Obtain the z coordinates using an astigmatism-based calibration method [15], in which differences in values of width parameters in x and y directions for the 2D Gaussian model are used for estimating z coordinate (for calibration procedure, *see* **Note 13**).

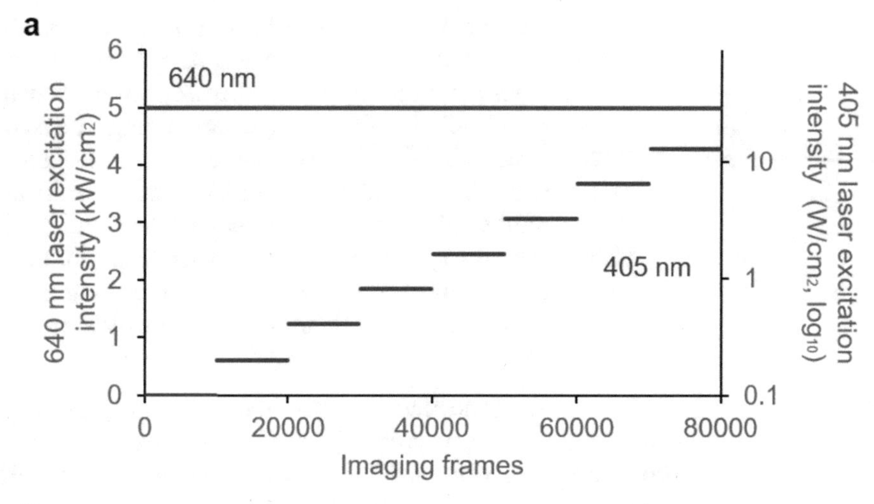

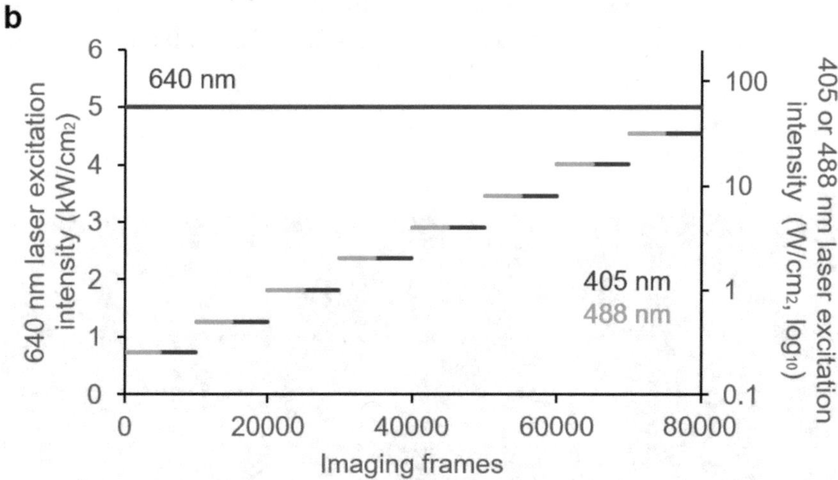

Fig. 2 STORM imaging acquisition protocols. (**a**) An acquisition protocol for single-color STORM imaging using Alexa-647. (**b**) An acquisition protocol for two-color STORM imaging using Alexa-647 with activator dyes (Alexa-488 and Alexa-405)

3. Correct sample drift during recording using fluorescent fiducial markers (*see* **Note 14**). Obtain *x*, *y*, and *z* coordinates of fluorescent beads using 2D Gaussian model fitting and astigmatism-based calibration as in Alexa-647 localizations. Use the 3D trajectory of coordinates of the fluorescent beads as "drift vector" after smoothing by taking moving median of 100 consecutive frames. Then, correct all Alexa-647 localization data sets by subtraction of the drift vector.
4. Reconstruct 3D-STORM images by pixelating the drift-corrected single-molecule localization data sets with a 10 nm binning (32-bit tiff images), and process a 3D Gaussian spatial filter with a radius of 10 nm. Confirm the sufficiency in the number of localizations per synapse by checking reproducibility of the reconstructed images (*see* **Note 15**).

5. For analysis of the spatial distribution of molecules within the active zone, rotate STORM images in 3D so that the active zone surface faces the front (*see* **Note 16**). In particular, rotate the cropped 3D-STORM image containing the synapse of interest manually along all the three axes (*y*-, *z*-, and then *x*-axis) one by one on the image display monitor. Initially, rotate data around *y*-axis to find the side view of the active zone (at the point where the apparent short axis of active zone is narrowest), then rotate around *z*-axis so that the apparent long axis of active zone is aligned horizontally, and finally rotate 90° around *x*-axis (*see* **Note 17**).

Examples of the reconstructed images are shown in Figs. 3 and 4 for single- and two-color STORM, respectively. In our imaging procedures, we obtained localization precisions (standard deviations of the localization distribution) of 9.0 nm in *x*-axis, 9.7 nm in *y*-axis, and 19.0 nm in *z*-axis (within 500 nm of the focal plane), respectively (*see* **Note 18**).

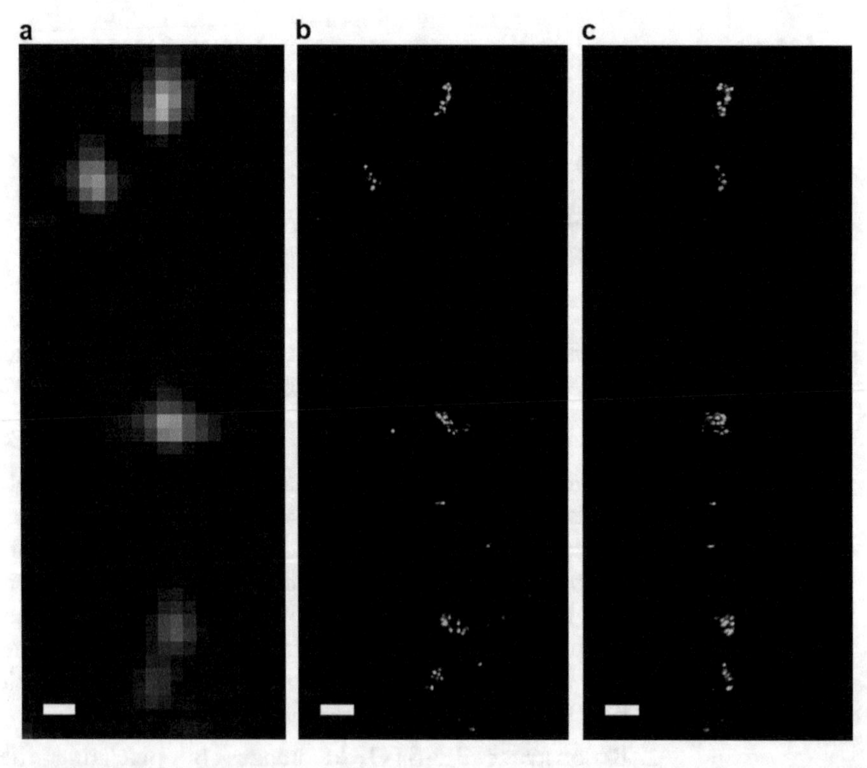

Fig. 3 Single-color 3D-STORM. (**a**) A wide-field immunofluorescence image of Munc13-1. (**b**) A STORM image corresponding to (**a**). The image shown is a projection of 800 nm in *z* direction (±400 nm from the focal plane). (**c**) A 90° rotated view of (**b**). Scale bar, 500 nm

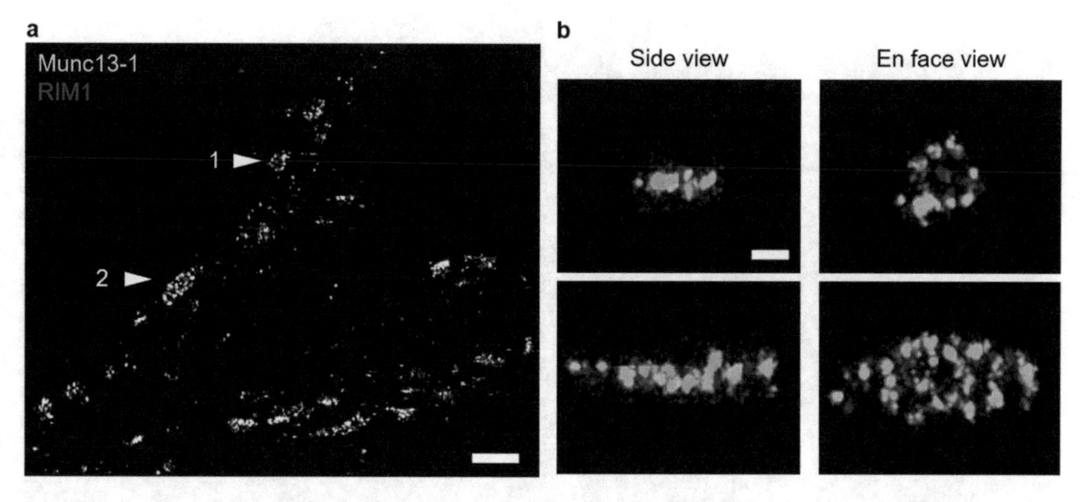

Fig. 4 Two-color 3D-STORM. (**a**) STORM images of Munc13-1 (yellow) RIM1 (magenta) immunostained with Alexa-647/405- and Alexa-647/488-labeled secondary antibodies, respectively. Scale bar, 1 μm. The image shown is a projection of 800 nm in *z* direction (±400 nm from the focal plane). (**b**) Side and en face views of Munc13-1 and RIM1 of synapses indicated by arrowheads in (**a**). Scale bar, 200 nm

4 Notes

1. Although other fluorescent dyes, such as a longer wave cyanine dye DyLight755, can be used for STORM imaging, Alexa-647 shows the best performance in photo-switching kinetics and emitting photon numbers [22]. We recommend Alexa-647 as the first-choice dye for STORM imaging.
2. Over-labeling of dyes leads to increase in nonspecific staining. Especially for Alexa-405, the dye labeling efficiency should not exceed 3.
3. The timing of plating is crucial for stable culture of neurons. We recommend that hippocampal cells are seeded onto glial cells just 1 h after plating of glial cells on coverslips. To accomplish this, we routinely perform plating of glial cells shortly after the hippocampi are dissected out.
4. To evaluate nanoscale structures with super-resolution microscopy, it is important to label most of the target proteins (epitopes) as much as possible to avoid misinterpretations due to under-labeling. Optimal concentrations of fixatives and reaction temperature should be determined to avoid epitope masking during fixation. Strong fixation sometimes leads to poor labeling of active zone proteins, presumably due to the masking of epitopes because of high protein concentrations in the active zone (Fig. 5).

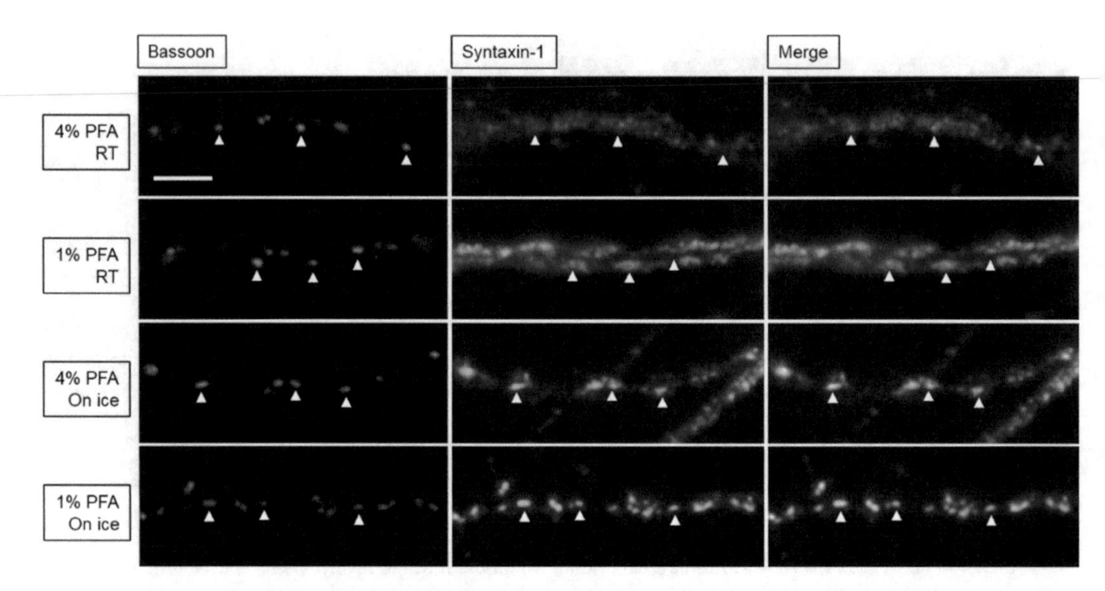

Fig. 5 Immunofluorescence of Bassoon and syntaxin-1 in various fixation conditions. Cultured hippocampal neurons were fixed with 4% PFA or 1% PFA at RT or on ice. Scale bar, 5 μm

5. When immunocytochemistry is performed with weak fixation conditions, saponin is recommended as a permeabilization agent, especially suitable for staining of membrane proteins.
6. In addition to optimization of fixation strength described in **Note 4**, the concentrations of primary antibodies should be titrated to determine the concentrations at which available epitopes are saturated with primary antibodies.
7. It is important to postfix the immunostained specimens for long-term preservation. Also, it prevents dissociation of bound antibodies during STORM imaging.
8. You need to completely replace the remaining water with the STORM imaging buffer to get the maximum performance of STORM imaging.
9. Sealed specimens should be used on the same day. Buffer pH might gradually decrease with oxidization of glucose and the concentration of free thiolate anions might drop, which adversely affects the photo-switching kinetics of Alexa-647.
10. Shortly after switching 640 nm laser to strong excitation intensity (~5 kW/cm^2), the number of activated Alexa-647 molecules is too large to image by an EMCCD camera with a high EM gain. You need to turn off the EM gain until the density of activated fluorophores becomes low to protect EMCCD camera. Within several minutes, the number of activated Alexa-647 molecules reaches low-level optimal for single-molecule imaging at synapses.

11. As described in **Note 1**, Alexa-647 shows the best performance in STORM imaging. Therefore, we also recommend using Alexa-647 with two activator dyes rather than using other fluorescent dyes in two-color imaging, although care needs to be taken to estimate the cross talk between two-color channels [14].
12. The cross talk fractions between two-color channels (Alexa-405 activated and Alexa-488 activated) are quantitatively estimated using different specimens singly labeled with either of the secondary antibodies under respective imaging conditions, and therefore the intensity of true signals can be estimated by simultaneous equations using the value of cross talk fractions and two-color channel data. In our imaging conditions, both of the crosstalk fractions across two-color channels (Alexa-488-activated signals to Alexa-405 channel and Alexa-405-activated signals to Alexa-488 channel) are below 8%. Although we do not know the exact reason, the crosstalk fraction between two-color channels becomes low when the Alexa-488 channel is measured first.
13. To construct an astigmatism-based calibration, red fluorescent beads nonspecifically bound on coverslips are measured under the same condition in STORM imaging when a cylindrical lens is inserted into the emission detection path. Series images with 10 nm steps (from −500 nm to +500 nm) are obtained using a piezo-positioning stage. Then, each fluorescent signal of beads is fitted with the 2D Gaussian model as with the localization procedures in the STORM. The relationship between set z-positions and differences in values of width parameters in x and y directions (the index for astigmatism) provides the calibration (Fig. 6).
14. We do not use an autofocusing system, which might introduce unexpected localization uncertainty. Drift correction using fluorescent beads sufficiently improves the localization errors due to sample drift.
15. One should be cautious when interpreting the observed nanostructures in the STORM images. If the number of data points is not enough, erroneous structures might appear in the STORM images. To verify the reproducibility of nanoscale structures in STORM images, we split the data set into early and late subsets and compare two independently reconstructed images from the subsets [19]. This verification helps to avoid misinterpretation of reconstructed images attributed to incomplete data acquisition for STORM image reconstruction.
16. The active zone planes randomly incline relative to the imaging (x–y) plane. The actual resolution on the 2D projected active zone plane is affected by localization uncertainty of both lateral (σ_{xy}) and axial (σ_z) depending on the degree of inclination of the synapse.

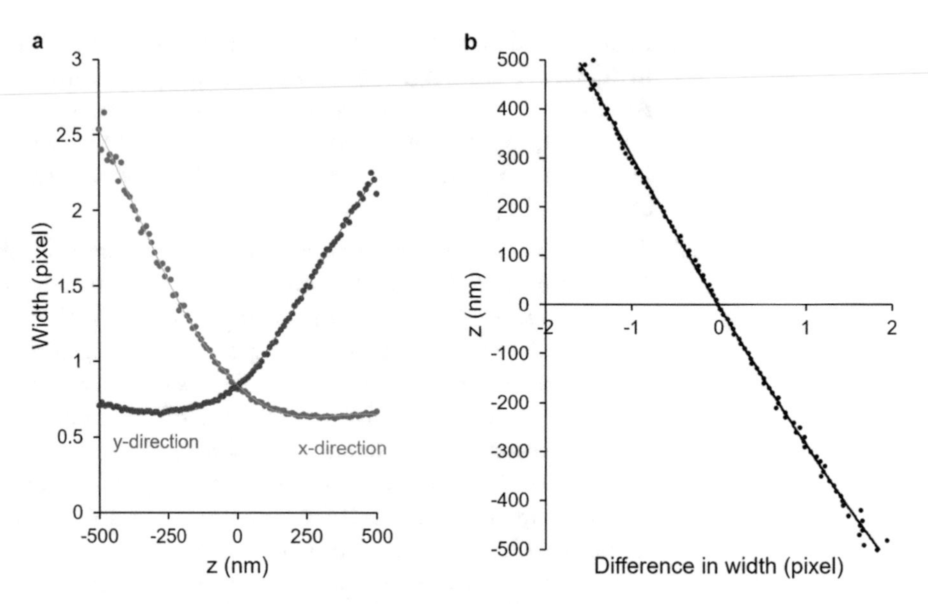

Fig. 6 An astigmatism-based calibration. (**a**) Widths of fluorescence spots of fluorescent beads in *x* and *y* directions plotted against *z*-positions. (**b**) The relation between *z*-positions and differences in width between *x* and *y* directions

17. The validity of rotation can be checked by comparing the manually obtained active zone planes with the planes numerically estimated by principal component analysis of localization data. In our experience, there were little or negligible differences in inclinations between the manually obtained planes by trained students and the numerically estimated planes.
18. In typical experiments, the number of detected photons per activated Alexa-647 molecule will be ~3000, providing the standard deviations of the localization distribution of ~10 nm in *x*–*y*-axes and ~20 nm in *z*-axis (within 500 nm of the focal plane), respectively. It is important to determine the actual localization uncertainty (the spatial resolution) in your imaging setup and analysis software for quantitative analysis of STORM imaging data. Statistical analysis of localizations of single-molecule fluorescence spots of Alexa-647-labeled antibodies which is nonspecifically bound on the fixed neuronal cell cultures would provide a practical estimation of the actual localization uncertainty.

Acknowledgments

This work was supported by JSPS KAKENHI (Grant nos. 17H04029 and 24116004 to K.H., 15K15048 to S.N.), AMED-SENTAN; Development of Advanced Measurement and Analysis Systems to K.H.; the Asahi Glass Foundation; and the Precise Measurement Technology Promotion Foundation (PMTP-F) to S.N.

References

1. Katz B (1969) The release of neural transmitter substances. Liverpool University Press, Liverpool
2. Heuser JE, Reese TS, Dennis MJ, Jan Y, Jan L, Evans L (1979) Synaptic vesicle exocytosis captured by quick freezing and correlated with quantal transmitter release. J Cell Biol 81:275–300
3. Watanabe S, Rost BR, Camacho-Perez M, Davis MW, Söhl-Kielczynski B, Rosenmund C, Jorgesen EM (2013) Ultrafast endocytosis at mouse hippocampal synapses. Nature 504:242–247
4. Peters A, Palay S, Webster H (1991) The fine structure of the nervous system: neurons and their supporting cells, 3rd edn. Oxford University Press, New York
5. Südhof TC (2012) The presynaptic active zone. Neuron 75:11–25
6. Frauke A, Clarissa LW, Craig CG (2015) Presynaptic active zones in invertebrates and vertebrates. EMBO Rep 16:923–938
7. Atwood HL, Karunanithi S (2002) Diversification of synaptic strength: presynaptic elements. Nat Rev Neurosci 3:497–516
8. Nancy AO, Nicholas CW, Kristina DM, Stephen JS (2012) Deep molecular diversity of mammalian synapses, why it matters and how to measure it. Nat Rev Neurosci 13:365–379
9. Stefan WH (2007) Far-field optical nanoscopy. Science 316:1153–1158
10. Huang B, Babcock H, Zhuang X (2010) Breaking the diffraction barrier: super-resolution imaging of cells. Cell 143:1047–1058
11. Rust MJ, Bates M, Zhuang X (2006) Sub-diffraction-limit imaging by stochastic optical reconstruction microscopy (STORM). Nat Methods 3:793–796
12. Heilemann M, van de Linde S, Schuettpelz M, Kasper R, Seefeldt B, Mukherjee A, Tinnefeld P, Sauer M (2008) Subdiffraction-resolution fluorescence imaging with conventional fluorescent probes. Angew Chem Int Ed 47:6172–6176
13. van de Linde S, Löschberger A, Klein T, Heidbreder M, Wolter S, Heilemann M, Sauer M (2011) Direct stochastic optical reconstruction microscopy with standard fluorescent probes. Nat Protoc 6:991–1009
14. Bates M, Huang B, Dempsey GT, Zhuang X (2007) Multicolor super-resolution imaging with photo-switchable fluorescent probes. Science 317:1749–1753
15. Huang B, Wang W, Bates M, Zhuang X (2008) Three-dimensional super-resolution imaging by stochastic optical reconstruction microscopy. Science 319:810–813
16. Dani A, Huang B, Bergan J, Dulac C, Zhuang X (2010) Superresolution imaging of chemical synapses in the brain. Neuron 68:843–856
17. Ehmann N, van de Linde S, Alon A, Ljaschenko D, Keung XZ, Holm T, Rings A, Di Antonio A, Hallermann S, Ashery U, Heckmann M, Sauer M, Kittel RJ (2014) Quantitative super-resolution imaging of Bruchpilot distinguishes active zone states. Nat Commun 5:4650
18. Tang A, Chen H, Li TP, Metzbower SR, MacGillavry HD, Blanpied TA (2016) A trans-synaptic nanocolumn aligns neurotransmitter release to receptors. Nature 536:210–214
19. Sakamoto H, Ariyoshi T, Kimpara N, Sugao K, Taiko I, Takikawa K, Asanuma D, Namiki S, Hirose K (2018) Synaptic weight set by Munc13-1 supramolecular assemblies. Nat Neurosci 21:41–49
20. Glebov OO, Jackson RE, Winterflood CM, Owen DM, Barker EA, Doherty P, Ewers H, Burrone J (2017) Nanoscale structural plasticity of the active zone matrix modulates presynaptic function. Cell Rep 18:2715–2728
21. Kaech S, Banker G (2006) Culturing hippocampal neurons. Nat Protoc 1:2406–2415
22. Dempsey GT, Vaughan JC, Chen KH, Bates M, Zhuang X (2011) Evaluation of fluorophores for optimal performance in localization-based super-resolution imaging. Nat Methods 8:1027–1036

Chapter 13

Three-Dimensional Super-Resolution Imaging of the Cytoskeleton in Hippocampal Neurons Using Selective Plane Illumination

Frances Camille M. Wu, Feby Wijaya Pratiwi, Chin-Yi Chen, Chieh-Han Lu, Wei-Chun Tang, Yen-Ting Liu, Bi-Chang Chen, and Peilin Chen

Abstract

Recent advances in single-molecule-based super-resolution imaging have been in the forefront of biological research for the visualization of the detail structures in cellular and molecular biology. A number of super-resolution optical microscopy techniques have been reported; however, several challenges such as the use of high-activation sources resulting in photobleaching and photodamage effects, restrictions in three-dimensional imaging, long data acquisition, and limited field of view remain unresolved. To address these concerns, a rapid, large-scale, and three-dimensional super-resolution fluorescence microscopy has been developed through the introduction of selective plane illumination microscopy based on scanning Bessel beam and a spontaneously blinking dye HMSiR as a reporter. This localization-based super-resolution microscope offers several advantages. namely minuscule levels of photodamage and phototoxicity effects due to low activation source, good optical sectioning suitable for three-dimensional imaging, large field of view, and fast data acquisition. In this chapter, protocols on the three-dimensional super-resolution imaging of neurons through the application of selective plane illumination technique are discussed.

Key words Super-resolution microscopy, Light-sheet microscopy, Cellular imaging, Hippocampal neurons, Three-dimensional imaging, Blinking fluorophore HMSiR

1 Introduction

To visualize the unseen complex and dynamic biological structures present in living cellular organisms, the quantitative measurement and integrated analysis of structural and molecular dynamics are necessary. For instance, in neuroscience, the intrinsic complexity of the brain along with the associated diversity of the behavior of neuron cells requires a more multifaceted quantitative approach toward a systematic investigation. To study the significant neuron interactions, wherein the majority of neurons are settling together and create circuits that are unlikely to act independently [1], a large-scale single-molecule localization microscopy is necessary.

Nobuhiko Yamamoto and Yasushi Okada (eds.), *Single Molecule Microscopy in Neurobiology*, Neuromethods, vol. 154,
https://doi.org/10.1007/978-1-0716-0532-5_13,

Indeed, the neurons primarily act on their respective roles such as a sensory motor transformation or a memory retrieval system [2, 3], involving a huge population of neurons.

Conventional fluorescent microscopy techniques enable the imaging of cellular structures and dynamics of biomolecules through the labeling of biomolecules with fluorophores [4, 5] or fluorescent proteins [6, 7] at a single-molecule level. However, it prohibits the observation of several biomolecules at subcellular structures within the diffraction-limited areas. To circumvent the barriers imposed by the diffraction limit, the super-resolution microscopy techniques are introduced. These super-resolution imaging techniques include the stimulated emission depletion (STED) microscopy [8], structured illumination microscopy (SIM) [9], photoactivated localization microscopy (PALM) [10], and stochastic optical reconstruction microscopy (STORM) [11]. In comparison, the mechanism of conventional fluorescent microscopy involves the excitation of all fluorophores at once, resulting in overlapped position of fluorophores in diffraction-limited areas which yields to a blurred image, while in super-resolution imaging, only a subset of fluorophores are excited at a given time and therefore localizing the precise location of a single fluorophore in diffraction-limited areas, resulting in a super-resolution image (*see* Fig. 1) [12]. The technological advancements of super-resolution microscopy techniques as well as high-powered

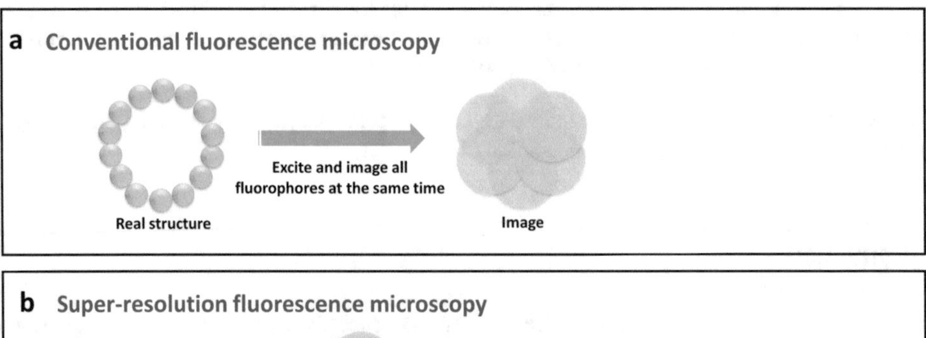

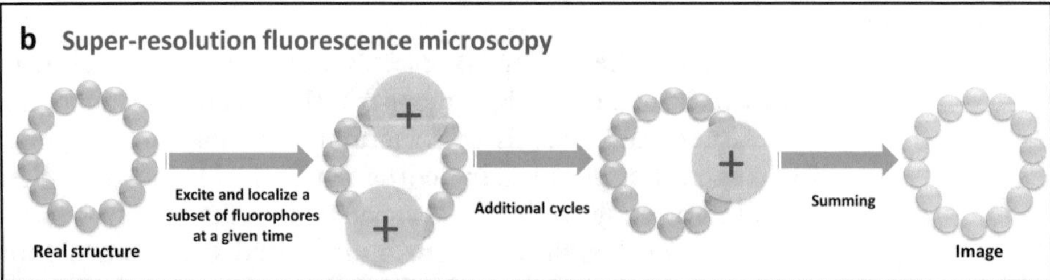

Fig. 1 The principle of conventional fluorescence microscopy vs. super-resolution fluorescence microscopy. (**a**) In conventional fluorescence microscopy, all the fluorophores are excited at once which results in the diffraction-limited areas positioned overlapped, yielding to a blurred image, while in (**b**) super-resolution fluorescence microscopy, only a small subset of fluorophores are excited at a given time and therefore localizing the precise location of a single fluorophore in diffraction-limited areas, resulting in a super-resolution image

lasers, rapid computational tools, and fast and robust digital image acquisition systems offer a unique opportunity to successfully probe locations of proteins in neurons and record the spatiotemporal information of neuron morphology and events with unprecedented resolution and speed [13, 14]. It permits the imaging of large neuron populations and precise location and morphology of neurons as compared with the standard electrical recordings which requires the insertion of the electrodes [15].

The earlier super-resolution microscopy techniques such as PALM and STORM are originally derived from single-molecule spectroscopy which allows the precise localization of a single molecule wherein the fluorescence emission signal is coming from one molecule within the diffraction-limited area at a specific acquisition period [16–19]. In a typical experimental setup, the separation of overlapped fluorophores in the axial direction is a huge challenge, and therefore the localization-based super-resolution microscopy remains circumscribed in two-dimensional imaging [20]. To overcome this challenge, optical astigmatism was introduced for the generation of three-dimensional super-resolution image of thin samples via the determination of the axial position [21]. However, it fails to provide a satisfying result for thick sample specimens due to the associated creation of unwanted noise and overlapping fluorescence signals in the axial position resulting in out-of-focus fluorescence signals. These issues toward a three-dimensional super-resolution technique were overcome with the optical sectioning techniques such as temporal focusing [22], multi-focus grating [23], and selective plane illumination microscopy (SPIM) [24–26].

In selective plane illumination microscopy (SPIM), a light sheet illuminates the sample specimen through a thin excitation plane (approximately half a micron) orthogonal to the detection axis, which improves the imaging optical sectioning capability, eliminates the out-of-focus fluorescence background, and reduces the unwanted photobleaching and photodamage effects [27, 28]. The combined good optical sectioning and parallelization permit a long-term imaging and rapid acquisition with minimal photobleaching effect suitable for the recording of a large population of neurons with a high spatiotemporal resolution [24–26]. In order to address the fundamental resolution limit of conventional light-sheet fluorescence microscopy based on Gaussian beam illumination, the Bessel beam plane illumination was introduced [29]. The non-diffracting characteristic of a Bessel beam created a significantly thinner light sheet as compared with the Gaussian beam, therefore providing a better axial resolution in the same field of view [30].

In this chapter, the application of Bessel beam plane illumination is used to perform single-molecule localization toward the three-dimensional super-resolution imaging of neurons. A spontaneously blinking dye, HMSiR, is used as a neuron probe for the effective reconstruction of a super-resolution image [31]. In contrary to the established super-resolution techniques such as the direct stochastic optical reconstruction microscopy (dSTORM) [32] and ground-state depletion microscopy followed by individual molecule return (GSDIM) [33] which require high-intensity laser activation sources and additives such as thiols or oxygen scavengers to induce the on-off switching of fluorophores [34–36], the self-blinking HMSiR fluorophore can be subsequently activated by a single-wavelength laser with a relative low power density without any additives [31, 37, 38]. The low activation threshold of the spontaneously blinking dye and the exceptional optical sectioning capability of the Bessel beam scanning light sheet could beneficially save the photon budget and extend the imaging volume beyond the limitation of the optical field of view of the detection optics. This technique allows the systematic investigations of the interaction among multiple neurons distributed in a vast area and the three-dimensional super-resolution imaging of neurons which opens new insights into our understanding on the complex nervous system behavior and development.

2 Materials and Methods

This section aims to provide a detailed protocol on the three-dimensional super-resolution imaging of hippocampal neurons via the selective plane illumination microscopy (SPIM) technique. A step-by-step procedure of the selective plane illumination microscope based on Bessel beam illumination setup, hippocampal neuron culture including the glial cell culture as conditional medium for neurons, neuronal culture and labeling of neurons with primary and secondary antibodies (anti-β-tubulin- and HMSiR-labeled goat anti-mouse, respectively), neuron mounting, and image processing and analysis are elaborated and discussed.

2.1 Selective Plane Illumination Microscopy Based on Bessel Beam

The super-resolution imaging of neuron cells is performed on a self-assembled selective plane illumination microscope based on Bessel beam illumination. The schematic setup of the microscope system is shown in Fig. 2. The selective plane illumination microscope is equipped with three excitation lasers ($\lambda = 488$, 567, 637 nm) and two dichroic filters (DF1, DF2), which can be openly combined for the optimum setting in individual fluorophores. To permit an effective wavelength selection in a controllable exposure time, an acousto-optical tunable filter (AOTF) is employed. After passing through the AOTF, the laser beam is expanded to a diameter of ~4 mm via the beam expanders (L7 and L8, L9 and L10), providing an equal Gaussian beam intensity distribution on the annular ring pattern of the custom-made annulus mask. The relay pairs of achromatic lenses (L11 and

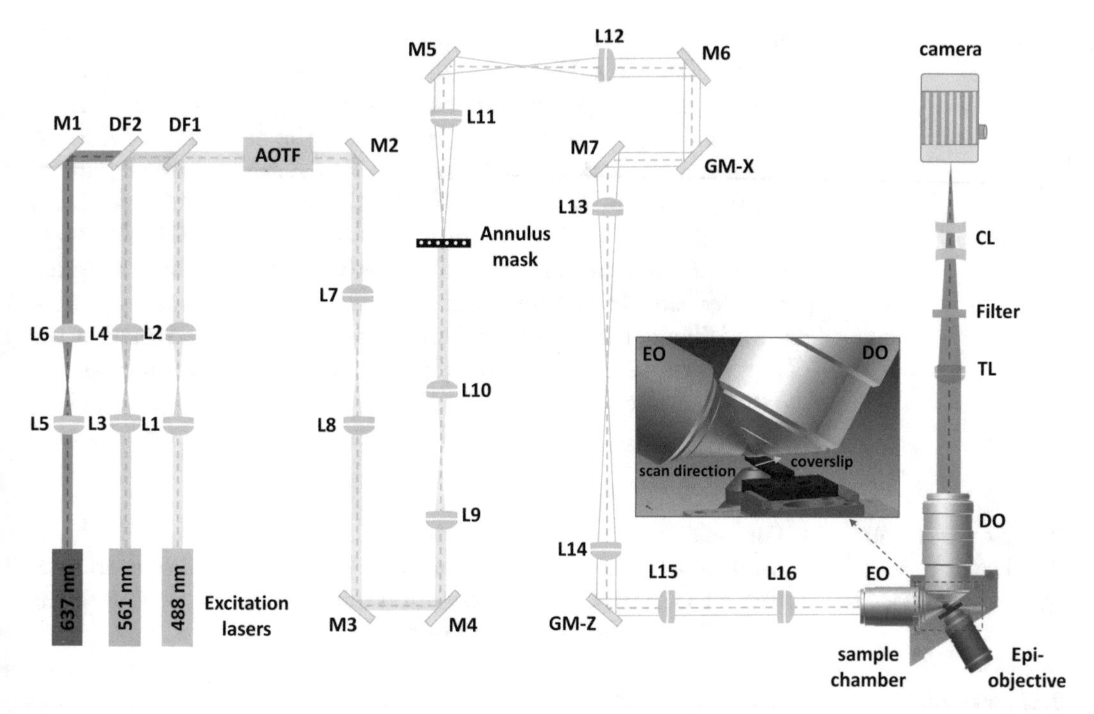

Fig. 2 Schematic diagram of the selective plane illumination microscope based on Bessel beam illumination setup for super-resolution imaging of neuron cells. (Abbreviations: *L* lens, *M* mirror, *DF* dichroic filter, *AOTF* acousto-optical tunable filter, *GM-X* galvanometer mirror in *x*, *GM-Z* galvanometer mirror in *z*, *EO* excitation objective lens, *DO* detection objective lens, *TL* tube lens, *CL* cylindrical lens.) Note: The schematic diagram is not drawn into scale

L12, L13 and L14) in a 4-F arrangement are used for the conjugation of the annular ring pattern beam to a set of galvanometer scanners, namely GM-X and GM-Y (Cambridge Technology, 6215H) for the scanning of *x*- and *y*-directions, respectively. Subsequently, after passing through the galvanometer scanning mirrors, the ring-pattern laser beam is then magnified by a pair of relay lens (L15 and L16) and projected to the rear focal plane of the excitation objective lens (special optics, NA = 0.66, WD = 3.74 mm). As a result, a self-reconstructive Bessel beam is created via the optical interference mechanism. The created Bessel beam confines the laser energy within the illumination plane while conserving an adequately long propagation length.

In the excitation arm setup, an appropriate numerical aperture (NA) of the excitation objective is used for the creation of a Bessel beam with an FWHM that is comparable with the thickness of the specimen to be observed. For neuron observations, the length of the generated light sheet is set to ~20 μm to allow an entire coverage of the cell intersection at an incident angle of 32.5° with respect to the sample coverslip. Hence, the thinness of the generated light sheet allows an effective background rejection and minimum levels of photobleaching. On the other hand, in the detection arm setup, a water-immersion objective (Nikon, CFI Apo LWD 25XW, NA = 1.1, WD = 2 mm) which is orthogonally oriented to the illumination plane and mounted on a piezo scanner (Physik Instrumente, P-726 PIFOC) is used for the collection of fluorescence signals, which is then imaged by a sCMOS camera (Hamamatsu, Orca Flash 4.0 v2 sCMOS) through a tube lens. By inserting a pair of cylindrical lens (Thorlabs, LK1002RM-A, LJ1516RM-A) between the tube lens and camera, a weak astigmatism for sub-diffraction imaging in the axial direction was introduced. Then, the associated distortion of the PSF is recorded by the imaging of fluorescent beads (Thermo Fisher) with a *z*-step size of ~40 nm and calibrated by the fitting with a defocusing model for the derivation of PSF shape dependence with respect to the axial direction.

2.2 Selective Plane Illumination Microscope Setup

The selective plane illumination microscope system is built up according to the standard procedure and protocol discussed elsewhere [39, 40]. Certain procedures are modified according to the availability of equipment and optics in the laboratory.

2.2.1 Materials and Equipment

1. Optical tables:
   - Optical table (Newport RS200).
2. Laser sources for linear excitation:
   - 488 nm laser (Oxxius PS-LBX).
   - 561 nm laser (Coherent, OBIS).
   - 637 nm laser (Coherent, OBIS).
3. Laser intensity modulation devices:
   - Acousto-optical tunable filter (AOTF, AA Opto-Electronic).
4. Laser, objective, and sample-scanning device:
   - Scanning galvanometer mirrors, protected silver (GM-X, GM-Z, Cambridge Technology).
   - Objective scan piezo, 400 μm range (Physik Instrumente, P-725).
   - Objective scan piezo controller (Physik Instrumente, E-665).
5. Objectives:
   - Excitation objective (special optics, NA = 0.66, WD = 3.74 mm).
   - Detection objective (Nikon, CFI Apo LWD 25XW, NA = 1.1, WD = 2 mm).
   - Wide-field objective (Olympus, UMPlanFLN 20X, NA = 0.5).

6. Detection camera:
   - sCMOS camera (Hamamatsu, Orca Flash 4.0 v2, higher speed, larger field of view).
   - Alignment inspection camera, wide-field camera (AVT Guppy F-146 Firewire).
7. Laser shaping devices:
   - Custom aluminum-coated quartz annulus mask.
8. System control devices:
   - Computer (Xeon CPU E5-2637 v3).
   - Signal generation board (National Instruments).
   - Scanning amplifier mainframe (Stanford Research Systems, SIM900).
   - Scanning amplifier (Stanford Research Systems, SIM983).
   - Control software written in Labview 2013 (National Instruments).
9. Filters and dichroic:
   - Lasermux laser combining filters (LM, Semrock).
   - Short-pass dichroic beam splitter (DCI, Semrock).
   - Multiband laser-flat dichroic beam splitter (DC2, Semrock).
   - Band-pass filters (Semrock).
   - Multiband band-pass filters (Semrock).
10. Other optical and sample components:
    - Achromatic lenses.
    - Optical mirrors, protected silver.
    - Sample chamber.
    - Sample stage and stage controller.

2.3 Microscope System Built-Up Procedure

1. Collimate the CW excitation laser beams ($\lambda = 637$, 561, 488 nm) through the beam-expanding components to provide the same beam diameter for all the beams and then combine the collimated CW laser beams into a single coincident beam before the acousto-optical tunable filter (AOTF). By using the 488 nm CW laser beam as reference, align the rest of the microscope yielding to one optical path.
2. Conjugate the galvanometer mirrors (GM-Z and GM-X) on the microscope. Adjust the power of GM-Z and GM-X galvos at midrange control voltages and then rotate the GM-Z on its axis until the light beam from GM-Z is on the center of GM-X and the light beam from GM-X is orthogonal to the annulus mask (AM).
3. Conjugate the GM-Z to GM-X. Place an inspection camera at the plane of GM-X, modulate the GM-Z angle at several Hz,

and simultaneously adjust the two relay lenses until the mirror of GM-Z is focus and clear when observed with the inspection camera.

4. Conjugate the annulus mask (AM) to GM-X. Place the inspection camera at the plane of AM and then repeat the procedure in **item 3** until the mirror of GM-X is focus and clear when observed with the inspection camera.
5. After the conjugation of the GM-Z, GM-X, and AM, place the AM at the second conjugation plane and create a Bessel beam of approximately ~600 nm in diameter and ~30 μm in length. Move the inspection camera at the side of AM opposite the GM-Z and GM-X galvos and then image the illuminated beam annulus onto the camera. Confirm that the illumination through the annulus does not vary when the galvo tilts are modulated by adjusting the relay lenses in the conjugation plane.
6. Conjugate the excitation objective (EO) to the annulus mask (AM). Design a sample chamber which is filled with water and has a small exit beam window opposite to the EO. View the annulus of light that emerges from the window on a white paper and place a mark at the point where the axis of the excitation objective intersects. Direct the annular beam at the center of the rear pupil of the EO by translating the *x* and *y* right-angle prism mirrors. Finally, adjust the axial position of the EO until the image of the annulus is focused on the white paper.
7. For the alignment of the detection objective (DO), fill the sample chamber with Alexa Fluor 488 hydrazide solution and move the AM laterally to the position wherein a Gaussian beam of at least 0.3 is created in the chamber. Set a maximum intensity output for the 488 nm laser, and midrange position for the GM-Z, GM-X, and detection objective piezo, and remove the detection tube lens (TL1). Place a white paper at approximately ~50 cm from the rear pupil of the DO and mark the point of intersection. Adjust the position of the DO axially until the beam of the fluorescence signal is collimated and laterally until it is in the central position with respect to the reference point.
8. For the alignment of the tube lens (TL1), mount the TL1 on the microscope and move the AM laterally which creates a Bessel beam with a diameter of ~600 nm and length of ~30 μm in the chamber. Adjust the axial position of the TL1 until the image of the Bessel beam is focused on the detection camera and adjust its lateral position until the image of the Bessel beam is at the center of the camera.
9. Precisely position the annular beam at the center of the rear pupil of EO. Confirm the symmetry of the two axial locations of the Bessel beam at the center of EO as observed from the detection camera. Adjust the *z*-axis position of the annular

beam entering the EO via the right-angle prism positioned before the EO until a symmetry of beam widths at both sides is observed. Adjust the *x*-axis position of the annular beam until the image of the Bessel beam is parallel to the *x*-axis on the detection camera.

10. Fine-tune the conjugation of EO, by placing the detection objective piezo on the top of the desired *z*-scan range. Adjust the control voltage of the GM-Z until the center of the Bessel beam is focused in the new position and repeat the process until the piezo moved to the bottom position of the desired *z*-scan range. Adjust the axial positions of the relay lenses between the AM and EO until the beam is focused along the entire length (from top to bottom extremities) of the *z*-scan range.
11. Calibrate the GM-Z and GM-X galvos. Place the detection objective piezo at 10 μm steps from −40 to 40 μm. In every 10 μm step position, adjust the control voltage of GM-Z until the Bessel beam is focused and then apply linear regression to GM-Z voltage as a function of the objective position for the determination of calibration constant *z* beam displacement per voltage applied. The correlation coefficient should be at least 0.999 for the Bessel beam to remain at the focal plane of DO across the entire *z* field of view. Next, regulate the control voltage to GM-X in nearly ten equivalent steps of amplitude such that the beam moves across the entire *x* field of view as observed on the detection camera. Using the camera images as reference, determine the beam position for each control voltage and apply linear regression to define the calibration constant of *x* beam displacement per voltage applied.
12. Confirm the conjugation and centration of the microscope. Verify that the light sheet created by Bessel beam remains in focus across the entire *xy* field of view for all the *z* planes. Repeat the **items 2–6** and **9–10** as needed to accomplish these conditions. When it is confirmed in focus, it completes the alignment of the linear excitation beam.
13. Mount the inspection camera on the microscope. Adjust the camera position to focus the annular light beam and center in the field of view (conjugated to AM and EO). Record and save the annular ring of light observed in the camera. If subsequent alignment of the system is necessary, adjust the position, focus, and conjugation of the excitation path as described in **items 2–6**.
14. Fine-tune and measure the wide-field detection PSF. Replace the Alexa Fluor 488 solution with water from the imaging chamber until no fluorescence is observed on the detection camera. Place the fluorescent bead sample on the specimen chamber, translate it to the center of the imaging field, position

a single fluorescent bead within the Bessel beam, and then measure the three-dimensional detection PSF. This could be carried out by fixing the GM-X galvo while moving the GM-Z galvo and detection objective piezo together to acquire a 3D image stack. Confirm if there is any asymmetry in the axial PSF by checking the *xz* view of the 3D stack. To achieve an axial symmetry, adjust the correction collar of DO and if the range of the collar is insufficient, fix the collar on its midrange position and then slightly iterate the axial positions of DO and TL1 until axial symmetry is attained. Confirm if there is asymmetry in the PSF by checking the *xz* and *yz* views of the 3D stack. To achieve lateral symmetry, iteratively adjust the lateral positions of DO and TL1 while maintaining the image of the bead at the same position in camera.

15. Characterize the cross-sectional intensity of the Bessel beam. Place a single fluorescent bead in focus at the common center of the imaging field and Bessel beam using the sample translational stages. Measure the cross section of the Bessel beam via the program utility system. Using the GM-Z and GM-X galvos, scan the excitation beam in 100 nm steps across the fluorescent bead with a 5 μm range in the *xz* plane while recording an image at each position. The measured fluorescence from the bead corresponds to the beam intensity at the *xz* position and thus the entire set of measurements represents the cross-sectional intensity of the Bessel beam. This image must be closely similar to an ideal Bessel beam with concentric and symmetric side lobes. If it is not symmetric, confirm if the intensity is uniform across the annulus through the rear pupil of the inspection camera and verify if the position remains constant as the GM-Z and GM-X galvos are scanned. Repeat the fine-tuning on the conjugation of all the elements and centration of the annular beam as described in **items 2–6** until a symmetric Bessel beam is achieved.

16. Characterize the Bessel beam swept-sheet PSF. Place the fluorescent beads within the volume to be imaged and then take a 3D image stack in the swept-sheet mode across an image volume of ~50 × 50 × 50 μm. The 3D image of each fluorescent bead resembles the Bessel sheet-scan PSF at the bead position. All the acquired PSFs must be identical and symmetric in the axial direction across the entire imaging field. Adjust the initial offset of the detection objective piezo to achieve symmetry if the PSFs are uniformly asymmetric in the same axial direction. Return to **items 10–12** if the PSFs are not uniform or are asymmetric in different directions. Record the image of an isolated bead taken in the swept-sheet mode to serve as the reference 3D PSF in deconvolution of image data.

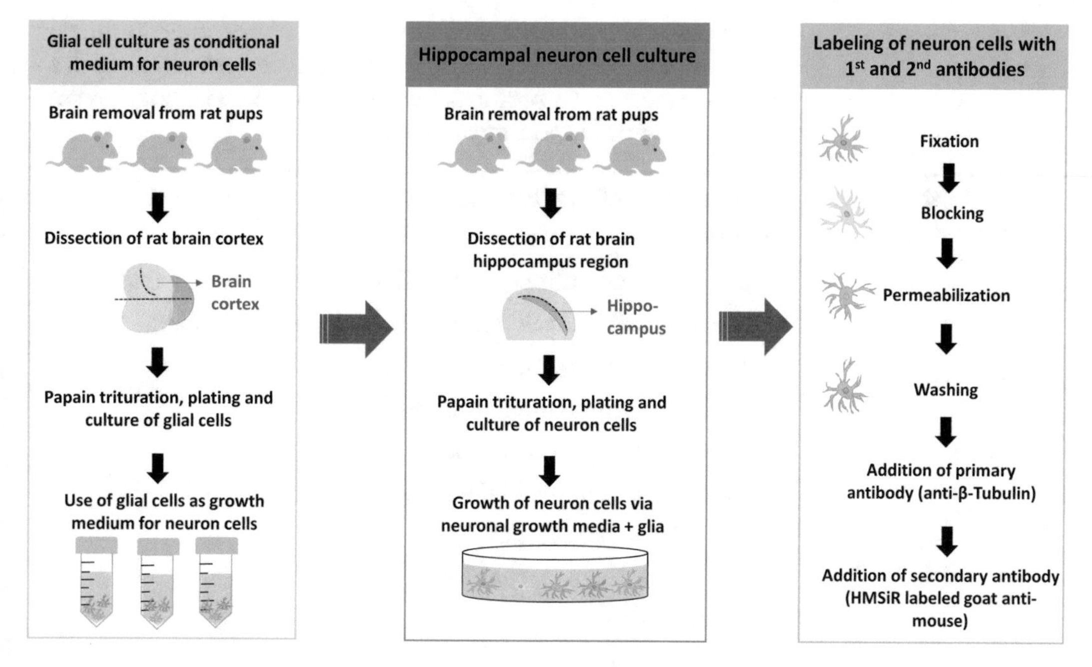

Fig. 3 Flow chart of experimental procedures outlining the glial cell culture as conditional medium for neurons, hippocampal neuron culture, and labeling of neurons with primary and secondary antibodies

2.4 Hippocampal Neuron Cell Culture

A detailed protocol and step-by-step procedure of the hippocampal neuron culture experiments are discussed in this section [41, 42]. Figure 3 illustrates a flow chart on the brief summary of experimental procedures outlining the glial cell culture as conditional medium for neurons, hippocampal neuron culture, and labeling of neurons with primary and secondary antibodies. In the glial and neuronal cultures, the glia and neurons are cultured via the papain-medium dissociation method. The papain-medium dissociation method has been chosen because it was found that papain is the best among the six tested proteases to isolate and dissect neurons from brain tissue. Moreover, it has been proven to be more effective and less destructive for survival of isolated cells after 4 days of culture [43]. The cultures are obtained from the brain of postnatal rat pups (P0–P1, postnatal days 0–1). The glial cell cultures are acquired from the brain cortex while the neuronal cultures are obtained from the hippocampus region of the rat pup brain. The confluent glial cells are induced into differentiation for the collection of secreted growth factors, the conditional medium, required for the hippocampal neuron growth medium and maturation. Subsequently, the hippocampal neuron immunofluorescence staining is performed after the synaptogenesis, which is on the 15th-day culture in vitro (15 *DIV*). To specifically immuno-label the microtubule of neurons, the β-tubulin-specific monoclonal primary antibody is used for the staining. Furthermore, the HMSiR-conjugated goat anti-mouse is used as secondary antibody to recognize the primary antibody-labeled microtubule in the neurons (*see* Subheading 2.3). The HMSiR fluorescent blinking dye is selected in order to perform the three-dimensional super-resolution imaging of hippocampal neurons.

2.4.1 Materials and Equipment

A standard cell culture laboratory equipment, including the following:

- Laminar flow cabinet, CO_2 cell incubator, inverted microscope.
- Hemocytometer cell counter, 0.4% trypan blue stain.
- 30 and 50 mL syringe.
- 0.22 μm Syringe filter (Millipore).
- 15 and 50 mL Falcon tubes.
- 5 and 10 mL individual wrapped pipettes.
- 2 × 10 cm Petri dishes.
- T75 flasks (canted, cent cap).
- 24-Well plate.
- 5 mm Coverslips.
- Centrifuge system.

2.4.2 Dissection Instruments

The instruments have to be cleaned and sterilized on a regular basis to preserve its functionality and avoid contamination.

- Large and small sharp dissecting scissors.
- Gritted curved tweezer.
- Curved-sharp tweezer.
- Spatula.
- Sharpened spatulas or scalpels.
- Whatman no. 5 filter paper (Sigma-Aldrich, WHA1005090).

2.5 Glial Cell Culture for the Collection of Conditional Medium for Neuron Cells

2.5.1 Reagents

- Minimum essential medium (MEM; Thermo Fisher Scientific, 11090-081) and GlutaMax (Thermo Fisher Scientific, 35050061).
- Heat-inactivated fetal bovine serum (FBS, Thermo Fisher Scientific).
- Papain solution (Sigma-Aldrich, P3125-100 mg).
- L-Cysteine hydrochloride (L-cysteine HCl, Sigma-Aldrich, C-1276-10G).
- Sodium hydroxide (1 N NaOH, Sigma-Aldrich, S2770).
- Neuronal growth medium (NGM, *see* below for details).
- Hanks' balanced salt solution (HBSS, Thermo Fisher Scientific, 141-70).

2.6 Preparation of Glial Medium (MGS: MEM, GlutaMax, and Serum)

2.6.1 Reagent Preparation

- 50 mL, Heat-inactivated fetal bovine serum (FBS, Thermo Fisher Scientific).
- 50 mL, GlutaMax at 20 °C (100×, Thermo Fisher Scientific, 35050061).
- Minimum essential medium (MEM; Thermo Fisher Scientific, 11090-081).

2.6.2 Procedures

1. In a 250 mL filter unit system, place a low-protein-binding membrane filter with a pore size of 0.22 μm (SteriCup, Millipore).
2. Separate from the filter vacuum line, add 50 mL of FBS and 5 mL of GlutaMax into the top compartment, and fill the filter unit with the MEM up to the 500 mL scale mark.
3. Filter and sterilize by vacuum and store the MGS at 4 °C before use.

2.7 Preparation of Neuronal Growth Medium (NGM)

2.7.1 Reagent Preparation

- 5 mL, B27 at 20 °C (50×, Thermo Fisher Scientific, 17504044).
- 2.5 mL GlutaMax at 20 °C (100×, Thermo Fisher Scientific, 35050061).
- 250 mL, Neurobasal A medium (Thermo Fisher Scientific, 10888022).

2.7.2 Procedures

1. In a 250 mL filter unit system, place a low-protein-binding membrane filter with a pore size of 0.22 μm (SteriCup, Millipore).
2. Separate from the filter vacuum line, add 5 mL of B27 and 2.5 mL of GlutaMax into the top compartment, and fill the filter unit with the Neurobasal A medium up to the 250 mL scale mark.
3. Attach the vacuum line and filter the solution from the top compartment to the lower compartment.
4. Unscrew the top compartment and discard, and then close the lower compartment with a cap.
5. Finally, cover the obtained neuronal growth media (NGM) with an aluminum foil since it is light sensitive.
6. Store the medium at 4 °C before use.

2.8 Dissection Procedure of Rat Brain Cortex for Glial Culture

This subsection aims to provide a detailed protocol on the dissection procedure of the rat pup brain cortex for the glial culture and growth. The confluent glial cells will be applied for the further collection of conditional glial medium. All the experimental procedures will be discussed next.

2.9 Preparation of the Fresh Solutions Prior to Dissection

2.9.1 Cysteine-Activated Papain Solution

1. Transfer 12.8 mg of cysteine into a 50 mL Falcon tube (for up to five rat pups). Cysteine is incorporated in the dissection process to enhance papain activity and reduce the time of incubation [44].
2. Add 40 mL of HBSS to the 12. 8 mg of cysteine. (The final concentration of cysteine is 0.32 mg/mL.) The color of the solution will turn to yellowish due to the intrinsic acidic nature of cysteine.
3. Gently swirl and mix the tube while adding few drops of 1 N NaOH until the color turned to orange or pink (~pH = 7.4).
4. Add 250 μL of papain solution (250 μL of papain solution for five pairs of cortex from five rat pups; the volume of the papain is dependent on the enzyme unit from batch to batch). The color will change into a milky-white solution. Continue the gentle swirling of the solution and then pipette it up and down to mix.
5. Place the tube-contained solution on ice prior to the dissection procedure.
6. During the dissection, warm up the solution in the 37 °C water bath.
7. After dissection, check the clearance and transparency of the solution and filter it through a 0.22 μm syringe filter disc.
8. Return the tube-contained solution back to the 37 °C water bath.

2.9.2 Hanks' Balanced Salt Solution (HBSS)

1. Divide the ice-cold HBSS into several aliquots in Falcon tubes and keep on ice.
2. Distribute the HBSS according to the following:
   - 30 mL of HBSS into a 50 mL Falcon tube for the dissection.
   - 10 mL of HBSS into a 50 mL Falcon tube for temporary placement of cortex.
   - 5 mL of FBS and 45 mL of HBSS into a 50 mL Falcon tube.
   - 50 mL of HBSS into a 50 mL Falcon tube for FBS rinsing.

2.9.3 Rat Pups (P0–P1)

- Pick up the rat pups for dissection. Remember to cover the rat pups with a piece of cloth or paper towel to protect them from the heat loss being away from their mother.

2.10 Dissection Procedure

All the rat experiments are carried out according to the standard experimental protocol and with the permission of Academia Sinica Institutional Animal Care and Utilization Committee.

1. Prepare the sterile dissection instruments in the laminar flow cabinet.

2. Put the prepared solutions and reagents on the ice bucket inside the laminar flow cabinet.
3. In a 10 cm Petri dish, use the main part for the tool holding and the lid part for the dissection. Place a Whatman no. 50 filter paper on the lid part of Petri dish and put a frozen metal block under it to keep the dissection procedure on ice-cold condition.
4. Add 14 mL of the ice-cold HBSS into the lid part of a Petri dish containing a Whatman no. 5 filter paper. Keep the temperature by placing a frozen ice pack or metal block under the dish.
5. Dissect the rat pup. First, decapitate the rat pup with a large and sharp dissecting scissor (*see* Fig. 4b) [45]. Second, spray a 70% ethanol into the head and slice the skin to reveal the skull. Third, cut the skull in an upward movement to prevent unwanted cuts in the brain cortex. Fourth, open the skull using the tweezers and remove the rat pup brain with a spatula. Fifth, place the obtained rat pup brain in the dissection dish. Finally, wipe out the excess blood in your gloves and spray a 70% ethanol.
6. For the brain cortex dissection: First, split the rat pup brain into two hemispheres with a sharpened spatula or scalpel (*see* Figs. 4a and 5a). Second, use two spatulas to remove the midbrain or telencephalon area which is located at the right side of the hippocampi (*see* Fig. 5b). Third, take out the meninges from the two hemispheres using curved fine tweezers (*see* Figs. 4b and 5c). Fourth, place the obtained cortex tissues

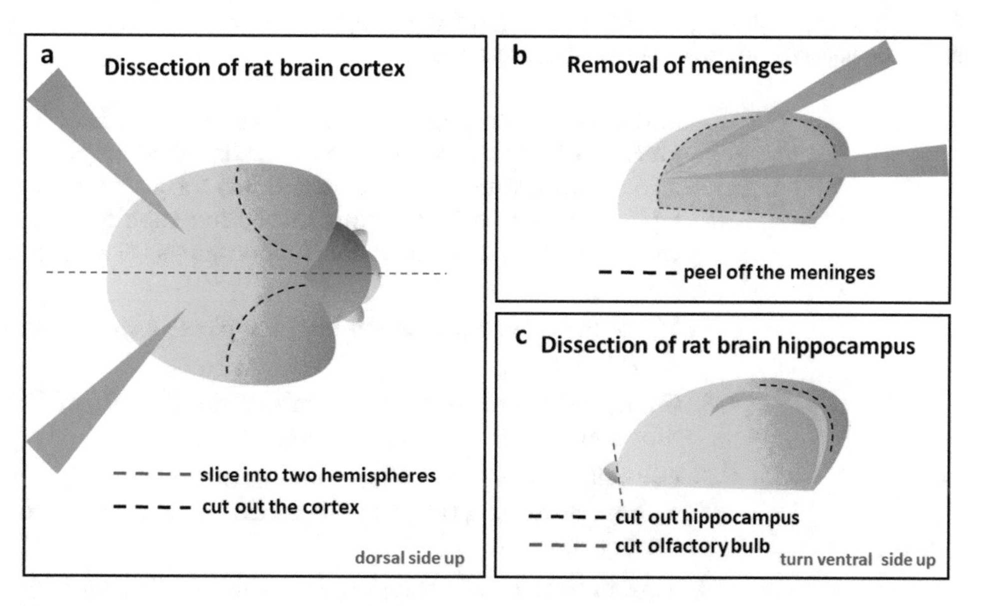

Fig. 4 Schematic diagram of rat cortex and hippocampus dissection: (**a**) dissection of cortex, (**b**) removal of meninges, and (**c**) dissection of hippocampus

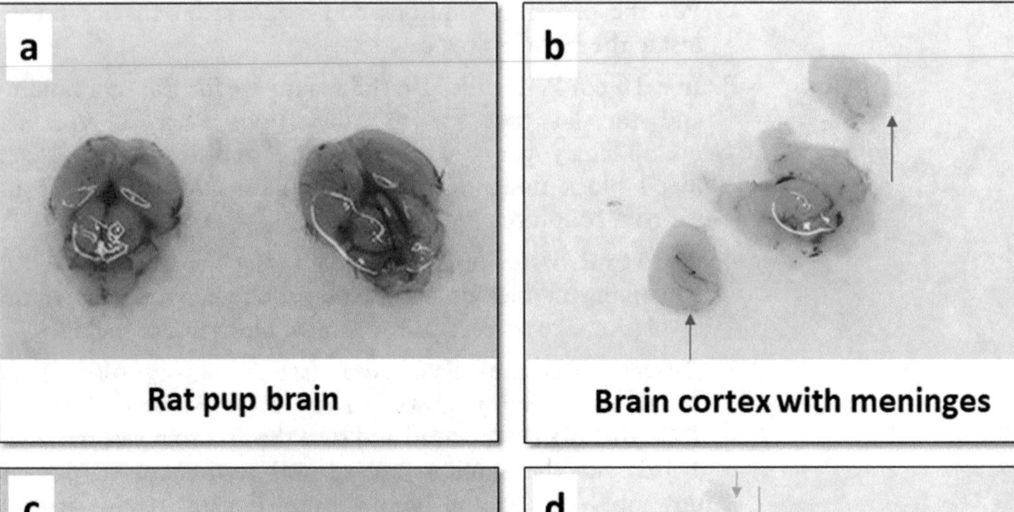

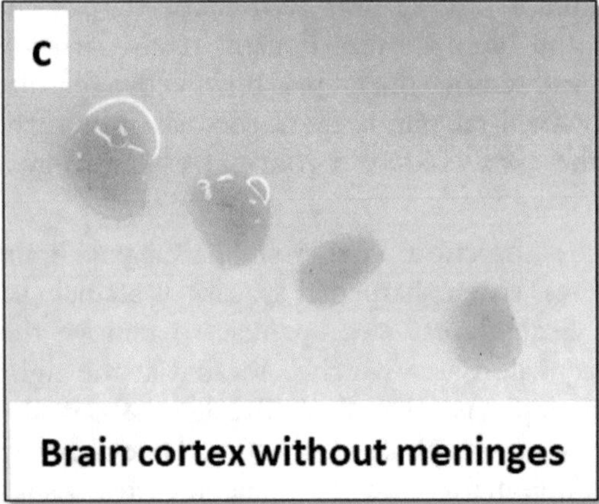

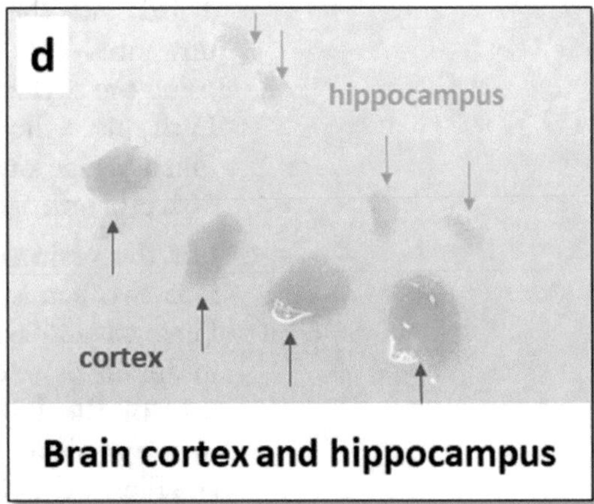

Fig. 5 Representative digital images of the dissected (**a**) rat pup brain, (**b**) brain cortex with meninges, (**c**) brain cortex without meninges, (**d**) brain cortex and hippocampus

into a new dissecting Petri dish lid with a 5 mL of HBSS. Finally, collect the cortex from all the dissected rat pup brains into a Falcon tube (*see* Figs. 4c and 5d) for glial culture. The hippocampus is collected into the Falcon tube for neuronal culture (*see* next section for hippocampus dissection and culture).

7. Mince the brain hemispheres into fine pieces using a sterilized razor blade.
8. Gently pour the solution-filled brain tissue into a 50 mL Falcon tube and place it on ice immediately.
9. Carefully remove the HBSS solution out of the Falcon tube without disturbing the collected minced brain cortex tissues at the bottom part.
10. Take out the papain solution from the 37 °C water bath and spray a 70% ethanol. Then, add 20 mL of the papain solution into the mince brain cortex tissues in the 50 mL Falcon tube.

11. Place and incubate the 50 mL Falcon tube containing papain solution and brain cortex tissues into a 37 °C water bath for 15 min.
12. After 15 min, take out the 50 mL Falcon tube containing solution from the 37 °C water bath and replace the papain solution with a new one and continue the incubation process for another 15 min at 37 °C water bath.
13. Remove the papain solution and quickly add a 20 mL of ice-cold HBSS/FBS (10% FBS in HBSS) solution, inverting the tube twice to mix. Wait until the brain cortex tissues settle down at the bottom part of the tube. Rinse with ice-cold HBSS/FBS again.
14. Take out the HBSS/FBS solution and quickly add a 20 mL of ice-cold HBSS solution, inverting the tube twice to mix. Wait until the brain cortex tissues settle down at the bottom part of the tube. Rinse again.
15. Carefully remove the HBSS solution out of the 50 mL Falcon tube containing brain cortex tissues.
16. Add a 10 mL of ice-cold MGS (*see* the above details for MGS preparation) into the 50 mL Falcon tube. Triturate the brain cortex tissues using a 10 mL pipette, resuspending in an upward and downward motions for up 30 times until the brain cortex tissues disappeared.
17. Place the 50 mL Falcon tube in an ice bucket for 3 min.
18. Carefully transfer 8–9 mL of the cell suspension into a 15 mL Falcon tube.
19. Centrifuge the cells at 67 × *g* (6000 rpm) for 5 min at 4 °C.
20. Take out the supernatant and resuspend the cells with a 10 mL of ice-cold MGS.
21. Count the number of cells: Mix 20 μL of cell suspension with 20 μL of trypan blue, measure 10 μL of the cell suspension in the hemocytometer, calculate the number of cells per mL, and distribute approximately 5×10^7 cells into the T75 flasks.
22. After plating the cells, add 17 mL of MGS for each T75 flask and incubate it overnight at 37 °C incubator.
23. The next day, check the cell attachment at the bottom part of the T75 flask and observe the cultured cell cells with an inverted microscope. To remove the debris, replace the MGS with new MGS. Replace the MGS by aspirating the old one and exchanging with the new one.
24. Check the cultured cell and replace the medium every 3 days until it reached the optimum confluency. It takes approximately 2 weeks for glial cells to reach confluency.

25. To collect the glial conditional medium for neuronal culture, replace the MGS with NGM to induce the glial differentiation when the cells reached the 95–100% confluency and then incubate the cells for 4 days.
26. Take out and collect the NGM with glial secreted growth factors and transfer into a 50 mL Falcon tube. Store the glial conditional medium at −80 °C for future neuronal cell culture.
27. Continuously harvest the glial conditional medium every 4 days until the cells start to develop the star protrusion (stage wherein glial cells begin to differentiate into astrocyte).

2.11 Neuronal Culture

2.11.1 Reagents

- Poly-D-lysine (BD Sciences, 354210, 20 mg/ea or Sigma-Aldrich, P1024, stored at 4 °C).
- Tissue culture-grade water (Sigma, W3500-1L).
- Neuronal growth medium glial (NGMG, *see* below for details).
- Glial conditional medium (preparation procedure described in Subheading 2.2.1).

All other reagents are included and described in glial cell culture (*see* Subheading 2.2.1).

2.11.2 Preparation of Neuronal Growth Medium Glial (NGMG)

1. Mix 200 mL of neuronal growth medium (NGM) with 50 mL of glial conditional medium to prepare NGMG.
2. Filter the obtained 250 mL of NGMG (with 20% glial conditional medium) by a 0.22 μm (SteriCup, Millipore) and store at 4 °C before use.

2.11.3 Dissection Procedure of the Rat Postnatal Hippocampal Neuronal Culture

This subsection aims to discuss the step-by-step dissection procedure of the rat pup brain hippocampus for the associated neuronal culture. For the neuronal culture and growth, the previously obtained glial conditional medium is mixed with neuronal growth medium (NGM), yielding to the neuronal growth media glial (NGMG). The cultured neurons will be used for the neuronal staining of primary and secondary antibodies labeled with HMSiR blinking dye for three-dimensional super-resolution imaging experiments.

2.12 Preparation of the Solution Prior to Dissection

2.12.1 Poly-D-Lysine-Coated Cover Glass

1. The coating of coverslip should be carried out a day before the rat pup sacrifice.
   - Prepare 0.5 M of borate buffer by dissolving 1.55 g of boric acid (Sigma-Aldrich, 185094) and 2.375 g of borax anhydrous (Sigma-Aldrich, 71997) in water, adjust the pH value to pH = 8.5, and then add water to a volume of 100 mL. Sterilize the 0.5 M borate buffer by autoclave and store at 4 °C before use.

- Prepare 10 mg/mL of poly-D-lysine stock solution by dissolving 0.05 g of poly-D-lysine (Sigma-Aldrich, P1024) in 5 mL of 0.5 M borate buffer. Aliquot into the Eppendorf and store at −20 °C. Avoid the repeated freezing and thawing.
- Prepare the coating poly-D-lysine working solution by mixing 7.6 mL of 0.5 M borate buffer (final concentration is 0.005 M) and 0.4 mg/mL of poly-D-lysine (final concentration is 0.1 mg/mL). Filter and sterilize the solution using 0.22 μm syringe filter.
- Place a 5 mm diameter coverslip into each well of the 24-well plate. Add 0.4 mL of poly-D-lysine working solution into each well, coating the coverslip overnight at 37 °C incubator.
- After incubation, remove the coating solution and rinse each well with distilled water twice.
- Add 0.4 mL of ice-cold NGM into each well and return back to the incubator at 37 °C until the step on neuron seeding.

2.12.2 Neuronal Growth Medium (NGM)

1. *See* the above discussion on the preparation of neuronal growth medium in Subheading 2.2.1.

2.12.3 Neuronal Growth Medium Glial (NGMG)

1. Prepare 200 mL of NGM, and then add 50 mL of glial conditional medium.
2. Store the medium on ice under the dark before use.

2.12.4 Cysteine-Activated Papain Solution

1. Transfer 3.2 mg of cysteine into a 15 mL Falcon tube (for up to five rat pups).
2. Add 10 mL of HBSS to 3.2 mg of cysteine (final concentration of cysteine is 0.32 mg/mL). *See* the above discussion for details on the preparation of cysteine-activated papain solution in Subheading 2.2.1.

2.12.5 Hanks' Balanced Salt Solution (HBSS)

1. *See* the above discussion for the details on the preparation of HBSS in Subheading 2.2.1.
   - Prepare 5 mL of HBSS into a 15 mL Falcon tube for the temporary placement of hippocampus.

2.12.6 Rat Pups (P0–P1)

1. *See* the above discussion for the details on rat pup preparation prior to dissection in Subheading 2.2.1.

2.12.7 Dissection Procedure

1. Follow exactly the same procedure in **items 1–6** in the dissection procedure for glial culture (Subheading 2.2.1).
2. For the hippocampus dissection: First, locate the C-shaped hippocampus along the cortical fold (*see* Fig. 4c). Second, sever the tissue at the junction of the hippocampus and cortex

using a spatula. Third, separate the cortex out by sliding the spatula into the deeper layer of the cortex and subsequently flip the cortex and hippocampal tissue onto the reversed side. Fourth, locate the C-shaped bump (back side of the hippocampus) visually and use the spatula to cut along the boundary (*see* Fig. 5d). Finally, transfer the C-shaped hippocampus into the temporary-holding Falcon tube (contains 5 mL HBSS in a 15 mL Falcon tube) immediately and place on ice until the dissection is finished (*see* Figs. 4 and 5).

3. After the dissection of all the hippocampi and placement on the temporary-holding Falcon tube, remove the HBSS out of the Falcon tube using a 5 mL pipette.
4. Take the papain solution out of the 37 °C water bath, spray with 70% ethanol, and filter it through the 0.22 μm syringe filter. Then, add 10 mL of papain solution into the hippocampi-containing tube and place it into the 37 °C water bath for 15 min.
5. After 15 min, take the papain and hippocampi tubes out of the water bath, replace the old papain solution by adding new 10 mL of papain solution into the hippocampi tubes, and then incubate again in 37 °C water bath for 15 min.
6. Remove the papain solution and quickly add 10 mL of ice-cold HBSS/FBS solution. Invert the tube up and down two times to mix.
7. When the hippocampi tissues are settled down at the bottom of the Falcon tube, repeat the rinsing with a new 10 mL of ice-cold HBSS/FBS solution.
8. Carefully remove most of the HBSS/FBS solution after the second rinsing procedure.
9. Add 5 mL of ice-cold neuronal growth medium (NGM) into the Falcon tube and triturate the hippocampi tissues using a 5 mL pipette, resuspending in upward and downward movement along the wall side of tube to prevent bubble formation. Perform the trituration 30 times until all the tissues disappeared and then put the tube on ice.
10. Carefully transfer 4.5 mL of the cell suspension from the top of the solution into a new Falcon tube.
11. Centrifuge the cells at $67 \times g$ (6000 rpm) for 5 min at 4 °C.
12. Take out the supernatant and resuspend the cells with a 5 mL of ice-cold NGM.
13. Count the number of cells: First, mix a 20 μL of cell suspension with a 20 μL of trypan blue. Second, perform cell counting using 10 μL of the solution into the hemocytometer. Third, calculate the number of cells per mL in every 5 mL suspensions.

14. Dilute the cells into the total volume of ice-cold NGM and plate into the dishes as described: use 0.4 mg/well on a 24-well plate (one pair of hippocampus could be seeded onto one 24-well plate, with a neuron density of 1×10^5 per well and final volume of 0.8 mL/well).
15. Place the 24-well plate in the 37 °C incubator and incubate the cells overnight.
16. Remove the NGM, and add 0.8 mL of room-temperature NGMG into each well. Culture and incubate the cells at 37 °C.
17. Every 4 days, replace the old NGMG with a new 0.4 mL of room-temperature NGMG. Culture and incubate the cells at 37 °C until the neuron maturation for 15 days. NGMG contains 80% of the fresh neuronal growth medium (NGM) and 20% of glial conditional medium (see Subheading 2.2.1).

2.13 Labeling of Neurons with Primary and Secondary Antibodies

2.13.1 Reagents

Fixation Solution

- 4% Paraformaldehyde (diluted from 16% paraformaldehyde aqueous solution, Electron Microscopy Sciences, 15710).
- 4% Sucrose (Sigma-Aldrich, S0389).
- 1× Phosphate-buffered saline (PBS; Sigma-Aldrich, P3813, freshly prepared and kept on ice under the dark; store for a maximum of 2 weeks).

Blocking Solution

- 2% Bovine serum albumin (BSA; Sigma-Aldrich A1933).
- 4% Normal goat serum (Thermo Fisher Scientific, 16210064).
- 1× Phosphate-buffered saline (PBS; Sigma-Aldrich, P3813, prepare in advance, then filter and sterilize by 0.22 μm syringe filter, store at 4 °C, warm up to room temperature before the blocking procedure).

Permeabilization Solution

- 0.1% Triton (Sigma-Aldrich, T8787) in blocking solution (freshly prepared, stored at room temperature).

Washing Solution

- 1× Phosphate-buffered saline (Sigma-Aldrich, P3813) or
- 1× PBS-MC (*see* detailed components next, prepare in advance, store one bottle at 4 °C and another at room temperature).

Imaging Solution

- 1× Phosphate-buffered saline (PBS; Sigma-Aldrich, P3813).

Antibodies

- Primary antibodies suitable for tubulin staining: either monoclonal anti-α-tubulin antibody produced in mouse (Sigma-Aldrich, T9026, clone DM1A) or monoclonal anti-β-tubulin antibody produced in mouse (Sigma-Aldrich, T8328, clone AA2).
- Secondary antibody: HMSiR-labeled goat anti-mouse whole IgG (Tebu-Bio, A202-01).

2.13.2 Preparation of the Solutions Prior to Neuronal Cell Staining

Preparation of the Fixation Solution

1. Paraformaldehyde (4%, 10 mL in 40 mL).
2. Sucrose (4%, 1.6 g in 40 mL).
3. PBS (1×, 4 mL in 40 mL):
   - Mix the reagents and dissolve all the components in a 50 mL Falcon tube.
   - Add deionized water for up to 40 mL and store at 4 °C.
4. Usage: 0.6 mL in a well of the 24-well plate.

Preparation of the Blocking Solution

1. BSA (2%, 2 g in 100 mL).
2. NGS (4%, 4 mL in 100 mL).
3. PBS (1×, 10 mL in 100 mL).
   - Dissolve all the reagents and components in a 100 mL tissue-grade water using a 150 mL beaker and a small stirrer bar.
   - Stir the solution until the BSA is completely dissolved in water. Filter and sterilize by 0.22 μm syringe filter. Store at 4 °C before use.
4. Usage: 0.6 mL in a well of the 24-well plate.

Preparation of the Permeabilization Solution

1. Blocking solution (10 mL).
2. Triton X-100 (final concentration is 0.1%, 1.0 μL).
   - Add 50 μL of 20% Triton X-100 into the 10 mL of blocking solution. The solution should be freshly prepared and stored at room temperature.
3. Usage: 0.6 mL in a well of the 24-well plate.

Preparation of the Washing Solution (with PBS-MC)

- PBS (1×).
- $MgCl_2$ (1 mM, add 1 mL of 1 M of $MgCl_2$ stock solution into 1 L of 1× PBS).

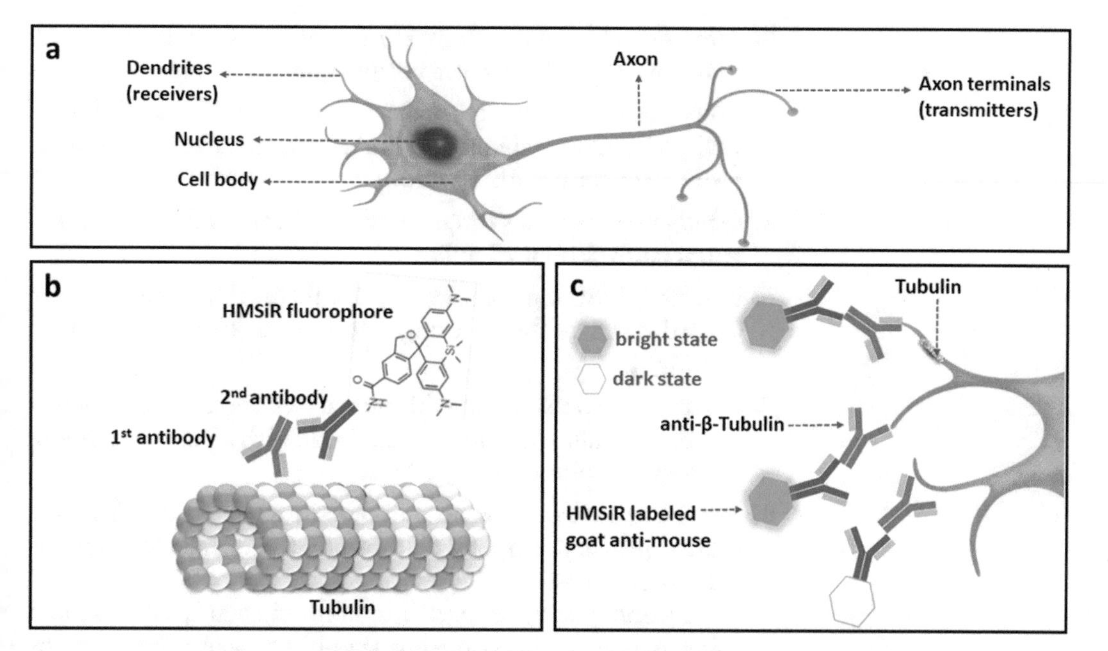

Fig. 6 Labeling of neurons with primary and secondary antibodies: anti-β-tubulin- and HMSiR-labeled goat anti-mouse, respectively. (**a**) Representative schematic diagram of a neuron with labeled parts, (**b**) neuronal staining via the immunochemistry methods, and (**c**) illustration of a neuron with labeled first antibody, second antibody, and HMSiR fluorescent blinking dye

- $CaCl_2$ (0.1 mM, add 1 mL of 0.1 M of $CaCl_2$ stock solution into 1 L of 1× PBS).
- Prepare 1 mM of $MgCl_2$ and 0.1 mM of $CaCl_2$ in 1× PBS in advance, and store one bottle at 4 °C and another at room temperature.
- Usage: 1 mL in a well of the 24-well plate.

2.14 Labeling of Neurons with Primary and Secondary Antibodies

A detailed protocol on the neuronal staining will be discussed in this section. The primary and secondary antibodies used in the neuronal staining are the mouse anti-β-tubulin- and HMSiR-labeled goat anti-mouse, respectively (*see* Fig. 6). The abovementioned studies are considered crucial for the effective labeling of tubulin found in neurons for three-dimensional super-resolution imaging.

2.14.1 Neuronal Staining Procedure

1. In the laminar flow cabinet, place the cultured 15 DIV (days in vitro) neuron plates on a metal tray and quickly rinse it three times with room-temperature PBS-MC.
2. For the fixation procedure, remove the PBS-MC and add 0.6 mL/well of the room-temperature fixation solution. Then, incubate the cells at room temperature for 20 min. (For the staining for other cytoskeleton or molecules, the PBS-MC and fixation solution should be ice-cold and the incubation should be on ice for 20 min.)

3. After the stipulated time, wash the neurons three times with 1 mL/well of PBS at room temperature for 5 min.
4. To permeabilize the neurons, take out the PBS, add 0.5 mL/well of the permeabilization buffer, and incubate the cells at room temperature for 15 min.
5. Gently wash the cells three times with 1 mL/well of the room-temperature PBS for 5 min.
6. For the blocking procedure, incubate the neurons with 0.5 mL/well of the blocking solution for 20 min or at 4 °C overnight.
7. For the neuron staining with primary antibody (mouse anti--β-tubulin), dilute the primary antibody with a minimal amount of blocking solution (1:1000 ratio).
8. For the neuron staining on the coverslips: First, drop a 10 μL of diluted primary antibody on a parafilm, creating different circular spots with a fixed distance at each spot.

   Second, place the coverslips on the circular spots and flip it using tweezers (the coverslips should be facing the antibody solution). Incubate the coverslips at room temperature within a humidified chamber for 2 h or overnight at 4 °C.
9. After incubation, pick up the coverslips with neuron-stained antibody and transfer into a 24-well plate filled with 1 mL/well of room-temperature PBS. Repeat the washing twice.
10. For the neuron staining with secondary antibody (HMSiR-labeled goat anti-mouse), dilute the primary antibody with a minimal amount of blocking solution (1:1000 ratio).
11. Repeat **item 8** and return it back to the humid chamber for incubation. Incubate the neuron-stained secondary antibody at room temperature for 2 h or overnight at 4 °C in the dark (place the tray on a box and cover it with aluminum foil).
12. Repeat **item 9** for the neuron-washing procedure.
13. Store the coverslips with neuron-stained primary and secondary antibodies in a 24-well plate with 1 mL/well of PBS until imaging experiments.

2.15 Image Acquisition

This subsection provides a detailed procedure on neuron sample mounting and image acquisition procedure. The 5 mm coverslips with neurons stained with primary and secondary antibodies are cultured as described above. A three-dimensional super-resolution imaging of microtubules is successfully achieved through selective plane illumination technique.

1. Mount and fix 5 mm coverslip containing neurons with stained primary and secondary antibodies on the sample chamber of the selective plane illumination microscope.

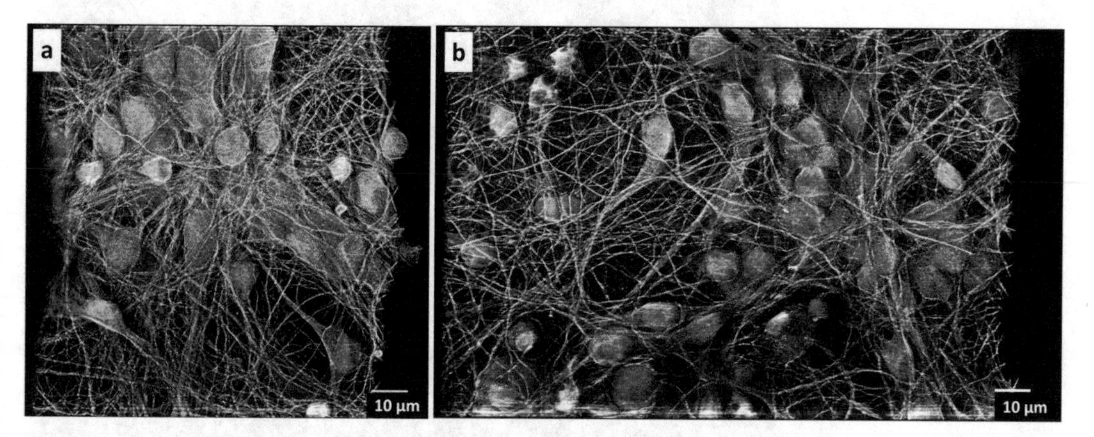

Fig. 7 Super-resolution images of primary cultured neuron cells in different area locations: (**a**) first area location, (**b**) second area location at disparate field of views. Scale bar is 10 μm

2. Switch on the epi-illumination mode to determine the precise location of the image of interest. In this configuration, the light-sheet excitation mode is switched off and epi-illumination is carried out via the epi-objective. Scan sample to define the precise imaging location.
3. When the location of the image of interest is determined, change the configuration to light-sheet excitation mode. In this configuration, the epi-illumination is switched off and the created excitation light sheet is illuminated onto the sample. A continuous light sheet scan is performed via the excitation objective (EO) and the optimum light sheet position is defined for super-resolution imaging.
4. To obtain a large-scale three-dimensional super-resolution image of microtubules, the sample scan strategy is carried out, wherein the sample chamber is mounted on an automatic 3D axis stage. The sample is moved across the light sheet and the detected fluorescence signals are recorded by sample scan mode, providing a series of 2D *z*-stack images of the sample. The image volume (Figs. 7 and 8) is constructed of 384 × 1024 pixels in the *XY* plane and 151 *z*-planes with a *z*-step interval of 0.5 μm.
5. Finally, the acquired 2D *z*-stack images are processed and analyzed (*see* Subheading 2.4 for image processing and analysis) to achieve a three-dimensional super-resolution image of individual microtubules present in neuron cells.

2.16 Image Processing and Analysis, Source Code, and References

A quick and easy program is designed for the image processing and analysis of three-dimensional super-resolution image of neurons. This section provides detailed information of the series of sequential events on 3D image processing and analysis (*see* Fig. 9). This customized program is created to efficiently run the program pipeline using the available resources at hand.

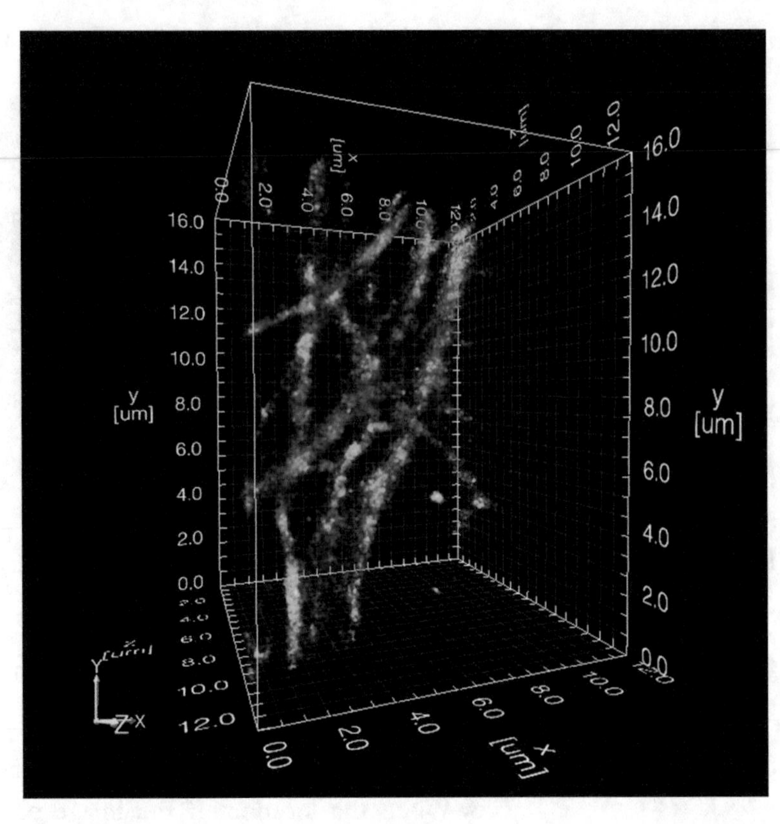

Fig. 8 Three-dimensional super-resolution image of primary cultured neurons

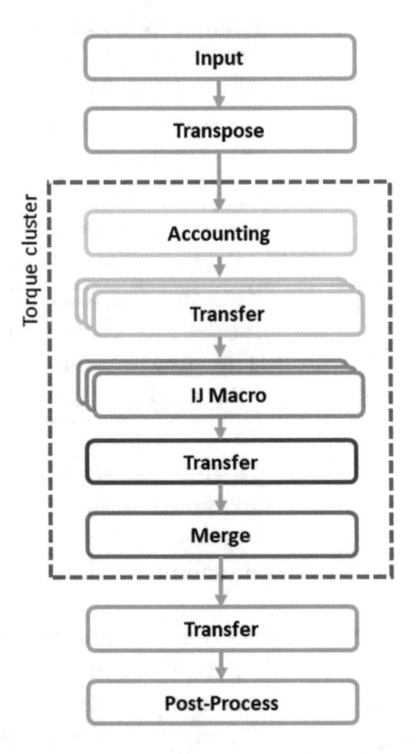

Fig. 9 Flow chart of the chronological sequence of events on image acquisition, processing, and analysis used for the three-dimensional imaging of neurons

1. Input:
   - For the efficient processing of massive dataset, the 3D TIFF image stacks are transferred to a remote Lustre [46] storage and a subsequent analysis of 2D processing per slice is performed using a four-node Torque cluster (Intel Xeon X5660 with 48 GB memory each, connected to the Lustre storage) [47].
2. Transpose:
   - To perform drift correction in a single pass, the XYZ image stacks are converted to time series XYT image stacks on the acquisition workstation post-experiment since the Thunder-STORM [48] requires the users to provide the time series 3D TIFF image stacks.
   - A straightforward C program would parse the file by libtiff [49] and rewrite them back in round-robin fashion. The files in process are not buffered in the memory first but directly concatenated to the new XYT stack for performance reason.
3. Accounting:
   - Accounting is performed on the head node of the Torque cluster. In this step, no data transfer is carried out since all the data are located on the local NAS storage.
   - After the acknowledgment of the file list, they are evenly distributed to multiple parts which are determined by the number of available worker nodes. Partial file lists are assigned to each worker node for data transfer.
4. Transfer:
   - The partial file list is fetched to the rsync [50] utility and pulled from the local NAS to a temporarily shared work-space (between worker nodes) created under home directory of the worker node, since /tmp space is not enough to store each batch of data.
   - Though the user could hide the transfer latency through dynamic job submission, it would increase the complexity and could hardly keep the script robust enough for general user.
   - Transfer and processing of data will not be terminated but rather ignored if some errors occur for certain files. Instead, the output will be recorded in a log file for later inspection and manual resubmission.
   - Additional wrapper scripts, IJ macro, and calibration file (required for 3D localization) are copied from the local NAS to worker nodes as well, in order to provide a mini-malistic way to update scripts on the cluster. The introduction of additional transfer time is negligible.

5. IJ Macro:
   - While IJ [51] is still buggy in headless mode, XVFB (X virtual framebuffer) [52], an in-memory display server, is created by each worker node to contain each IJ instance.
   - IJ reads the user preference file during initialization and creates one when it does not exist.
   - This will cause race condition and put IJ in an unpredictable state; therefore the user preference file is redirected to each workspace using JVM command-line arguments.
   - Temporary directory under /tmp is created for the CSV results to avoid excessive small-batch data transfer. It is removed upon termination of the JVM instance by shell script trap.
   - The macro detects the existence of calibration file and calls the respective 2D/3D configuration for ThunderSTORM and executes them for all the files in the workspace. The CSV files are dumped in per file basis; hence, each Z slice will have a respective output.
6. Transfer:
   - The CSV files are rsync back to the shared workspace for merging.
7. Merge:
   - The first worker, determined randomly by head node during resource request, would execute a Python script to merge the CSV files.
   - A SQLite [53] database is created on the fly to hold all the CSV files of interest.
   - For the 2D localization, the sequential z information is deduced by z step size and source file, since each z slice is contained in a single 3D stack (*see* "Transpose").
   - NaN filtering and geometry manipulation, required by data acquired through sample scan, are also performed here.
   - A new CSV file is generated by the final content in the database.
8. Transfer:
   - Prior to the transfer, the discrete CSV files and the database are removed.
   - The resulting CSV is rsync back to the local NAS storage, under the source directory for further processing.
9. Post-process:
   - Additional processing happened here, i.e., ASH plot and FRC evaluation [54].

Note: Detailed information on image processing and analysis is provided in the source code [55].

2.17 Anticipated Results

For the super-resolution imaging of neurons, selective plane illumination technique is used to obtain a high spatiotemporal resolution over a large imaging volume. Here, the optical sectioning is determined by the full width at half maximum (FWHM) of the Bessel beam central lobe, corresponding to approximately ~0.5 μm. The precise location of each fluorophore within the optical sectioned layer is successfully determined by the introduction of optical astigmatism. In order to demonstrate the abovementioned super-resolution localization microscopy, the primary cultured neurons are labeled with monoclonal anti-β-tubulin as primary antibody and subsequently stained with HMSiR-labeled goat anti-mouse as secondary antibody. The neurons are cultured on 5 mm diameter coverslips. After the staining of microtubules, the neurons are imaged in phosphate-buffered saline at pH 7.4 wherein the optimum environment for HMSiR blinking fluorophore is efficiently achieved for localization microscopy.

In conventional super-resolution optical microscopy, a high-resolution image relies on the reconstruction of thousands of frames and a longer acquisition rate, resulting in associated photobleaching and photodamage effects. This is the reason why a thin-plane illumination by the scanning Bessel beam is used in our optical configuration, yielding to the detection of densely populated fluorophores at a remarkably low background level. Furthermore, the HMSiR blinking fluorophore is activated by a low laser intensity which precludes the intense photobleaching and photodamage effects commonly observed in localization-based microscopy. Figure 7 shows the super-resolution images of primary cultured neurons at different area locations. It is clearly demonstrated that a larger field of view is achieved by the selective plane illumination technique as compared with conventional super-resolution optical microscopy techniques. By labeling the microtubules with HMSiR blinking fluorophore, the visible structures such as the cell body or soma, dendrites, axons, and axon terminals are conspicuously resolved. This super-resolution technique is highly beneficial for the study of neuron interaction and activity in a large scale. Furthermore, the activity of individual neurons could be clearly observed, creating neuron circuits that act as a whole.

Figure 8 illustrates a three-dimensional (3D) super-resolution image of β-tubulin labeled with the HMSiR blinking fluorophore. A 3D image of individual fluorophores is observed using selective plane illumination technique, revealing the structure of tubulin filaments which comprise interconnected and heavily overlapped neurons. By extracting the line profiles of certain locations of tubulin filaments, it is observed that a fine structure with a dimension of approximately ~100 nm is resolved by the sleeve plane illumination microscope system.

3 Conclusion

Three-dimensional super-resolution imaging of microtubules in neurons of postnatal rat pup brain is successfully achieved by the introduction of light-sheet microscopy based on selective plane illumination and a spontaneously blinking fluorophore, HMSiR, as reporter crucial for localization microscopy. The developed localization-based microscopy opens new insights toward the discovery of super-resolution imaging with good optical sectioning for three-dimensional imaging, large field of view for huge specimen volume, fast data acquisition, and low activation source which minimizes the effects of photobleaching and photodamage. Hence, this technique could be of great advantage for the super-resolution imaging of live-cell sample specimens.

Acknowledgments

The control software of the selective plane illumination microscope was licensed by Howard Hughes Medical Institute, Janelia Farm Research Campus. P.C. and B.-C.C. acknowledge the partial support of the Ministry of Science and Technology of Taiwan under contract numbers 106-2119-M-001-023 and 105-2119-M-001-026-MY2. P.C. is grateful for the support of the Thematic project of Academia Sinica.

References

1. Ok Kyu P, Jina K, Yoo Jung J, Young Ho K, Hyun-Seok H, Byung Joon H, Seung-Hae K, Yun K (2015) 3D light-sheet fluorescence microscopy of cranial neurons and vasculature during zebrafish embryogenesis. Mol Cells 38 (11):975–981
2. Pulvermüller F, Garagnani M (2014) From sensorimotor learning to memory cells in prefrontal and temporal association cortex: a neurocomputational study of disembodiment. Cortex 57:1–21. https://doi.org/10.1016/j.cortex.2014.02.015
3. Peyrache A, Schieferstein N, Buzsáki G (2017) Transformation of the head-direction signal into a spatial code. Nat Commun 8(1):1752. https://doi.org/10.1038/s41467-017-01908-3
4. Giepmans BNG, Adams SR, Ellisman MH, Tsien RY (2006) The fluorescent toolbox for assessing protein location and function. Science 312(5771):217–224. https://doi.org/10.1126/science.1124618
5. Lippincott-Schwartz J, Patterson GH (2009) Photoactivatable fluorescent proteins for diffraction-limited and super-resolution imaging. Trends Cell Biol 19(11):555–565. https://doi.org/10.1016/j.tcb.2009.09.003
6. Noji H, Yasuda R, Yoshida M, Kinosita K Jr (1997) Direct observation of the rotation of F1-ATPase. Nature 386:299. https://doi.org/10.1038/386299a0
7. Rief M, Rock RS, Mehta AD, Mooseker MS, Cheney RE, Spudich JA (2000) Myosin-V stepping kinetics: a molecular model for processivity. Proc Natl Acad Sci U S A 97 (17):9482–9486. https://doi.org/10.1073/pnas.97.17.9482
8. Hell SW, Wichmann J (1994) Breaking the diffraction resolution limit by stimulated emission: stimulated-emission-depletion

fluorescence microscopy. Opt Lett 19 (11):780–782. https://doi.org/10.1364/OL.19.000780

9. Gustafsson MGL (2000) Surpassing the lateral resolution limit by a factor of two using structured illumination microscopy. J Microsc 198(2):82–87. https://doi.org/10.1046/j.1365-2818.2000.00710.x
10. Betzig E (2006) Imaging intracellular fluorescent proteins at nanometer resolution. Science 313:1642–1645
11. Huang B, Jones SA, Brandenburg B, Zhuang X (2008) Whole-cell 3D STORM reveals interactions between cellular structures with nanometer-scale resolution. Nat Methods 5:1047–1052
12. Chiu S-W, Leake MC (2011) Functioning nanomachines seen in real-time in living bacteria using single-molecule and super-resolution fluorescence imaging. Int J Mol Sci 12 (4):2518–2542. https://doi.org/10.3390/ijms12042518
13. Amat F, Lemon W, Mossing DP, McDole K, Wan Y, Branson K, Myers EW, Keller PJ (2014) Fast, accurate reconstruction of cell lineages from large-scale fluorescence microscopy data. Nat Methods 11:951. https://doi.org/10.1038/nmeth.3036. https://www.nature.com/articles/nmeth.3036#supplementary-information
14. Tomer R, Khairy K, Amat F, Keller PJ (2012) Quantitative high-speed imaging of entire developing embryos with simultaneous multiview light-sheet microscopy. Nat Methods 9:755. https://doi.org/10.1038/nmeth.2062. https://www.nature.com/articles/nmeth.2062#supplementary-information
15. Zhang J, Liu X, Xu W, Luo W, Li M, Chu F, Xu L, Cao A, Guan J, Tang S, Duan X (2018) Stretchable transparent electrode arrays for simultaneous electrical and optical interrogation of neural circuits in vivo. Nano Lett 18:2903. https://doi.org/10.1021/acs.nanolett.8b00087
16. Szymborska A, de Marco A, Daigle N, Cordes VC, Briggs JAG, Ellenberg J (2013) Nuclear pore scaffold structure analyzed by super-resolution microscopy and particle averaging. Science 341(6146):655–658. https://doi.org/10.1126/science.1240672
17. Shim S-H, Xia C, Zhong G, Babcock HP, Vaughan JC, Huang B, Wang X, Xu C, Bi G-Q, Zhuang X (2012) Super-resolution fluorescence imaging of organelles in live cells with photoswitchable membrane probes. Proc Natl Acad Sci U S A 109(35):13978–13983. https://doi.org/10.1073/pnas.1201882109
18. Fernández-Suárez M, Ting AY (2008) Fluorescent probes for super-resolution imaging in living cells. Nat Rev Mol Cell Biol 9:929. https://doi.org/10.1038/nrm2531
19. Jones SA, Shim S-H, He J, Zhuang X (2011) Fast, three-dimensional super-resolution imaging of live cells. Nat Methods 8:499. https://doi.org/10.1038/nmeth.1605. https://www.nature.com/articles/nmeth.1605#supplementary-information
20. French JB, Jones SA, Deng H, Pedley AM, Kim D, Chan CY, Hu H, Pugh RJ, Zhao H, Zhang Y, Huang TJ, Fang Y, Zhuang X, Benkovic SJ (2016) Spatial colocalization and functional link of purinosomes with mitochondria. Science 351(6274):733–737. https://doi.org/10.1126/science.aac6054
21. Huang B, Wang W, Bates M, Zhuang X (2008) Three-dimensional super-resolution imaging by stochastic optical reconstruction microscopy. Science 319(5864):810–813. https://doi.org/10.1126/science.1153529
22. Vaziri A, Tang J, Shroff H, Shank CV (2008) Multilayer three-dimensional super resolution imaging of thick biological samples. Proc Natl Acad Sci U S A 105:20221–20226
23. Abrahamsson S, Chen J, Hajj B, Stallinga S, Katsov AY, Wisniewski J, Mizuguchi G, Soule P, Mueller F, Darzacq CD, Darzacq X, Wu C, Bargmann CI, Agard DA, Dahan M, Gustafsson MGL (2013) Fast multicolor 3D imaging using aberration-corrected multifocus microscopy. Nat Methods 10(1):60–63. https://doi.org/10.1038/nmeth.2277. http://www.nature.com/nmeth/journal/v10/n1/abs/nmeth.2277.html#supplementary-information
24. Cella Zanacchi F (2011) Live-cell 3D super-resolution imaging in thick biological samples. Nat Methods 8:1047–1049
25. Legant WR, Shao L, Grimm JB, Brown TA, Milkie DE, Avants BB, Lavis LD, Betzig E (2016) High-density three-dimensional localization microscopy across large volumes. Nat Methods 13(4):359–365. https://doi.org/10.1038/nmeth.3797. http://www.nature.com/nmeth/journal/v13/n4/abs/nmeth.3797.html#supplementary-information
26. Cella Zanacchi F, Lavagnino Z, Perrone Donnorso M, Del Bue A, Furia L, Faretta M, Diaspro A (2011) Live-cell 3D super-resolution imaging in thick biological samples. Nat Methods 8(12):1047–1049. http://www.nature.com/nmeth/journal/v8/n12/abs/nmeth.1744.html#supplementary-information
27. Schmid B, Shah G, Scherf N, Weber M, Thierbach K, Campos CP, Roeder I,

Aanstad P, Huisken J (2013) High-speed panoramic light-sheet microscopy reveals global endodermal cell dynamics. Nat Commun 4:2207. https://doi.org/10.1038/ncomms3207. https://www.nature.com/articles/ncomms3207#supplementary-information

28. Silvestri L, Paciscopi M, Soda P, Biamonte F, Iannello G, Frasconi P, Pavone FS (2015) Quantitative neuroanatomy of all Purkinje cells with light sheet microscopy and high-throughput image analysis. Front Neuroanat 9:68. https://doi.org/10.3389/fnana.2015.00068
29. Planchon TA, Gao L, Milkie DE, Davidson MW, Galbraith JA, Galbraith CG, Betzig E (2011) Rapid three-dimensional isotropic imaging of living cells using Bessel beam plane illumination. Nat Methods 8:417. https://doi.org/10.1038/nmeth.1586. https://www.nature.com/articles/nmeth.1586#supplementary-information
30. Chen B-C, Legant WR, Wang K, Shao L, Milkie DE, Davidson MW, Janetopoulos C, Wu XS, Hammer JA, Liu Z, English BP, Mimori-Kiyosue Y, Romero DP, Ritter AT, Lippincott-Schwartz J, Fritz-Laylin L, Mullins RD, Mitchell DM, Bembenek JN, Reymann A-C, Böhme R, Grill SW, Wang JT, Seydoux G, Tulu US, Kiehart DP, Betzig E (2014) Lattice light-sheet microscopy: imaging molecules to embryos at high spatiotemporal resolution. Science 346(6208):1257998. https://doi.org/10.1126/science.1257998
31. Uno S-N, Kamiya M, Yoshihara T, Sugawara K, Okabe K, Tarhan MC, Fujita H, Funatsu T, Okada Y, Tobita S, Urano Y (2014) A spontaneously blinking fluorophore based on intramolecular spirocyclization for live-cell super-resolution imaging. Nat Chem 6(8):681–689. https://doi.org/10.1038/nchem.2002. http://www.nature.com/nchem/journal/v6/n8/abs/nchem.2002.html#supplementary-information
32. van de Linde S, Löschberger A, Klein T, Heidbreder M, Wolter S, Heilemann M, Sauer M (2011) Direct stochastic optical reconstruction microscopy with standard fluorescent probes. Nat Protoc 6(7):991–1009
33. Folling J, Bossi M, Bock H, Medda R, Wurm CA, Hein B, Jakobs S, Eggeling C, Hell SW (2008) Fluorescence nanoscopy by ground-state depletion and single-molecule return. Nat Methods 5(11):943–945. http://www.nature.com/nmeth/journal/v5/n11/suppinfo/nmeth.1257_S1.html
34. Ha T, Tinnefeld P (2012) Photophysics of fluorescence probes for single molecule biophysics and super-resolution imaging. Annu Rev Phys Chem 63:595–617. https://doi.org/10.1146/annurev-physchem-032210-103340
35. Dempsey GT, Vaughan JC, Chen KH, Bates M, Zhuang X (2011) Evaluation of fluorophores for optimal performance in localization-based super-resolution imaging. Nat Methods 8(12):1027–1036. https://doi.org/10.1038/nmeth.1768
36. Bates M, Huang B, Dempsey GT, Zhuang X (2007) Multicolor super-resolution imaging with photo-switchable fluorescent probes. Science 317(5845):1749–1753. https://doi.org/10.1126/science.1146598
37. Fölling J, Bossi M, Bock H, Medda R, Wurm CA, Hein B, Jakobs S, Eggeling C, Hell SW (2008) Fluorescence nanoscopy by ground-state depletion and single-molecule return. Nat Methods 5:943. https://doi.org/10.1038/nmeth.1257. https://www.nature.com/articles/nmeth.1257#supplementary-information
38. Takakura H, Zhang Y, Erdmann RS, Thompson AD, Lin Y, McNellis B, Rivera-Molina F, Uno S-N, Kamiya M, Urano Y, Rothman JE, Bewersdorf J, Schepartz A, Toomre D (2017) Long time-lapse nanoscopy with spontaneously blinking membrane probes. Nat Biotechnol 35:773. https://doi.org/10.1038/nbt.3876. https://www.nature.com/articles/nbt.3876#supplementary-information
39. Gao L, Shao L, Chen B-C, Betzig E (2014) 3D live fluorescence imaging of cellular dynamics using Bessel beam plane illumination microscopy. Nat Protoc 9:1083. https://doi.org/10.1038/nprot.2014.087. https://www.nature.com/articles/nprot.2014.087#supplementary-information
40. Huisken J, Stainier DYR (2009) Selective plane illumination microscopy techniques in developmental biology. Development 136(12):1963–1975. https://doi.org/10.1242/dev.022426
41. Fath T, Ke YD, Gunning P, Götz J, Ittner LM (2008) Primary support cultures of hippocampal and substantia nigra neurons. Nat Protoc 4:78. https://doi.org/10.1038/nprot.2008.199
42. Chen CY, Chen YT, Wang JY, Huang YS, Tai CY (2017) Postsynaptic Y654 dephosphorylation of β-catenin modulates presynaptic vesicle turnover through increased n-cadherin-mediated transsynaptic signaling. Dev Neurobiol 77(1):61–74. https://doi.org/10.1002/dneu.22411
43. Brewer GJ (1997) Isolation and culture of adult rat hippocampal neurons. J Neurosci

Methods 71(2):143–155. https://doi.org/10.1016/S0165-0270(96)00136-7

44. Homaei AA, Sajedi RH, Sariri R, Seyfzadeh S, Stevanato R (2010) Cysteine enhances activity and stability of immobilized papain. Amino Acids 38(3):937–942. https://doi.org/10.1007/s00726-009-0302-3
45. Lautenschläger J, Mosharov EV, Kanter E, Sulzer D, Kaminski Schierle GS (2018) An easy-to-implement protocol for preparing postnatal ventral mesencephalic cultures. Front Cell Neurosci 12:44. https://doi.org/10.3389/fncel.2018.00044
46. Lustre. http://lustre.org/
47. Adaptive Computing Torque Resource Manager. http://www.adaptivecomputing.com/products/open-source/torque/
48. Oxford Academic ThunderSTORM: a comprehensive ImageJ plug-in for PALM and STORM data analysis and super-resolution imaging. https://academic.oup.com/bioinformatics/article/30/16/2389/2748167
49. LibTIFF – TIFF Library and Utilities. http://www.simplesystems.org/libtiff/
50. rsync. https://rsync.samba.org/
51. BMC Bioinformatics ImageJ2: ImageJ for the next generation of scientific image data. https://bmcbioinformatics.biomedcentral.com/articles/10.1186/s12859-017-1934-z
52. XVFB Xvfb – virtual framebuffer X server for X Version 11. https://www.x.org/releases/X11R7.7/doc/man/man1/Xvfb.1.xhtml
53. SQLite. https://www.sqlite.org/index.html
54. Europe PMC Fourier ring correlation as a resolution criterion for super-resolution microscopy. http://europepmc.org/abstract/MED/23684965
55. Bitbucket Torque and Thunderstorm. https://bitbucket.org/account/user/cbc-group/projects/LOC

Chapter 14

A Protocol for Single-Molecule Translation Imaging in *Xenopus* Retinal Ganglion Cells

Florian Ströhl, Julie Qiaojin Lin, Francesca W. van Tartwijk, Hovy Ho-Wai Wong, Christine E. Holt, and Clemens F. Kaminski

Abstract

Single-molecule translation imaging (SMTI) is a straightforward technique for the direct quantification of local protein synthesis. The protein of interest is fused to a fast-folding and fast-bleaching fluorescent protein, allowing one to monitor the appearance of individual fluorescence events after photobleaching of pre-existing proteins in the cell under investigation. The translation of individual molecules is then indicated by photon bursts of sub-second length that appear over a dark background. The method thus shares attributes with fluorescence recovery after photobleaching (FRAP) microscopy. Resulting datasets are similar to those generated by localization-based super-resolution microscopy techniques and can be used both to generate density maps of local protein production and to quantify the kinetics of local synthesis. The detailed protocol described in this chapter uses a Venus-β-actin fusion construct to visualize and measure the *β-actin* mRNA translational activity in *Xenopus* retinal ganglion cell growth cones upon Netrin-1 stimulation, which can be readily adapted for detecting translation events of other mRNAs in various cell types.

Key words Local translation, Localization microscopy, β-Actin, Venus, Axonal navigation, Guidance cues, Netrin-1, Retinal ganglion cells

1 Introduction

1.1 Motivation

Translation is the process in which ribosomes synthesize specific polypeptide chains according to a DNA-encoded sequence blueprint, the messenger RNA (mRNA). Folding of the nascent polypeptide is necessary for it to function as an active protein. In neurons, ribosomes are found in the cell body, but they are also present in more distal compartments like dendrites and axons, where they synthesize proteins locally [1, 2].

Local translation can impart a certain regulatory autonomy to subcellular compartments, which is of great importance for axonal

Florian Ströhl and Julie Qiaojin Lin contributed equally to this work.

Nobuhiko Yamamoto and Yasushi Okada (eds.), *Single Molecule Microscopy in Neurobiology*, Neuromethods, vol. 154, https://doi.org/10.1007/978-1-0716-0532-5_14, © Springer Science+Business Media, LLC, part of Springer Nature 2020

guidance during neurodevelopment. The tip of a growing axon (the growth cone) often navigates to a remote target region, for instance from the retina to the optic tectum/superior colliculus in the brain, which requires it to respond to extracellular guidance cues by turning, advancing, or pausing in a timely manner [3–5]. Given the complexity of this task and the significant travel distance relative to the cell body length, growth cones must possess a high degree of autonomy to react sufficiently rapidly to spatially restricted cues [6]. Previous studies have provided evidence that local protein synthesis contributes to this fast and local response [7–12]. For instance, mRNA of the structural protein β-actin has long been known to be present in growth cones [13]. Local translation of *β-actin* mRNA occurs within 10 min upon the application of a gradient of the extrinsic cue Netrin-1, causing the protein to accumulate on the gradient's near-side, as demonstrated by immunocytochemistry and using a translation reporter [8, 12].

The localized regulation of cue-mediated local translation represents a response mechanism distinct from the cell-wide shifts in gene expression, which are mediated by signaling pathways regulating transcription factor activity. Therefore, investigating the precise spatiotemporal dynamics of local translation activity at the single-molecule level could provide valuable new insights. For example, it is not known precisely where in the growth cone new proteins are synthesized, how fast synthesis occurs in response to a cue, and whether it occurs repetitively in the same spot or singly in diverse sites (indicative of translation patterns of polysomes (complexes of single-mRNA molecules and multiple ribosomes) or monosomes, respectively). To obtain intracytosolic localization data for the de novo synthesis of a protein of interest, fluorescence microscopy approaches are most suitable, as photo depletion of pre-existing fluorescence allows event registration with minimal background within the native environment.

1.2 Imaging Translation Events

Two approaches to visualize translation have been developed recently, termed single-molecule translation imaging (SMTI) and nascent polypeptide fluorescent tagging method. The latter uses fluorescence amplification systems via multi-epitope tagging of tandem repeats of a peptide-binding site linked to the protein of interest by site-specific proteins fused to fluorescent proteins [14–16] or conjugated to dyes [17]. This chapter focuses on SMTI, which can provide a direct readout of translation [1, 2, 18].

An SMTI protocol involves several sample preparation and imaging steps. The fast-folding and fast-bleaching fluorescent protein Venus [19] is fused to a protein of interest, and the DNA encoding the fusion construct introduced into the specimen, for instance via chemical transfection or electroporation. The imaging protocol begins with illumination of the sample at high intensity for a short time period to reduce initial fluorescence and remove signal

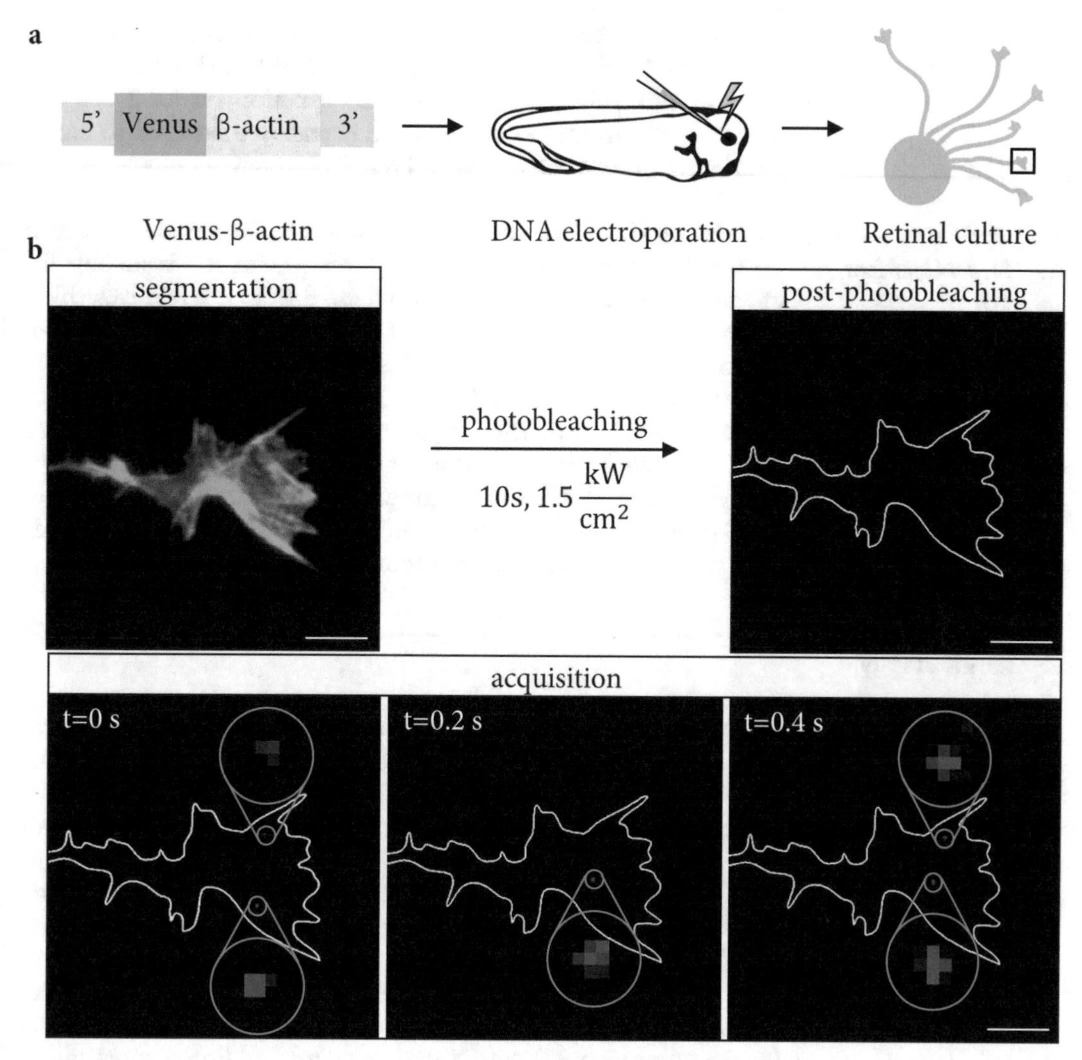

Fig. 1 (**a**) A reporter construct, coding for, e.g., Venus and β-actin, is electroporated into *Xenopus* eye primordia, which are dissected and cultured. (**b**) During imaging, a fluorescent image of a growth cone is acquired to segment the outline for later processing. Existing fluorescence is then photobleached using a brief pulse of laser irradiation. Subsequently, the translation of individual β-actin is recorded as individual Venus molecule emission. Scale bars are 5 μm. (Figure adapted from [1])

background caused by pre-existing protein. Subsequent imaging is performed at low illumination intensity. Each newly synthesized Venus protein is then detected as a short burst of fluorescence before becoming bleached, thus maintaining a background-free environment for high-precision readouts. A burst in fluorescence photons therefore signifies an individual translation event, which can be localized spatially via single-molecule localization procedures and temporally within sequentially acquired imaging sequences (Fig. 1b). Localizations of individual translation events are retrieved using specialized fitting software and appropriate filtering of detected events. Filtering can be performed using thresholds on the number of photons, and the duration, shape, and size of candidate events. The information on locations and frames of occurrence of all detected events is then used to create translation rate density maps and translation rate graphs.

1.3 *Xenopus* as a Model Organism for SMTI

Xenopus laevis (African clawed frog) is a common model organism for the investigation of embryonic development and is suitable for imaging studies on neurodevelopment due to its relatively high tissue transparency. Retinal cultures produced from *Xenopus* are especially suitable for SMTI studies into axon guidance: the transparent retinal ganglion cell (RGC) axons are the only neuronal protrusions that exit the eye; they grow flatly onto the culture dish, and can be kept at room temperature. In this chapter, we detail a protocol for SMTI imaging of *β-actin* mRNA translation in RGC growth cones and the investigation of the effect of the guidance cue Netrin-1 on localized translation rates [1].

2 Methodology

2.1 Venus-β-Actin Plasmid Construction

Construction of a plasmid encoding the fusion protein of interest with appropriate coding regions (CDSs) and untranslated regions (UTRs) can be achieved through amplification of separate DNA fragments and the subsequent insertion into a suitable plasmid (also *see* **Note 1**).

Fragment A (*β-actin* 5′UTR and Venus) can be obtained by PCR amplification from the monomeric Venus plasmid (Addgene #27793) using a forward primer containing a BamHI site and *β-actin* 5′UTR sequence (underlined region in the forward primer) followed by 16 nucleotides complementary to the 5′ end of Venus (start codon in bold), and a reverse primer adding a short linker sequence (AAGCTTGAATTCAAA) containing an EcoRI site.

Primers for fragment A (*β-actin* 5'UTR and Venus):

Forward: 5′-TACTCGGATCCGGCTCAGTGACCCGCCCGCATAGAAAGGAGACAGTCTGTGTGCGTCCAACCCTCAGATCACA**ATG**GTTAGTAAGGGCG-3′.

Reverse: 5′-GTATGAATTCAAGCTTTTTGTAAAGTTCATCC-3′.

Subsequently, fragment B (*β-actin* CDS and 3′UTR) can be obtained using the *Xenopus laevis* cDNA library as the template. Coding regions of interest for plasmid construction, namely the sequences of the *β-actin* mRNA CDS and 3′UTR, can be obtained from a suitable *Xenopus* cDNA library. The library can be constructed by extracting total mRNA from Stage 32 *Xenopus laevis* embryos using RNeasy Mini Kit (QIAGEN) and reverse transcribing using the SuperScript III First-Strand Synthesis System

(Thermo Fisher Scientific) with the Oligo(dT) primer. The forward primer used in the PCR reaction contains the linker sequence with the EcoRI site and 16 nucleotides complementary to the 5′-end of *β-actin* CDS, while the reverse primer includes 16 nucleotides complementary to the 3′-end of *β-actin* 3′UTR and an XbaI site.

Primers for fragment B (*β-actin* CDS and 3′UTR):

Forward: 5′-GCTTGAATTCAAAATGGAAGACGATATTG-3′.

Reverse: 5′-CGTATCTAGAGTGAAACAACATAAGT-3′.

The two fragments of the fusion construct can be inserted into a plasmid. As the backbone for plasmid construction, the pCS2+ vector (Addgene) was used, which contains a strong simian CMV promoter, followed by a multiple cloning site and an SV40 late polyadenylation site. Fragment A can be inserted into a linearized pCS2+ vector using BamHI and EcoRI restriction enzymes. Subsequently, fragment B can be inserted into the fragment A-containing pCS2+ plasmid using the EcoRI and XbaI restriction sites. The resulting Venus-β-actin construct consists of the following elements cloned into pCS2+ vector: 5′-(*β-actin* 5′UTR, Venus, a short linker (KLEFK), *β-actin* CDS, and *β-actin* 3′UTR)-3′ (Fig. 1a).

2.2 *Xenopus laevis* Embryos

Xenopus laevis eggs were fertilized in vitro and embryos were raised in 0.1× modified Barth's saline (MBS; 8.8 mM NaCl, 0.1 mM KCl, 0.24 mM $NaHCO_3$, 0.1 mM HEPES, 82 μM $MgSO_4$, 33 μM $Ca(NO_3)_2$, 41 μM $CaCl_2$) at 14–20 °C and staged according to the tables of Nieuwkoop and Faber [20]. All animal experiments were approved by the University of Cambridge Ethical Review Committee in compliance with the University of Cambridge Animal Welfare Policy. This research has been regulated under the Animals (Scientific Procedures) Act 1986 Amendment Regulations 2012, following ethical review by the University of Cambridge Animal Welfare and Ethical Review Body (AWERB).

2.3 Targeted Eye Electroporation

Targeted eye electroporation can be performed as previously described [21, 22]. A "†" shape electroporation chamber suitable for stage 26–30 embryos can be modified from a Sylgard dish (Sigma) as described [21]. To anesthetize embryos during electroporation, 0.4 mg/ml tricaine methanesulfonate (MS-222) (Sigma) in 1× MBS, pH 7.5, is used. An anesthetized embryo is positioned along the longitudinal channel of the electroporation chamber, with its head positioned at the cross of the longitudinal and transverse channels. A pair of flat-ended platinum electrodes (Sigma) is held in place by a manual micromanipulator (World Precision Instruments) at the ends of the transverse channel. A glass capillary with a fine tip containing plasmid solution is inserted into the eye primordium of stage 26–30 embryos to inject 8 × 5–8 nl doses of

2 μg/μl Venus-β-actin plasmid (also *see* **Note 2**) driven by an air-pressured injector, such as a Picospritzer (Parker Hannifin). Immediately following the plasmid injection, eight electric pulses of 50-ms duration at 1000-ms intervals are delivered at 18 V by a square wave generator, such as the TSS20 OVODYNE electroporator (Intracel). The embryos are recovered and raised in 0.1× MBS until they reach stage 32–35 as required for retinal cultures.

2.4 Xenopus Retinal Culture

For the retinal culture, dishes must be appropriately pretreated to minimize background fluorescence. Glass-bottom dishes (MatTek) are pretreated with 5 M KOH (Sigma) for 1 h, followed by 5–8 rinses with deionized water (Sigma), and are then left to dry in a hood (all culturing procedures should be performed in a laminar flow hood or a microbiological safety cabinet). They are next coated with 10 μg/ml poly-L-lysine (Sigma) and left overnight at room temperature. On the following day, excess poly-L-lysine is discarded from the dishes, which are then rinsed three times with double-distilled water and left in the hood until dry. Subsequently, the dishes are coated with 10 μg/ml laminin (Sigma) in L15 for 1–3 h (also *see* **Note 3**). Finally, the laminin solution is replaced with culture medium (60% (v/v) of L15 (GIBCO), 1% (v/v) antibiotic–antimycotic (100×), in double-distilled water, pH 7.6–7.8, sterilized with 0.22 μm pore-size filters), in which dissected eyes can be placed.

Embryos should be screened and washed before the retinal culture is commenced. Electroporated embryos at stage 32–35 should be screened for Venus fluorescence to check for successful electroporation. Successfully electroporated embryos are then rinsed three times in the embryo wash solution (0.1× MBS with 1% (v/v) antibiotic–antimycotic (100×) (Thermo Fisher Scientific), pH 7.5, sterilized with 0.22 μm pore-size filters) and anesthetized in MS-222 solution (0.04% (w/v) MS-222 and 1% (v/v) antibiotic–antimycotic (100×) in 1× MBS, pH 7.5, filtered with 0.22 μm filters). After transfer of the embryos to the Sylgard dish, the electroporated eye primordia are dissected out and washed three times in culture medium. A stereomicroscope, 0.1–0.2 mm minutien pins (Fine Science Tools), pin holders (Fine Science Tools), forceps, and a Sylgard dish (Sigma) are needed for the eye dissection. Finally, the dissected eye primordia are placed in the center of the dish, and the cultures incubated at room temperature overnight. During this period, explanted eyes maintain their anatomical integrity in culture and only RGC axons exit the eye and grow on coverslips [23].

2.5 Optical Setup

An SMTI optical setup consists of a standard wide-field fluorescence microscope equipped with high-power lasers and a sensitive camera suitable for single-molecule detection. Implementation is relatively straightforward if built around commercially available inverted microscope frames (e.g., Olympus IX73) [1]. For imaging of Venus, an illumination laser wavelength of 488 nm (Coherent Sapphire) can be used in combination with a 525/45 emission filter (Semrock) and a dichroic beam splitter (Chroma ZT405/488/561/640rpc). The laser beam must be circularly polarized by a quarter-wave plate (AQWP05M-600, Thorlabs) to result in homogeneous excitation of fluorescent proteins (irrespective of their orientation). To collect data, an EMCCD camera (Andor iXon Ultra 897) is required with an effective optical pixel size near the optimum for single-molecule localization (ca. 120 nm) [24]. This can be achieved through use of a 100× magnification, 1.49 NA, oil-immersion TIRF objective (Olympus UAPON100XOTIRF) in combination with an additional 1.3× magnification optical relay after the tube lens (e.g., via a TwinCam (CAIRN)) for a camera with physical pixel size of 16 μm (standard for EMCCD cameras).

The SMTI setup can be adapted to facilitate dual-color imaging. For dual-color imaging of mRFP as well as Venus, a 561 nm laser (Cobolt Jive) must be added in combination with a 600/37 emission filter (Semrock). This configuration allows easier identification and clearer visualization of transfected growth cones and is needed when bright-field identification is not feasible, e.g., when imaging in whole organisms [25]. Where simultaneous imaging is required, two identical EMCCD cameras (Andor iXon Ultra 897) can be used in conjunction with a TwinCam housing a dichroic beam splitter (Chroma T565spxr). The beam splitter must contain a 525/45 emission filter (Semrock) in the transmission direction to capture Venus fluorescence and a 600/37 emission filter (Semrock) in reflection to capture mRFP. Alterative fluorophore combinations are also possible, such as of Venus and Cy5. To capture Cy5 fluorescence, a 647 nm laser (MPB VFL-P-200-647) can be used in combination with a 680/42 emission filter (Semrock).

2.6 Imaging Protocol

For SMTI, fluorescent axons emanating from the dissected eye are identified and traced to find their growth cones. An outgrowing fluorescent growth cone is selected and, prior to the bleaching step, imaged with low irradiance (<2 W/cm^2) in both fluorescence and bright-field mode. These images can later be used to generate an outline image and to track growth cone health by comparison to an image taken once the SMTI acquisition sequence is completed. The growth cone is then photobleached for 10 s with an irradiance of 1.5 kW/cm^2. If pharmacological treatment or addition of a guidance cue is needed, they can be administered before or immediately after the photobleaching step. Afterwards, the flash-like recovery of Venus fluorescence is recorded with an exposure time of 200 ms at 5 Hz for 60 s to determine a baseline (non-stimulated) translation rate. Typically, a reduced intensity of around 0.3 kW/cm^2 is used to ensure survival of the axons while simultaneously bleaching newly synthesized Venus. To study the effect of guidance cues, Netrin-1 (final concentration: 600 ng/ml, Sigma) or the same volume of culture medium as a control can then be added to the dish. As the dish is uncovered during drug application, the laser has to be switched off or appropriate laser safety goggles have to be used if the running acquisition should not be interrupted. After total SMTI acquisition period of 180 s, the second bright-field image is taken to check for growth cone health; retracted growth cones should be excluded from analysis. All imaging steps can be performed under epifluorescence as well as HILO (highly inclined and laminated optical sheet) or TIRF illumination. An EM gain of about 200 should be used on the EMCCD camera to ensure single-molecule sensitivity. Furthermore, the field of illumination should be bigger than the size of the imaged field of view to bleach diffusing or transported fluorescent proteins from outside the growth cone prior to their entering the field of view (also *see* **Note 4**).

2.7 Data Processing and Analysis Software

Localizations of individual translation events are retrieved using maximum likelihood estimation with a Gaussian model fit, for example using the rapidSTORM software package [26] (also *see* **Note 5**). rapidSTORM provides information on locations and frames of occurrence of all detected events, which can be used to create translation rate density maps and translation rate graphs in custom-written MATLAB scripts [1]. It is useful to filter out potentially non-translation events through appropriate thresholding; for instance, in most experimental conditions a threshold of ~500 photons per localization is appropriate to filter out noise and blinking events that do not stem from Venus.

To examine the spatial distribution of the translational events, a Sholl analysis [27] can be performed: each growth cone is divided into n evenly spaced concentric arcs. For $n = 5$ arcs, a labeling of A1–A5 from central to peripheral domains can be used. Arc A5 is then set to circumscribe the outermost part of the growth cone (Fig. 3d). Performing this analysis on control and treated samples visualizes and quantifies the translocation of translation sites, for example as they move from central domains to the periphery [21].

3 Notes

1. The advantage of fusing the Venus sequence to the 5′-end of the candidate gene CDS is that the mRNA sequence encoding Venus will be translated first, which shortens the time between the initiation of polypeptide synthesis and signal detection. Therefore, N-terminal Venus fusion is the preferable design, unless expression levels are found to be severely impaired for such constructs.

The untranslated regions of an mRNA are often crucial for its localization, stability, and translation [28, 29]. If overexpression of the full-length protein impairs cellular physiology, a Venus sequence flanked by the 5′ and 3′ untranslated regions of the mRNA of interest can also be used in place of the fusion construct to investigate translational regulation.

2. The expression level and ubiquity of Venus fluorescence can be influenced in several ways. The former should be carefully titrated, as high expression is associated with significant phototoxicity due to the prolonged photobleaching step and/or the increased levels of reactive oxygen species resulting upon fluorophore excitation. Electroporation efficiency is known to decrease with increasing embryonic stage [21]. Therefore, eye-targeted electroporation is usually more efficient when performed at around stage 26 rather than in older embryos. In addition, choosing wider electrodes for electroporation can increase the number of fluorescence-positive cells.
3. Screening and washing of the embryos can be carried out in parallel with incubation of the dishes with laminin solution.
4. An adaption of the SMTI protocol in the form of "translation FRAP with low-power illumination" can also be used in vivo [25] and has, for instance, been employed for longer-term imaging of the interplay between ribosome/RNA trafficking and regulation of local protein synthesis during axon arborization. With this methodological variant, the challenges posed by the high intrinsic fluorescence background due to high-power illumination and to some extent out-of-focus light in whole organisms can be overcome. Thus the docking of RNA granules at branching positions in conjunction with spatially and temporally resolved local translation could be correlated [25]. Recovery of fluorescence is in this case observed analogously to fluorescence recovery after photobleaching, FRAP, but used to indicate protein synthesis.

   When highly mobile specimens like growth cones are imaged, it is useful to count all events in a small area even around and outside of the growth cone (e.g., a rectangular window that tightly crops the growth cones), to account for the high mobility of the growth cone filopodia.
5. The threshold used in the filtering step is found by manual selection of Venus flashes and determination of the *average* photon budget of a single emitting Venus molecule. The threshold value as well as the size of the emission point spread function (PSF) can be used as a parameter setting in rapidSTORM [26]. Further optimization of the parameters can be

performed on manually selected flashes in order to recognize as many real flashes as possible while at the same time keeping the number of false positives as low as possible. For validation, the manual selection process can be repeated blindly by multiple independent researchers to minimize any bias occurring during the manual selection process [1]. The tracking option of rapidSTORM can be used to recombine photons emanating from the same fluorescent protein during a translation event over multiple frames to increase the number of photons per localization, but also to determine the *on-time* of the Venus flashes before bleaching. Another optimization parameter is the survival fraction as the high intensities required for SMTI can cause cell death due to phototoxicity. In this case, the employed illumination intensities can be reduced, but correspondingly longer durations are needed for the bleaching step.

Various examples for measurements of these parameters are displayed in Fig. 2. A measured *on-time* of about 400 ms is displayed, which provides an indication of the average duration of translation flashes. Changing of the intensity of the irradiation light can be used to tune this duration to some degree, but care must be taken to avoid lethal dosages (compare panel (d)). The photon budget of a translation event can be calculated from the sum value of pixel values belonging to single events, for instance by using a 5 × 5 pixel area around the brightest pixels. When the conversion factor of the camera is known, this

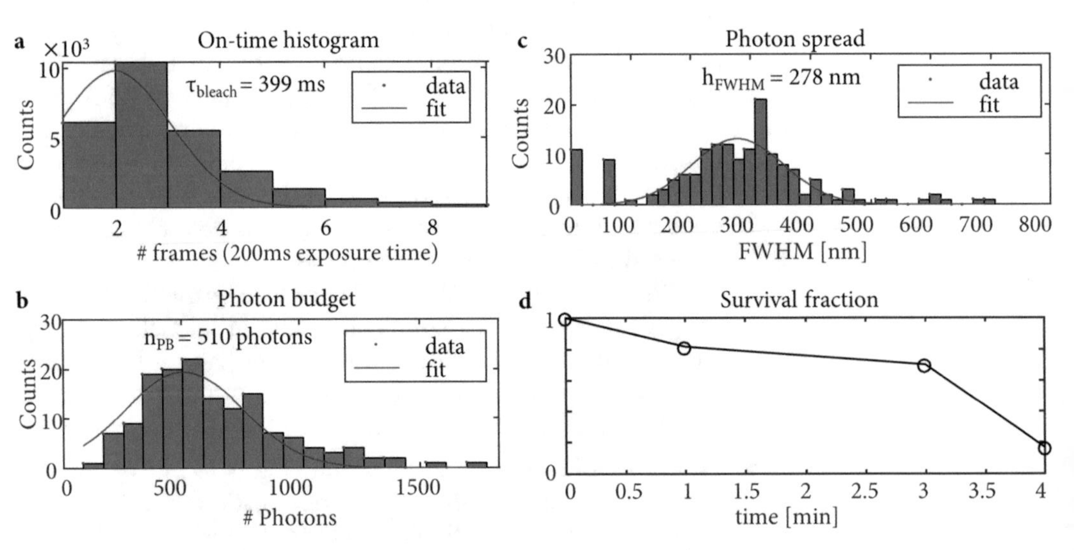

Fig. 2 (**a**) The measured *on-time*, i.e., duration of flashes, of the average Venus-β-actin molecule when using a laser irradiation of 0.3 kW/cm^2. (**b**) The average photon budget of a Venus-β-actin molecule as calculated from hundreds of manually selected flashes, which were subjectively classified as stemming from Venus emission. (**c**) Width of the PSF. The broad distribution stems from flashes in different focal planes. (**d**) After 3 min of continuous imaging with 0.3 kW/cm^2, about 30% of axons started to retract, while 4 min of imaging caused most axons to retract. Forty-six axons were observed for this plot. (Figure adapted from [1])

value can be translated directly into the number of photons. As shown in panel (b), an average photon budget of around 500 photons was found in this example dataset, which indicates that each event is caused by the emission of single Venus molecules. As another control measure, the width of the PSF can be used to validate the diffraction-limited performance of the microscope (panel (c)).

4 Example Results

Using the protocol described above, we studied the effect of Netrin-1 on *β-actin* mRNA translation rates and translocation of translation sites in navigating RGC growth cones.

Under no-treatment baseline conditions, single-molecule translation events were detected at an average rate of ~10 events/min, which were located predominantly in the central domain of the growth cones (regions A1 and A2 of the Sholl analysis depicted in Fig. 3d). Translation in RGC growth cones was sporadic, with limited events reoccurring at the same location. This was in contrast to HEK293T cells, which exhibited hot spots of repetitive Venus-*β-actin* translation [1], or to hippocampal dendrites, in which the activity-regulated cytoskeletal-associated (Arc) protein and the fragile X mental retardation protein (FMRP) undergo similar burst-like translation [18].

After translation was quantified for the baseline condition, the effect of the chemical guidance cue Netrin-1 was investigated. As displayed in the translation density maps in Fig. 3a, the event rate was up to 0.5 events/s. The same dataset also provides even more detailed temporal information, in the form of a curve of instantaneous rate as function of time (Fig. 3b). Remarkably, Netrin-1 stimulation led to a burst of *β-actin* translation starting 20 s post-treatment and lasting for 30 s before gradually declining to lower but still above baseline levels. The decline likely reflects a rapid desensitization of Netrin-1 receptors on the growth cone surface, which is known to occur within 1–2 min [30]. The initial increase in translation is clearly visible in a cumulative translation event plot, with the baseline translation of around 15 events/min and the event number doubled within the first minute after Netrin-1 application (Fig. 3c). To examine the spatial distribution of the translational events, a Sholl analysis [27] was performed. The percentage of *β-actin* translation events located in the central arcs (A1 and A2) decreased upon Netrin-1 stimulation, whereas translational events located at the growth cone periphery (A3–A5) increased (Fig. 3d). The Netrin-1-stimulated increase in fluorescence events was attenuated by preincubation of the samples with the translation inhibitor puromycin, indicating that it was truly reflecting an increase in local *β-actin* translation rate. Incubation with puromycin also

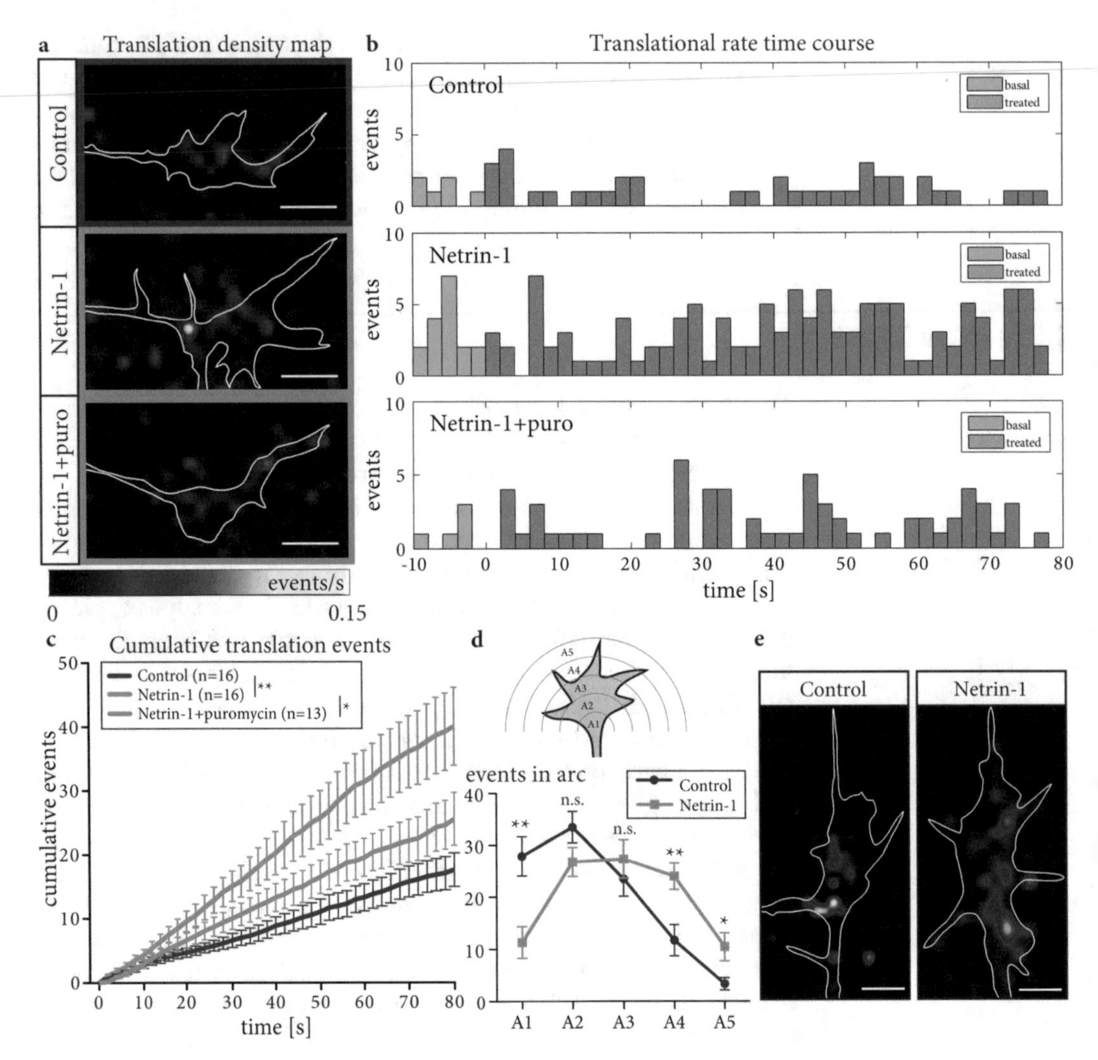

Fig. 3 (**a**) Translation density maps for RGC growth cones in different conditions. (**b**) Respective translation rate time courses. The difference in the pretreatment rates can be attributed to biological variability; the average pretreatment rate between the two groups ($n = 16$) was not significantly different ($p = 0.63$). (**c**) Cumulative event rates per growth cone $^{*}p < 0.05$; $^{**}p < 0.001$; two-way ANOVA. (**d**) Sholl analysis on culture medium- or Netrin-1-treated growth cones shows the spatial distribution of translational events. The center of five concentric circles is located at the base of the growth cone; the outermost circle is tangential to the growth cone tip. The radii are equidistant. Intracellular events within each arc in %. $^{**}p < 0.001$; Mann-Whitney Houston test; scale bars are 5 μm. Error bars indicate standard error of the mean. In (**c**), n is the number of growth cones. (Figure adapted from [1])

reduced the translation rate in HEK293T cells ($p = 0.011$) but had no significant effect on growth cones not stimulated with Netrin-1, which could be a result of the already low baseline translation overshadowing the effect of puromycin ($p = 0.57$).

5 Summary

SMTI is a technique that can be used to quantify the spatiotemporal dynamics of local translation, for instance to measure the effects of Netrin-1 on local *β-actin* mRNA translation in growth cones. Using SMTI, it has been shown that Netrin-1 transiently increases translation and that it causes relocation of potentially monosomal translation sites to the growth cone periphery. This result is in line with a previously described shift in translation activity inside of growth cones upon exposure to an external gradient of guidance cues [8]. A rapid stimulating effect of Netrin-1 on *β-actin* local translation in RGC growth cones was observed in real time and the concomitant shift of its translation toward the periphery of the growth cones measured. This translation localization pattern is consistent with the recently reported *β-actin* mRNA trafficking modes in axons and within growth cones [31, 32].

The SMTI protocol can be further developed to yield even richer insights. For example, the use of fluorescent proteins featuring shorter folding times and higher quantum yields will lead to improvements in temporal resolution and detection sensitivity. As the method has been shown to be capable of measurements of local translation rates at the single-molecule level in axonal growth cones, it holds great potential for application to other model systems. The method is thus a powerful new tool for investigating signal-driven subcellular responses affecting protein synthesis.

References

1. Ströhl F, Lin JQ, Laine RF et al (2017) Single molecule translation imaging visualizes the dynamics of local β-actin synthesis in retinal axons. Sci Rep 7:709. https://doi.org/10.1038/s41598-017-00695-7
2. Ifrim MF, Williams KR, Bassell GJ (2015) Single-molecule imaging of PSD-95 mRNA translation in dendrites and its dysregulation in a mouse model of fragile X syndrome. J Neurosci 35:7116–7130. https://doi.org/10.1523/JNEUROSCI.2802-14.2015
3. Dickson BJ (2002) Molecular mechanisms of axon guidance. Science 298:1959–1964. https://doi.org/10.1126/science.1072165
4. O'Donnell M, Chance RK, Bashaw GJ (2009) Axon growth and guidance: receptor regulation and signal transduction. Annu Rev Neurosci 32:383–412. https://doi.org/10.1146/annurev.neuro.051508.135614
5. Kolodkin AL, Tessier-Lavigne M (2011) Mechanisms and molecules of neuronal wiring: a primer. Cold Spring Harb Perspect Biol 3:1–14. https://doi.org/10.1101/cshperspect.a001727
6. Harris WA, Holt CE, Bonhoeffer F (1987) Retinal axons with and without their somata, growing to and arborizing in the tectum of Xenopus embryos: a time-lapse video study of single fibres in vivo. Development 101:123–133
7. Campbell DS, Holt CE (2001) Chemotropic responses of retinal growth cones mediated by rapid local protein synthesis and degradation. Neuron 32:1013–1026. https://doi.org/10.1016/S0896-6273(01)00551-7
8. Leung K-M, van Horck FPG, Lin AC et al (2006) Asymmetrical beta-actin mRNA translation in growth cones mediates attractive turning to netrin-1. Nat Neurosci 9:1247–1256. https://doi.org/10.1038/nn1775
9. Piper M, Anderson R, Dwivedy A et al (2006) Signaling mechanisms underlying Slit2-induced collapse of Xenopus retinal growth

cones. Neuron 49:215–228. https://doi.org/10.1016/j.neuron.2005.12.008

10. Welshhans K, Bassell GJ (2011) Netrin-1-induced local beta-actin synthesis and growth cone guidance requires zipcode binding protein 1. J Neurosci 31:9800–9813. https://doi.org/10.1523/JNEUROSCI.0166-11.2011
11. Wu KY, Hengst U, Cox LJ et al (2005) Local translation of RhoA regulates growth cone collapse. Nature 436:1020–1024. https://doi.org/10.1038/nature03885
12. Yao J, Sasaki Y, Wen Z et al (2006) An essential role for beta-actin mRNA localization and translation in Ca^{2+}-dependent growth cone guidance. Nat Neurosci 9:1265–1273. https://doi.org/10.1038/nn1773
13. Bassell GJ, Zhang H, Byrd AL et al (1998) Sorting of beta-actin mRNA and protein to neurites and growth cones in culture. J Neurosci 18:251–265
14. Wang C, Han B, Zhou R, Zhuang X (2016) Real-time imaging of translation on single mRNA transcripts in live cells. Cell 165:990–1001. https://doi.org/10.1016/j.cell.2016.04.040
15. Yan X, Hoek TA, Vale RD, Tanenbaum ME (2016) Dynamics of translation of single mRNA molecules in vivo. Cell 165:976–989. https://doi.org/10.1016/j.cell.2016.04.034
16. Wu B, Eliscovich C, Yoon YJ, Singer RH (2016) Translation dynamics of single mRNAs in live cells and neurons. Science 352:1430–1435. https://doi.org/10.1126/science.aaf1084
17. Morisaki T, Lyon K, DeLuca KF et al (2016) Real-time quantification of single RNA translation dynamics in living cells. Science 352:1425–1429. https://doi.org/10.1126/science.aaf0899
18. Tatavarty V, Ifrim MF, Levin M et al (2012) Single-molecule imaging of translational output from individual RNA granules in neurons. Mol Biol Cell 23:918–929. https://doi.org/10.1091/mbc.E11-07-0622
19. Nagai T, Ibata K, Park ES et al (2002) A variant of yellow fluorescent protein with fast and efficient maturation for cell-biological applications. Nat Biotechnol 20:87–90. https://doi.org/10.1038/nbt0102-87
20. Nieuwkoop PD, Faber J (1994) Normal table of Xenopus laevis (Daudin): a systematical and chronological survey of the development from the fertilized egg till the end of metamorphosis. Garland Publishing, New York
21. Falk J, Drinjakovic J, Leung KM et al (2007) Electroporation of cDNA/Morpholinos to targeted areas of embryonic CNS in Xenopus. BMC Dev Biol 7:107. https://doi.org/10.1186/1471-213X-7-107
22. Wong HH-W, Holt CE (2018) Targeted electroporation in the CNS in Xenopus embryos. Methods Mol Biol 1865:119–131. https://doi.org/10.1007/978-1-4939-8784-9_9
23. Cagnetta R, Frese CK, Shigeoka T et al (2018) Rapid cue-specific remodeling of the nascent axonal proteome. Neuron 99:29–46.e4. https://doi.org/10.1016/j.neuron.2018.06.004
24. Thompson RE, Larson DR, Webb WW (2002) Precise nanometer localization analysis for individual fluorescent probes. Biophys J 82:2775–2783. https://doi.org/10.1016/S0006-3495(02)75618-X
25. Wong HH-W, Lin JQ, Ströhl F et al (2017) RNA docking and local translation regulate site-specific axon remodeling in vivo. Neuron 95:852–868.e8. https://doi.org/10.1016/j.neuron.2017.07.016
26. Wolter S, Löschberger A, Holm T et al (2012) rapidSTORM: accurate, fast open-source software for localization microscopy. Nat Methods 9:1040–1041. https://doi.org/10.1038/nmeth.2224
27. Sholl DA (1953) Dendritic organization in the neurons of the visual and motor cortices of the cat. J Anat 87:387–406
28. Xing L, Bassell GJ (2013) MRNA localization: an orchestration of assembly, traffic and synthesis. Traffic 14:2–14. https://doi.org/10.1111/tra.12004
29. Chabanon H, Mickleburgh I, Hesketh J (2004) Zipcodes and postage stamps: mRNA localisation signals and their trans-acting binding proteins. Brief Funct Genomic Proteomic 3:240–256. https://doi.org/10.1093/bfgp/3.3.240
30. Piper M, Salih S, Weinl C et al (2005) Endocytosis-dependent desensitization and protein synthesis-dependent resensitization in retinal growth cone adaptation. Nat Neurosci 8:179–186. https://doi.org/10.1038/nn1380
31. Turner-Bridger B, Jakobs M, Muresan L et al (2018) Single-molecule analysis of endogenous β-actin mRNA trafficking reveals a mechanism for compartmentalized mRNA localization in axons. Proc Natl Acad Sci U S A 115:E9697–E9706. https://doi.org/10.1073/pnas.1806189115
32. Leung K-M, Lu B, Wong HH-W et al (2018) Cue-polarized transport of β-actin mRNA depends on 3′UTR and microtubules in live growth cones. Front Cell Neurosci 12:1–19. https://doi.org/10.3389/fncel.2018.00300

Chapter 15

Investigating Molecular Diffusion Inside Small Neuronal Compartments with Two-Photon Fluorescence Correlation Spectroscopy

Kazuki Obashi and Shigeo Okabe

Abstract

Molecular mobility within cells is regulated by cytoplasmic structures and cell geometries. Conversely, intricate cellular functions may depend on heterogeneity and compartment-specific regulation of molecular dynamics. Precise measurements of molecular mobility within neuronal cells are challenging due to their highly complex morphology and heterogeneity in the intracellular environment. Fluorescence correlation spectroscopy (FCS) is a sensitive method for measuring diffusion coefficients of fluorescent molecules. Since FCS relies on molecular motion across a detection volume defined by the optical resolution (<1 fL), FCS is an appropriate method for measurements inside small cellular compartments. Here, we describe basic equipment for two-photon FCS and procedures of measurement and data analysis in cells expressing fluorescent proteins. We also show the data from two-photon FCS measurements inside dendritic shafts or spines of cultured hippocampal neurons and discuss the effect of cellular geometry on measurements of small neuronal compartments.

Key words Fluorescence correlation spectroscopy (FCS), Two-photon microscopy, Live-cell imaging, Intracellular molecular diffusion, Hippocampal neuron

1 Introduction

1.1 Molecular Diffusion Inside Neurons

For proper cellular function, localization of molecules should be controlled spatially and temporally. Neurons are highly polarized cells extending elaborate processes specialized for propagation of electrical and biochemical signals. To achieve proper synaptic transmission and subsequent intracellular signaling, macromolecular localization inside specific compartments, such as the cell body, neurites, and synapses, should be regulated precisely. Localization of macromolecules depends on coordination between active transport and passive transport (diffusion). Active transport process is efficient for long-distance movement. In turn, diffusion is effective only over short distances. Active and passive processes of molecular translocation operate independently or in combination to support

Nobuhiko Yamamoto and Yasushi Okada (eds.), *Single Molecule Microscopy in Neurobiology*, Neuromethods, vol. 154, https://doi.org/10.1007/978-1-0716-0532-5_15, © Springer Science+Business Media, LLC, part of Springer Nature 2020

many key functions of neurons. For example, cytoplasmic diffusion filter at the axon initial segment is important for maintenance of the segregation of axonal and somatodendritic compartments [1]. This filter impedes both diffusion of macromolecules and active transport of vesicular carriers into the axon. Another example is diffusion of molecules activated at a subset of synapses along dendrites. This phenomenon has been shown to be important for hetero-synaptic plasticity [2]. In this situation, the size of dendritic segments that express hetero-synaptic plasticity is determined by the distance which signaling molecules can diffuse before they enter the inactivation step.

Regulation of molecular mobility is thought to be important for intracellular signaling from several reasons [3]. First, suppression of molecular mobility is required for compartmentalized biochemical reactions. In the case of dendritic spines, the spine necks serve as diffusion barrier [4] and spines can function as a biochemical compartment [5]. Second, lower molecular mobility reduces an encounter rate of signaling molecules. Dendritic spines are known to be enriched with F-actin. Higher concentration of cytoskeletal polymers inside spines may reduce the rate of molecular interaction. These two factors, confinement and reduction of mobility, may play complex roles during the process of synaptic plasticity, which is associated with spine structural changes. Experimental evidences indicate that plasticity-related molecules are temporally controlled for their localization and biochemical signaling [6, 7]. One of the underlying mechanisms of this process is temporally regulated balance between spine-shaft exchange rate and diffusion-restricted intermolecular interaction inside spines [8].

Molecular mobility in cellular compartments deviates from the data obtained from observation in dilute solutions. For example, translational diffusion coefficient of EGFP in the cytoplasm is 3–4 times slower than that in solution [9, 10]. This is because the composition, geometry, and solvent properties of the intracellular environment determine the characteristics of molecular diffusion [11, 12]. Due to the heterogeneity in these critical parameters, molecular mobility in different cellular compartments should be measured independently. In reverse, information about the cytoplasmic structure inside each cellular compartment can be extracted from the mobility data of biologically inert molecules [13].

1.2 Optical Methods for Measuring Fast Diffusion

To measure the molecular diffusion inside cells, single-particle tracking (SPT), fluorescence recovery after photobleaching (FRAP), and fluorescence correlation spectroscopy (FCS) are widely used. SPT is the most direct method of obtaining information about molecular diffusion [14]. SPT tracks molecular motion directly and can detect short pause or transient binding to intracellular structures. In other methods, these irregular molecular behaviors can be detected only indirectly as deviation from physical models based on Brownian diffusion. Currently a major limitation of SPT is the detection speed of molecules. The upper limit of the detection speed is determined by available photons from single molecules within a time frame, speed of photobleaching, and maximal frame rate of detectors. From the limitation in the detection speed, SPT is generally applied for slow two-dimensional movements of membrane molecules on the cell surface or on the membrane organelles.

Instead, FRAP and FCS are widely used to measure the average characteristics of faster moving molecules. In FRAP measurements, fluorescent molecules in a small region of the cell are irreversibly photobleached and subsequent movement of surrounding non-bleached fluorescent molecules into the photobleached area is monitored. From the recovery kinetics, molecular mobility and immobile fraction of molecules can be estimated [15]. To achieve rigorous measurement of fast-diffusing molecules, instantaneous bleaching of most of the fluorescent molecules and subsequent monitoring of fluorescence recovery without any time lag should be performed. Practically, sufficient bleaching requires the duration of intense light exposure in the order of 10–100 ms. In this condition, contamination of signals from fast-moving molecules cannot be ignored in the initial phase of the fluorescence recovery curve, limiting accurate measurements of diffusion coefficients only for relatively slow-moving molecules. Therefore, FRAP is suitable to measure the molecular replacements that take place in a time range of seconds to hours. Recently, measurements of the fluorescence decay after photoactivation of monomeric photoconvertible fluorescent proteins overcome this temporal limitation [16, 17].

FCS is a convenient and suitable method to measure fast molecular mobility inside cells. FCS is often referred to as a single-molecule method because it analyzes the fluorescence fluctuations caused by movement of single-fluorescent molecules. However, FCS does not use trajectories of single molecules, which were the key information of molecular dynamics in SPT. In FCS experiments, the fluctuation of fluorescence intensity from the detection volume fixed at a specific intracellular position is recorded as a function of time (Fig. 1a, b). Since the fluorescence intensity fluctuates as the molecules enter and leave the fixed detection volume, the characteristics of intensity fluctuation essentially contain information about local diffusion speed and concentration of molecules. When the molecules diffuse fast, fluorescence intensity fluctuates fast. Higher concentration of the molecules within detection volume leads to reduction in amplitude of fluorescence fluctuations. To analyze the fluctuation pattern and derive the information about molecular mobility and concentration, recorded intensity fluctuations are transformed to a time-dependent autocorrelation function [$G(\tau)$] (Fig. 1c). Autocorrelation analysis provides a measure of the self-similarity of the temporal signal so that it

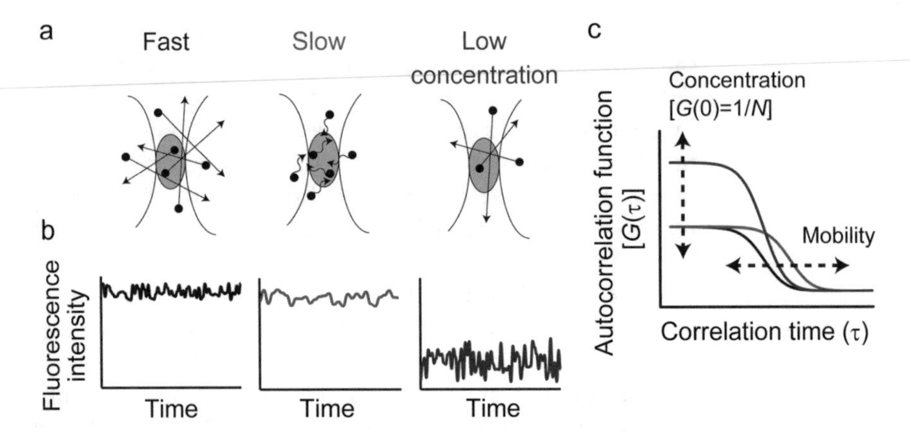

Fig. 1 Fluorescence fluctuations and autocorrelation analysis. In FCS experiments, the fluctuation of fluorescence intensity from the detection volume fixed at a specific intracellular position is recorded as a function of time. Since the fluorescence intensity fluctuates as the molecules enter and leave the fixed detection volume (**a**), the characteristics of intensity fluctuation are affected by diffusion speed and concentration of molecules (**b**). To analyze the fluctuation pattern and derive the information about molecular mobility and concentration, recorded intensity fluctuations are transformed to a time-dependent autocorrelation function [$G(\tau)$] (**c**)

allows us to extract the mobility information related to the intrinsic properties of the molecules and the cytoplasmic environment. As the diffusion becomes faster, autocorrelation curve shifts left with shorter correlation time. And as the concentration becomes higher, amplitude of autocorrelation curve [$G(0)$] becomes lower. By fitting the autocorrelation function to an appropriate physical model, information about molecular mobility, local concentration, states of aggregation, and molecular interactions can be estimated [18, 19]. Because FCS relies on molecular motion across a detection volume, slowly moving and immobile molecules cannot be properly measured.

1.3 Two-Photon FCS

FCS is usually combined with scanning microscopic systems, including confocal and two-photon microscopy, although the focal point of the excitation laser is fixed to a specific cellular domain during measurements. This configuration is advantageous in reducing detection volume, enhancing detection efficiency, and suppressing background noise. The two-photon excitation depends on square of photon flux and local confinement of fluorophore excitation can be easily achieved by focusing of the laser beam by an objective lens. Due to this excitation characteristics, two-photon FCS has properties different from confocal FCS [20]. First, a pinhole is not needed due to inherent depth discrimination of two-photon excitation. This makes the optical setup simpler and pinhole adjustment unnecessary. For confocal FCS, pinhole size and position are related to the shape of the detection volume [21] and an appropriate model should be selected based on the shape of the detection volume. Two-photon excitation takes a nearly three-dimensional Gaussian shape and a standard model can be applied. Second, fluorophore pairs with different emission spectra can be excited by a laser with the same wavelength. Therefore, fluorescence cross-correlation spectroscopy (FCCS) can be implemented without alignment of two laser lines [22]. Third, out-of-focus photobleaching is reduced due to spatially confined excitation. This enables a prolonged measurement in the cytoplasm and decreases phototoxicity. Although two-photon FCS has these advantages, it is not widely used for measuring intracellular molecular mobility, mainly due to relatively low levels of FCS signals. Brightness of fluorophores achieved by two-photon excitation is generally lower than one-photon excitation due to saturation [23] and photobleaching [24]. Although longer excitation wavelength of two-photon FCS contributes to less nonspecific background induced by cellular components [25], calculated signal-to-noise ratios are generally low, compared to one-photon excitation. Recently available detectors with high quantum efficiency and low noise such as gallium arsenide phosphide (GaAsP) photomultiplier tubes (PMT) or hybrid photodetectors are expected to compensate this disadvantage. Therefore, two-photon FCS is a promising technique with potential applications to a variety of cell types and cellular compartments.

In the following sections, we describe how to design and perform two-photon FCS experiments inside neurons. We first describe the setup and general consideration for live-cell FCS measurements. Then, we discuss the points and modification for two-photon FCS analysis inside neuronal compartments.

2 Materials and Equipment

Two-photon FCS setup is similar to a standard two-photon laser scanning microscopy for imaging experiments. A femtosecond pulse laser is focused on the sample through a high numerical aperture (NA) objective. Emitted fluorescence is collected by the same objective, reflected by a dichroic mirror and detected by a photon counting detector with high sensitivity. The detected fluorescence signal is sent to a digital correlator for real-time data processing, and then analyzed by an analysis software.

2.1 Laser

Fluorophore is excited by a mode-locked femtosecond pulse Ti: sapphire laser tuned to optimum wavelength and power. The excitation laser power is modulated by a glan laser polarizer and neutral-density filters. Other than wavelength and intensity of excitation laser, temporal and spectral pulse widths affect two-photon excitation efficiency [26, 27]. The molecular brightness is higher for shorter temporal width and wider spectral width of the excitation pulses at the same laser power. However, the fluorophore properties, such as bleaching and saturation, are reported to be major factors that limit maximum achievable molecular brightness [28]. Therefore, careful control of pulse width is not essential for most biological experiments. Positions of the fluorescence excitation can be controlled by either beam steering or stage movement. The former configuration is more popular, as built-in galvanometer mirrors in commercially available systems can be utilized. In the beam steering configuration, measurement position should be near the center of the scanning field because lateral position affects the result mainly due to optical aberrations [29].

2.2 Objective Lens

To reduce focal volume, objectives with high NA (>0.9) are used and the excitation laser beam is expanded to slightly overfill the back aperture of the objective to create diffraction-limited focal volume. For reliable FCS measurements, a shape of detection volume should be close to a theoretical model, namely a three-dimensional Gaussian. Aberrations can be induced by coverslip thickness and its tilt to the optical axis or by refractive index mismatch between immersion and imaging media. As these factors distort and elongate the size and shape of focal volume, objectives with adjustable correction collar are recommended. Although oil or silicon oil immersion objectives with higher NA are available [30], water immersion objectives are superior to these objectives for FCS measurements inside living samples in aqueous environment since the effect of aberration inside thick specimens is lower for water immersion objectives when focusing more than a few micrometers from a coverslip. We use 1.1 NA water dipping objective with adjustable correction collar (LUMFI 60×, Olympus).

2.3 Photon Counting Detector

An avalanche photodiode (APD) is a highly sensitive photon counting detector and widely used for FCS. The two-photon excitation efficiency depends on the square of the excitation laser intensity. Due to this nonlinearity, two-photon microscopy bears intrinsic optical sectioning ability and does not need confocal detection. Therefore, direct (non-descanned) detection is widely used. Its simple design and high efficiency are advantageous. Since direct detection needs large detection area, GaAsP PMT is used instead of APD that has small detection area. We use H7421-40 (Hamamatsu) as a detector. This GaAsP PMT module has a thermoelectric cooler, which reduces thermal noise generated from the photocathode and enables FCS measurements with better signal-to-background ratio. It should be noted that the GaAsP photocathode can be damaged by the excessive light input.

For accurate FCS measurements, background signal should be kept low. Direct detection without a pinhole is sensitive to stray light of the room, compared with a confocal system. Therefore, it is important to find light sources and shield the setup by monitoring readout of the detector while operating the system without fluorescent samples.

2.4 Correlator and Data Analysis Software

Both a hardware correlator and a software correlator are widely used. Real-time data processing helps for adjustment of experimental setup and procedures. We use a multiple tau hardware correlator (ALV-7004/USB, ALV GmbH). Using this correlator, raw counting trace cannot be saved. Therefore, if it is desirable to calculate correlation after selecting the time window or correcting the trace, a software correlator is more appropriate. Once correlation curves are obtained, they are evaluated by fitting to an optimal physical model using any analysis software (Origin, MATLAB).

2.5 Fluorophores

For selection of fluorophores and labeling methods, general considerations for fluorescence imaging should also be applied, such as size and physical properties of fluorophores, position of labeling, and ratio of endogenous molecules and probes. These parameters should be set to preserve protein localization and cellular functions. For FCS, the following points should be controlled more rigorously than conventional fluorescence imaging. First, biological inertness to intracellular structures is important. Even weak intermolecular interaction, association to intracellular structures, and aggregation change the mobility of probes and then lead to incorrect interpretation. Second, size of probes should also be selected carefully. The cytoplasm is crowded with small solutes, soluble macromolecules, cytoskeletal filaments, and membranes [31]. Mainly due to collision to these intracellular components, mobility for larger molecules is suppressed more prominently [3, 32]. Therefore, on the basal decrease of diffusion coefficient given by the Stokes-Einstein relationship, additional size-dependent suppression of mobility can occur inside cells. Although genetically encoded fluorescent proteins are convenient for protein labeling, their size (27 kDa for EGFP) is not negligible when attached to small proteins. Finally, photostability of fluorophores is a principal parameter for reliable FCS measurements. Molecular brightness of fluorophores is determined not only by absorption cross section and quantum yield but also by photostability, because practically the excitation power is limited by the condition that prevents irreversible photobleaching under a given excitation intensity and measurement duration. Photobleaching increases apparent diffusion coefficient or changes amplitude and shape of autocorrelation function [33]. Therefore, the effect of photobleaching on FCS should be minimized.

2.6 Cells

We use dissociated moue hippocampal neurons on a glass-bottom dish [8, 34]. Fluorescent proteins or probes fused with fluorescent proteins are expressed through Ca^{2+} phosphate transfection [35] or adenovirus infection. Expression level of proteins is controlled to be kept low. During FCS measurements, neurons are kept in HEPES-buffered imaging solution. If you use the inverted microscopes, the thickness of coverslips should be kept in the range for objective's compensation.

3 Methods

3.1 Basic Theory of FCS

With FCS, the mobility and concentration of fluorescent molecules are estimated by the autocorrelation analysis of fluorescence fluctuations (Fig. 1). The details of FCS theory can be found in previous publications [36–38]. Here, we overview a basic principle. The autocorrelation function [$G(\tau)$] is defined as

$$G(\tau) = \frac{< F(t)F(t+\tau) >}{< F(t)>^2} - 1 = \frac{\sum_n F(n\Delta t)F(n\Delta t + \tau)}{\left[\sum_n F(n\Delta t)\right]^2} - 1 \quad (1)$$

where $F(t)$ is the fluorescence intensity observed from the two-photon excitation volume at the time t, τ is the correlation time, and Δt is the sampling period. Brackets denote averaging over all time values t [25]. Through the analysis of this autocorrelation function using an appropriate physical model, mobility information [diffusion coefficient (D)] and average number of molecules (N) can be estimated (Fig. 1c). As movement of molecules is not directed, its speed is described by the diffusion coefficient (μm^2/s) which describes how far a molecule may reach due to its diffusive motion. Considering a single molecular species that freely diffuses in three dimensions and a three-dimensional Gaussian profile as a detection volume [39], the autocorrelation function is

$$G(\tau) = \frac{1}{N}\left(1 + \frac{\tau}{\tau_{\text{diff}}}\right)^{-1}\left(1 + \frac{\tau}{\left(z_0/w_0\right)^2 \tau_{\text{diff}}}\right)^{-1/2} \quad (2)$$

where N is the number of fluorescent molecules in the detection volume, τ_{diff} is the diffusion time which means the average lateral transit time of molecules through the focal volume, and w_0 and z_0 are lateral and axial $1/e^2$ excitation beam waists (Fig. 2). The ratio of lateral and axial beam radius is called structure parameter ($s = z_0/w_0$) which characterizes the shape of detection volume. Faster moving molecules (with higher D value) pass the detection volume with shorter durations (with shorter diffusion time). Therefore, the diffusion time is related to the diffusion coefficient (D) and the lateral radius of excitation intensity distribution (w_0):

$$w_0^2 = 8D \times \tau_{\text{diff}} \quad (3)$$

3.2 Optical Alignment

Optical alignment of the system follows the same principles of imaging experiments. Using fluorescent beads and solution of fluorescent dyes (or a fluorescent reference slide), position and angle of optical components are corrected to optimize collimation, position, and angle of excitation laser and correction efficiency of emitted fluorescence.

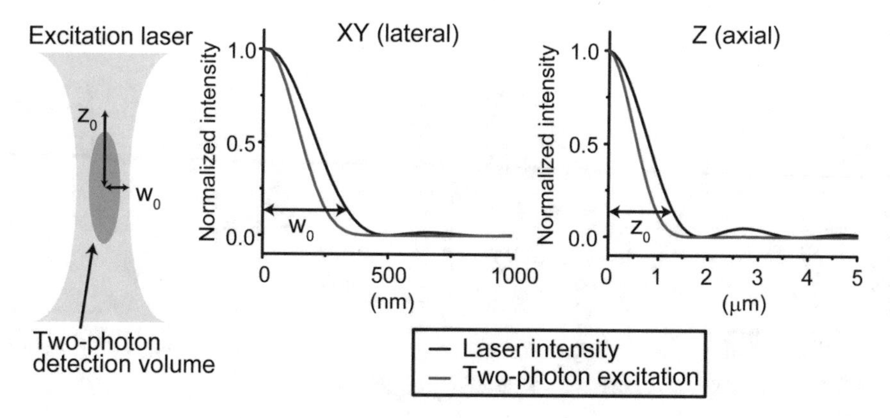

Fig. 2 Two-photon excitation profile. A two-photon excitation profile shows focal excitation volume smaller than the one-photon excitation profiles. This effective spatial confinement is derived from the fact that the photon efflux rate of two-photon excitation is proportional to the square of the local laser intensity. Laser intensity profiles were generated by PSF Generator [56] using a scalar-based diffraction model with $\lambda = 880$ nm and NA 1.1. The one-photon excitation profile was approximated by a three-dimensional Gaussian, which is decayed to $1/e^2$ at w_0 in lateral direction and z_0 in axial direction

3.3 Calibration with Rhodamine 6G Solution

After general alignment, position of correction collar is determined. FCS measurements are performed inside rhodamine 6G (Rh6G) solution dissolved in HEPES-buffered imaging solution at 10^{-6}–10^{-9} M. Correction collar position should be determined in the same refractive index solution for live-cell experiments. At the beginning, correction collar is set to the center position and laser power is set to obtain an appropriate molecular brightness [counts per molecules per seconds (CPM)] (above several kHz). Molecular brightness is calculated as measured count rate divided by molecular number [$N = 1/G(0)$] which is estimated by fitting the autocorrelation curve with Eq. 2. Count rate should be within the linear response range of the detector (for H7421-40, 1.5×10^6/s). If the count rate exceeds the limit, concentration of Rh6G solution is reduced. High molecular brightness indicates that the laser beam is focused well. By determining molecular brightness at different correction collar positions, correction collar position is optimized to maximize molecular brightness (Fig. 3a).

Next, maximum laser power is determined. Laser power should be controlled to avoid both photobleaching and excitation saturation in keeping with acquiring sufficient signals. The latter is caused by a depletion of the ground-state population. Saturation of the fluorescence excitation can alter the efficient size and profile of the focal volume due to uneven excitation profile [23]. Therefore, excitation power for two-photon FCS measurements should be adjusted to be in the range of squared power dependence. The relationship between laser power and molecular brightness is measured (Fig. 3b). Under our measurement condition, squared

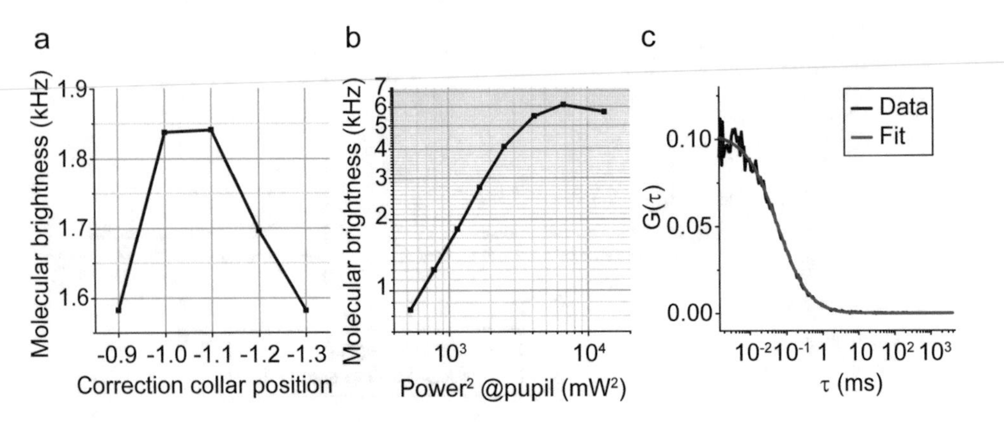

Fig. 3 Two-photon FCS measurements of rhodamine 6G in solution. Two-photon FCS measurements were performed with 60×, NA 1.1 objective and $\lambda = 880$ nm excitation laser. Rhodamine 6G was dissolved in HEPES-buffered imaging solution. (**a**) Relative change of molecular brightness against collection collar position of the objective. (**b**) Excitation power dependence of molecular brightness. Excitation power was measured at the pupil of the objective and represented as squared values. (**c**) An autocorrelation function of rhodamine 6G in solution (black). The autocorrelation function was fitted by the model considering the single molecular species that freely diffuses in three dimensions (Eq. 2, red)

power dependence was observed up to 50 mW laser power at objective pupil and molecular brightness of 4 kHz for Rh6G.

Finally, after setting correction collar position and excitation laser power, a measurement in standard solution, namely Rh6G solution, is performed to characterize focal volume size (Fig. 3c). An obtained autocorrelation curve is fitted by the single-component three-dimensional diffusion model (Eq. 2). From the fitting, the structure parameter (s) and the diffusion time ($\tau_{\mathrm{diff,Rh}}$) can be derived. A diffusion coefficient of Rh6G (D_{Rh}) has been reported as 414 $\mu m^2/s$ [40]. Using this reported value, the lateral radius of excitation intensity profile (w_0) can be determined following Eq. 3. Axial width (z_0) is also determined by combining w_0 and s. Because the two-photon excitation efficiency depends on the square of the excitation laser intensity, effective measurement volume (V_{eff}) approximated as a cylindrical shape [39] is calculated from the following equation:

$$V_{\mathrm{eff}} = \pi^{3/2} \times \left(\frac{w_0}{\sqrt{2}}\right)^2 \times \frac{z_0}{\sqrt{2}} = \left(\frac{\pi}{2}\right)^{3/2} \times w_0{}^2 \times z_0 \qquad (4)$$

In our two-photon setup equipped with NA 1.1 objective at $\lambda = 880$ nm, $\tau_{\mathrm{diff,Rh}}$ was 55 μs and s was 3.5. From these parameters, lateral width and effective measurement volume of focal spot were determined as 300 nm (equivalent FWHM = 355 nm) and 0.53 fL, respectively.

To estimate the diffusion coefficient of a target molecule (D_{target}), the structure parameter is used as fixed value for fitting. From Eq. 3, D_{target} can be estimated from a diffusion time ($\tau_{\text{diff,target}}$) with the following equation:

$$\frac{D_{\text{target}}}{D_{\text{Rh}}} = \frac{\tau_{\text{diff,Rh}}}{\tau_{\text{diff,target}}} \tag{5}$$

3.4 Adjustment of Measurement Position

FCS measurement is performed at a single spot inside solution or a cell. FCS measurement position is selected from a scanned image and laser position is kept stationary by stalling galvanometer mirrors during the measurement.

To construct images, excitation beam is raster-scanned across imaging region. Positions of galvanometer mirrors are determined by the command voltage output from the control device (i.e., a DAQ board). Because the galvanometer mirrors have inertia, their positions during scanning and stalling are not identical with the same command voltages. Temporal lag between the command voltage and physical position of the beam during scanning should be determined before FCS measurements. We measure the lag using a fluorescence reference slide. After initial image scanning, a fluorescence reference slide is photobleached at specified pixel positions. Subsequently an image at the same region is obtained. The temporal lag of mirror positions is translated into spatial mismatch between the bleached spot in the image and the position specified by the voltage command. Either offset of command voltage or pixel position can be adjusted. Until the lag become small enough, repeat this adjustment process. As the degree of lag depends on scanning parameters, the scanning speed and pixel size for imaging are recommended to be fixed.

3.5 Measurement

As in the case of in vitro FCS measurement of Rh6G, intracellular FCS requires prior information about relationship between laser power and molecular brightness of a fluorophore. Because properties of fluorescent proteins are dependent on physical and chemical parameters such as pH, which may differ in vitro and in the cytoplasm, it is important to check the power-brightness relationship directly inside cells. For intracellular measurements, care should be taken to avoid the photodamage. Furthermore, due to small volume of the cells, FCS in the cytoplasm is prone to photobleaching. Therefore, laser power should be reduced to avoid maximizing molecular brightness. Instead, laser power should be adjusted to obtain FCS signal just above 1 kHz of molecular brightness, which is sufficient for reliable fitting [20]. As mentioned in the calibration section, power should be in a range of squared power dependence.

Laser power should be controlled for each experimental condition, as its optimum value may change in different intracellular compartments and with mobility of molecules.

For intracellular measurements, aggregation of fluorescent molecules or movement of organelles frequently disturbs the measurements. Repetition of short measurements, instead of a single continuous measurement, may help reducing such effects. Practically, measurement duration is at least about three orders of magnitude longer than the slowest time component to be resolved [20].

3.6 Data Analysis

3.6.1 Sorting

As mentioned in the previous section, intracellular FCS measurements are distorted often by several factors such as photobleaching and large fluorescence fluctuations due to aggregations or movement of organelles. For example, large fluorescence spikes induce long time correlations and change the shape of autocorrelation function. To minimize these effects, sorting the traces after multiple short measurements is recommended. Single long measurements are more likely to be contaminated with large spikes, which affect the correlation analysis. Multiple short measurements can be sorted by the presence of large spikes and the runs without large spikes can be used for analysis. Short duration is also beneficial for reducing the effect of photobleaching. Because total fluorescence decline during data acquisition becomes small, the effect of photobleaching on the shape of correlation function is reduced [41]. Only runs without prominent photobleaching or spikes are used for further analysis. The criteria for sorting short runs of FCS are reported in the previous paper [20]. If your system can access the counting data before calculating the autocorrelation function, correction methods can be applied [42, 43].

3.6.2 Physical Models

After run sorting, an autocorrelation function is fitted with an appropriate physical model [44]. For measurements inside solution, standard three-dimensional diffusion model (Eq. 2) is appropriate (Fig. 3c). However, mobility inside cells often differs from that inside solutions even if molecules are biologically inert. The actual examples of FCS measurements inside neuronal compartments are shown in the next section.

Photophysical events, such as blinking or triplet state, are detected in an autocorrelation function at microsecond timescales. For intracellular measurements, correlation from these photophysical events and that from molecular mobility can be easily separated based on correlation time. In the fitting process, range of correlation time for fitting is determined by a visual examination or experimentally determined values. For fast diffusion molecules, separation is difficult and models considering these photophysical events are used.

4 Application Examples

4.1 Cell Body

Molecular mobility inside cells often differs from that inside solutions. This difference is derived from the fact that intracellular environment is structurally more complicated. Further, biological molecules interact with other molecules or intracellular structures. Therefore, an autocorrelation function from intracellular measurements is difficult to be evaluated by the standard diffusion model. With increase of obstacles in the environment, molecular mobility is reduced by collisions and binding to obstacles. In this case, mean squared displacement (MSD) does not increase linearly with time but shows a power law (anomalous diffusion, MSD $= \Gamma t^{\alpha}$, where Γ is constant and α is anomaly parameter). EGFP mobility inside cells has been described well by the anomalous diffusion model [45]:

$$G(\tau) = \frac{1}{N}\left(1 + \left(\frac{\tau}{\tau_{\text{diff}}}\right)^{\alpha}\right)^{-1}\left(1 + \frac{1}{s^2}\left(\frac{\tau}{\tau_{\text{diff}}}\right)^{\alpha}\right)^{-1/2} \quad (6)$$

where $\alpha < 1$ for anomalous subdiffusion. In fact, an autocorrelation function from a measurement inside the cell body of an EGFP-expressing dissociated hippocampal neuron was described well by the anomalous diffusion model (Fig. 4).

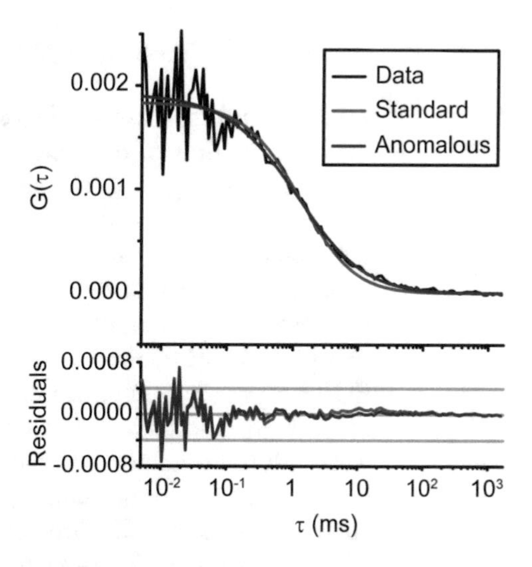

Fig. 4 Two-photon FCS measurement inside a neuronal cell body. Two-photon FCS measurement was performed inside a cell body of an EGFP-expressing dissociated hippocampal neuron at 19 days in vitro. An autocorrelation function (black) was fitted by the standard diffusion model (Eq. 2, red) and the anomalous diffusion model (Eq. 6, blue). The lower panel shows the residuals, which represent the deviations of the data from the fitted curve. The autocorrelation function was well described by the anomalous diffusion model

4.2 Dendritic Shaft

For FCS measurements inside small cellular compartments, boundary of focal volume will be located close to cellular boundaries such as plasma membranes and intracellular membranes. In these situations, cellular boundaries affect the molecular diffusion pattern around excitation volume. Further, if cellular boundaries cross the boundary of excitation volume, the shape of detection volume is affected. Therefore, FCS measurements inside small cellular compartments should be evaluated by the models considering boundary effects [46, 47]. Neurons are highly polarized and extend long axons and dendrites. Because the diameters of axons and dendrites are comparable to the axial dimension of focal excitation volume, results of FCS measurements inside axons and dendrites should be evaluated considering their geometry. Gennerich and Schild [48] proposed the modified FCS model including the confinement of the detection volume:

$$G(\tau) = \frac{1}{\pi {w_0}^2 d_z \langle C \rangle} \times \frac{1}{\sqrt{1 + \tau/\tau_D}} \times g_\Upsilon(\tau) \tag{7}$$

with

$$g_\Upsilon(\tau) = \frac{\sqrt{\pi}}{\Upsilon}\left[1 + \left(\frac{\Upsilon}{\sqrt{\pi}} \times \frac{\mathrm{erf}(\Upsilon)}{\mathrm{erf}^2(\Upsilon/\sqrt{2})} - 1\right) \times \frac{\exp\left[-k(\Upsilon)(\pi/\Upsilon)^2 \tau/\tau_D\right]}{\sqrt{1 + \tau/\tau_D}}\right] \tag{8}$$

$$k(i) = 0.689 + 0.34 \times \exp\left[-0.37 \times (i - 0.5)^2\right] \tag{9}$$

where a dendrite is assumed to extend in the direction of the x-axis, the cross section of dendrite is assumed to be rectangle with the width d_y and height d_z, C is the concentration of fluorescent molecules, and $\Upsilon\,(= d_y/w_0)$ is the confinement parameter. The authors demonstrated that autocorrelation function obtained from FCS measurements inside dendritic shafts of cultured neurons can be evaluated by this modified model. In Fig. 5, we performed a two-photon FCS measurement inside a dendritic shaft of an EGFP-expressing hippocampal dissociated neuron. A single-component anomalous diffusion model (Eq. 6) was largely fitted well with the obtained correlation curve but an apparent deviation was observed at longer correlation time (Fig. 5c). To evaluate the curve using the modified model (Eq. 7), we first measured the size of dendritic shafts. Line-scanning profile of dendritic shafts is the convolution product of the two-photon detection volume and cross section of dendritic shafts. Width and height of dendritic shafts were determined by fitting to the theoretical model under the assumption of rectangular cross section (Fig. 5b). The modified model that incorporated the shape of the dendritic shafts that additionally considered anomalous diffusion provided more reasonable description of the data (Fig. 5c). Therefore, molecular

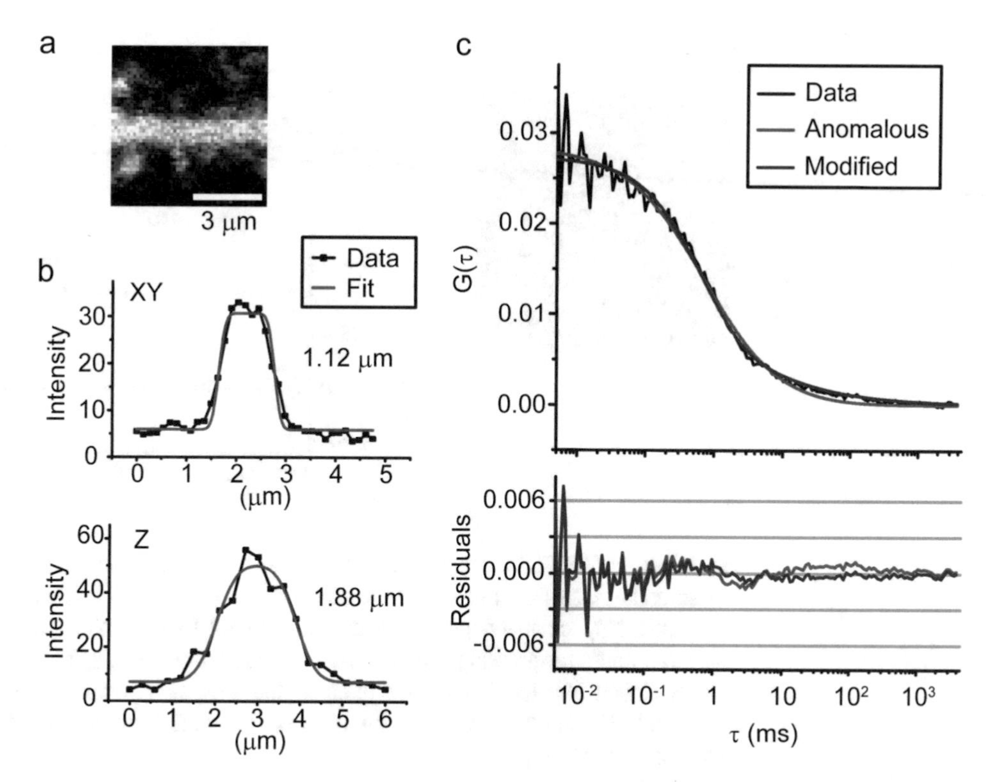

Fig. 5 Two-photon FCS measurement inside a dendritic shaft. Two-photon FCS measurement was performed inside a dendritic shaft of an EGFP-expressing dissociated hippocampal neuron at 19 days in vitro. (**a**) Two-photon image of a measured neuron expressing EGFP. (**b**) Lateral (upper) and axial (bottom) line scanning profiles of the dendritic shaft (black). The profiles were fitted to the theoretical model based on the convolution product of the two-photon detection volume and cross section of a dendritic shaft assuming rectangular shape (red). From the fit, width and height of the dendritic shaft were determined. (**c**) An autocorrelation function (black) was fitted by the anomalous diffusion model (Eq. 3, red) and the modified model considering the confined effect induced by the shape of the dendritic shaft (Eq. 7 with anomalous diffusion, blue). The lower panel shows the residuals. The autocorrelation function was well described by the modified model

mobility inside dendritic shafts can be quantitatively evaluated by FCS with the confined-diffusion model.

4.3 Dendritic Spine

Dendritic spines are tiny protrusions, with a roughly spherical head (0.5–1.5 μm in diameter) connected by a narrow neck (<0.5 μm in diameter) to a dendritic shaft. Although two-photon FCS inside spines has been reported [8, 49], quantitative measurements are still challenging due to their small size and unique shape. As the volumes of dendritic spines are 0.01–0.3 fL [50] and their cross sections are smaller than those of dendritic shafts, boundary effect will be more prominent. Further, the alignment of a laser focus to the center of spine is more difficult in spines with smaller heads. These two factors, the spine head volume and the lateral displacement of the laser focus from the spine center, affect FCS results [8]. Although a physical model considering actual dendritic

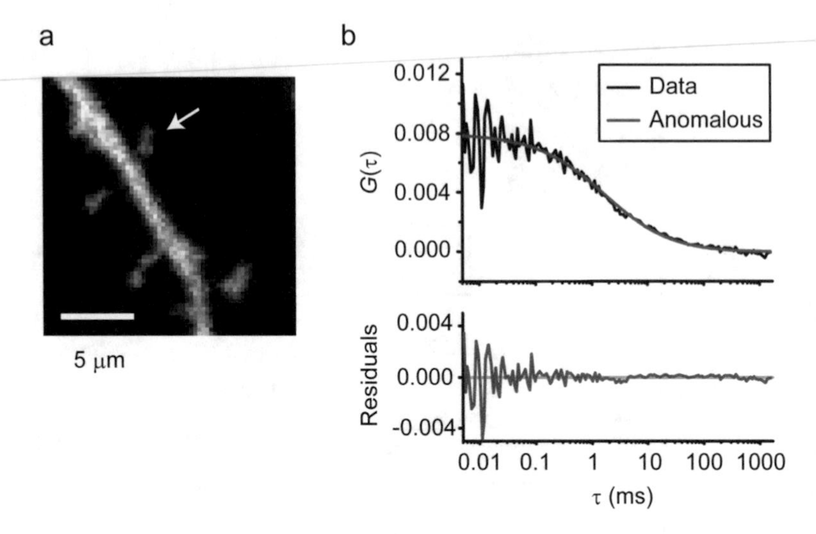

Fig. 6 Two-photon FCS measurement inside a dendritic spine. Two-photon FCS measurement was performed inside a dendritic spine of an EGFP-expressing dissociated hippocampal neuron at 19 days in vitro. (**a**) Two-photon image of a measured neuron expressing EGFP. Arrow indicates a measured spine. (**b**) An autocorrelation function (black) was fitted by the anomalous diffusion model (Eq. 6, red). The lower panel shows the residuals (figure is taken and modified from Obashi et al. [8])

geometry and laser focus position is needed for the establishment of a quantitative analysis, obtaining these factors is difficult due to small size and highly dynamic nature of dendritic spines [51]. To obtain the diffusion times, we used the anomalous diffusion model as an alternative. In Fig. 6, we performed a two-photon FCS measurement inside a dendritic spine of an EGFP-expressing hippocampal dissociated neuron. The single-component anomalous diffusion model (Eq. 6) was largely fitted well with the obtained correlation curve and deviation of a spine measurement was similar to that of a shaft measurement. Therefore, diffusion inside spines can be analyzed semiquantitatively by using the anomalous diffusion model. To analyze more quantitatively, a physical model considering complex nanostructure of individual spines should be developed.

Molecules can enter and leave dendritic spines only through thin spine necks. Therefore, mobility pattern inside dendritic spines will be modulated by spine shape. Furthermore, spine necks act as a diffusion barrier [4]. Small spine volume and confinement induced by thin spine necks accelerate photobleaching, which induces artifacts in FCS. Scanning FCS or raster image correlation spectroscopy (RICS) may be useful in reducing artifacts derived from photobleaching [8, 52].

5 Future Direction

To perform reliable FCS inside small cellular compartments, reducing the detection volume is a straightforward solution. FCS combined with STED microscopy (STED-FCS) is a promising approach. However, original STED-FCS in 3D is difficult due to a significant nonspecific background signal. This background signal changes the amplitude of autocorrelation function and precludes accurate FCS measurements [53]. Recently, this problem has been solved by background determination and subsequent subtraction from the readout [54] or by combining photon separation and fluorescence lifetime correlation spectroscopy [55]. In these reports, STED-FCS has been successfully performed inside living cells. Thus, combination of newly developed microscopic techniques and appropriate physical models will make it possible to perform accurate FCS measurements inside small cellular compartments such as dendritic spines.

Acknowledgments

This study was supported by Grants-in-Aid for Scientific Research (17H01387 and 18H04727), Core Research for Evolutional Science and Technology from the Japan Science and Technology Agency (JPMJCR14W2), the Project for Elucidating and Controlling Mechanisms of Aging and Longevity from the Japan Agency for Medical Research and Development (17gm5010003), the Tokyo Center for Integrative Science of Human Behavior (CiSHuB) to S.O., and a Grant-in-Aid for JSPS Fellows (13J03845) to K.O.

References

1. Song AH, Wang D, Chen G, Li Y, Luo J, Duan S, Poo MM (2009) A selective filter for cytoplasmic transport at the axon initial segment. Cell 136:1148–1160. https://doi.org/10.1016/j.cell.2009.01.016
2. Yasuda R (2017) Biophysics of biochemical signaling in dendritic spines: implications in synaptic plasticity. Biophys J 113 (10):2152–2159. https://doi.org/10.1016/j.bpj.2017.07.029
3. Verkman AS (2002) Solute and macromolecule diffusion in cellular aqueous compartments. Trends Biochem Sci 27:27–33. https://doi.org/10.1016/S0968-0004(01)02003-5
4. Bloodgood BL, Sabatini BL (2005) Neuronal activity regulates diffusion across the neck of dendritic spines. Science 310:866–869. https://doi.org/10.1126/science.1114816
5. Newpher TM, Ehlers MD (2009) Spine microdomains for postsynaptic signaling and plasticity. Trends Cell Biol 19:218–227. https://doi.org/10.1016/j.tcb.2009.02.004
6. Murakoshi H, Yasuda R (2012) Postsynaptic signaling during plasticity of dendritic spines. Trends Neurosci 35:135–143. https://doi.org/10.1016/j.tins.2011.12.002
7. Bosch M, Castro J, Saneyoshi T, Matsuno H, Sur M, Hayashi Y (2014) Structural and molecular remodeling of dendritic spine substructures during long-term potentiation. Neuron 82:299–304. https://doi.org/10.1016/j.neuron.2014.03.021

8. Obashi K, Matsuda A, Inoue Y, Okabe S (2019) Precise temporal regulation of molecular diffusion within dendritic spines by actin polymers during structural plasticity. Cell Rep 27:1503–1515.e8. https://doi.org/10.1016/j.celrep.2019.04.006
9. Swaminathan R, Hoang CP, Verkman AS (1997) Photobleaching recovery and anisotropy decay of green fluorescent protein GFP-S65T in solution and cells: cytoplasmic viscosity probed by green fluorescent protein translational and rotational diffusion. Biophys J 72:299–304. https://doi.org/10.1016/S0006-3495(97)78835-0
10. Pack C, Saito K, Tamura M, Kinjo M (2006) Microenvironment and effect of energy depletion in the nucleus analyzed by mobility of multiple oligomeric EGFPs. Biophys J 91:299–304. https://doi.org/10.1529/biophysj.105.079467
11. Luby-Phelps K (2000) Cytoarchitecture and physical properties of cytoplasm: volume, viscosity, diffusion, intracellular surface area. Int Rev Cytol 192:299–304. https://doi.org/10.1016/S0074-7696(08)60527-6
12. Lin YC, Phua SC, Lin B, Inoue T (2013) Visualizing molecular diffusion through passive permeability barriers in cells: conventional and novel approaches. Curr Opin Chem Biol 17:299–304. https://doi.org/10.1016/j.cbpa.2013.04.027
13. Baum M, Erdel F, Wachsmuth M, Rippe K (2014) Retrieving the intracellular topology from multi-scale protein mobility mapping in living cells. Nat Commun 5:299–304. https://doi.org/10.1038/ncomms5494
14. Moerner WE, Shechtman Y, Wang Q (2015) Single-molecule spectroscopy and imaging over the decades. Faraday Discuss 184:9–36. https://doi.org/10.1039/c5fd00149h
15. Lippincott-Schwartz J, Snapp E, Kenworthy A (2001) Studying protein dynamics in living cells. Nat Rev Mol Cell Biol 2:299–304. https://doi.org/10.1038/35073068
16. Matsuda T, Miyawaki A, Nagai T (2008) Direct measurement of protein dynamics inside cells using a rationally designed photoconvertible protein. Nat Methods 5:339–345. https://doi.org/10.1038/nmeth.1193
17. Gura Sadovsky R, Brielle S, Kaganovich D, England JL (2017) Measurement of rapid protein diffusion in the cytoplasm by photoconverted intensity profile expansion. Cell Rep 18:2795–2806. https://doi.org/10.1016/j.celrep.2017.02.063
18. Kinjo M, Sakata H, Mikuni S (2011) First steps for fluorescence correlation spectroscopy of living cells. Cold Spring Harb Protoc 2011:1185–1189. https://doi.org/10.1101/pdb.top065920
19. Bacia K, Haustein E, Schwille P (2014) Fluorescence correlation spectroscopy: principles and applications. Cold Spring Harb Protoc 2014:299–304. https://doi.org/10.1101/pdb.top081802
20. Kim SA, Heinze KG, Schwille P (2007) Fluorescence correlation spectroscopy in living cells. Nat Methods 4:963–973. https://doi.org/10.1038/nmeth1104
21. Hess ST, Webb WW (2002) Focal volume optics and experimental artifacts in confocal fluorescence correlation spectroscopy. Biophys J 83:2300–2317. https://doi.org/10.1016/S0006-3495(02)73990-8
22. Heinze KG, Koltermann A, Schwille P (2000) Simultaneous two-photon excitation of distinct labels for dual-color fluorescence cross-correlation analysis. Proc Natl Acad Sci U S A 97:10377–10382. https://doi.org/10.1073/pnas.180317197
23. Berland K, Shen G (2003) Excitation saturation in two-photon fluorescence correlation spectroscopy. Appl Opt 42:5566–5576. https://doi.org/10.1364/AO.42.005566
24. Dittrich PS, Schwille P (2001) Photobleaching and stabilization of fluorophores used for single-molecule analysis with one- and two-photon excitation. Appl Phys B Lasers Opt 73:829–837. https://doi.org/10.1007/s003400100737
25. Schwille P, Haupts U, Maiti S, Webb WW (1999) Molecular dynamics in living cells observed by fluorescence correlation spectroscopy with one- and two-photon excitation. Biophys J 77:2251–2265. https://doi.org/10.1016/S0006-3495(99)77065-7
26. Xu C, Webb WW (1996) Measurement of two-photon excitation cross sections of molecular fluorophores with data from 690 to 1050 nm. J Opt Soc Am B 13:481. https://doi.org/10.1364/JOSAB.13.000481
27. Helmchen F, Denk W (2005) Deep tissue two-photon microscopy. Nat Methods 2:299–304. https://doi.org/10.1038/nmeth818
28. Mütze J, Petrášek Z, Schwille P (2007) Independence of maximum single molecule fluorescence count rate on the temporal and spectral laser pulse width in two-photon FCS. J Fluoresc 17:805–810. https://doi.org/10.1007/s10895-007-0246-5
29. Dross N, Spriet C, Zwerger M, Müller G, Waldeck W, Langowski J (2009) Mapping eGFP oligomer mobility in living cell nuclei.

PLoS One 4:1–13. https://doi.org/10.1371/journal.pone.0005041

30. Banachowicz E, Patkowski A, Meier G, Klamecka K, Gapiński J (2014) Successful FCS experiment in nonstandard conditions. Langmuir 30:299–304. https://doi.org/10.1021/la5015708
31. Medalia O, Weber I, Frangakis AS, Nicastro D, Gerisch G, Baumeister W (2002) Macromolecular architecture in eukaryotic cells visualized by cryoelectron tomography. Science 298:299–304. https://doi.org/10.1126/science.1076184
32. Luby-Phelps K, Taylor DL, Lanni F (1986) Probing the structure of cytoplasm. J Cell Biol 102:2015–2022. https://doi.org/10.1083/jcb.102.6.2015
33. Bacia K, Schwille P (2007) Practical guidelines for dual-color fluorescence cross-correlation spectroscopy. Nat Protoc 2:2842–2856. https://doi.org/10.1038/nprot.2007.410
34. Okabe S, Kim HD, Miwa A, Kuriu T, Okado H (1999) Continual remodeling of postsynaptic density and its regulation by synaptic activity. Nat Neurosci 2:804–811. https://doi.org/10.1038/12175
35. Jiang M, Chen G (2006) High Ca^{2+}-phosphate transfection efficiency in low-density neuronal cultures. Nat Protoc 1:695–700. https://doi.org/10.1038/nprot.2006.86
36. Lakowicz JR (2006) Fluorescence correlation spectroscopy. In: Principles of fluorescence spectroscopy, 3rd edn. Springer, New York
37. Elson EL (2011) Fluorescence correlation spectroscopy: past, present, future. Biophys J 101:2855–2870. https://doi.org/10.1016/j.bpj.2011.11.012
38. Schwille P (2001) Fluorescence correlation spectroscopy and its potential for intracellular applications. Cell Biochem Biophys 34:383–408. https://doi.org/10.1385/CBB:34:3:383
39. Rigler R, Mets Ü, Widengren J, Kask P (1993) Fluorescence correlation spectroscopy with high count rate and low background: analysis of translational diffusion. Eur Biophys J 22:299–304. https://doi.org/10.1007/BF00185777
40. Müller CB, Loman A, Pacheco V, Koberling F, Willbold D, Richtering W, Enderlein J (2008) Precise measurement of diffusion by multi-color dual-focus fluorescence correlation spectroscopy. Europhys Lett 83:299–304. https://doi.org/10.1209/0295-5075/83/46001
41. Wachsmuth M, Conrad C, Bulkescher J, Koch B, Mahen R, Isokane M, Pepperkok R, Ellenberg J (2015) High-throughput fluorescence correlation spectroscopy enables analysis of proteome dynamics in living cells. Nat Biotechnol 33:299–304. https://doi.org/10.1038/nbt.3146
42. Delon A, Usson Y, Derouard J, Biben T, Souchier C (2004) Photobleaching, mobility, and compartmentalisation: inferences in fluorescence correlation spectroscopy. J Fluoresc 14:255–267. https://doi.org/10.1023/B:JOFL.0000024557.73246.f9
43. Ries J, Bayer M, Csúcs G, Dirkx R, Solimena M, Ewers H, Schwille P (2010) Automated suppression of sample-related artifacts in fluorescence correlation spectroscopy. Opt Express 18:11073. https://doi.org/10.1364/OE.18.011073
44. Vukojević V, Pramanik A, Yakovleva T, Rigler R, Terenius L, Bakalkin G (2005) Study of molecular events in cells by fluorescence correlation spectroscopy. Cell Mol Life Sci 62:535–550. https://doi.org/10.1007/s00018-004-4305-7
45. Wachsmuth M, Waldeck W, Langowski J (2000) Anomalous diffusion of fluorescent probes inside living cell nuclei investigated by spatially-resolved fluorescence correlation spectroscopy. J Mol Biol 298:677–689. https://doi.org/10.1006/jmbi.2000.3692
46. Fradin C, Abu-Arish A, Granek R, Elbaum M (2003) Fluorescence correlation spectroscopy close to a fluctuating membrane. Biophys J 84:2005–2020. https://doi.org/10.1016/S0006-3495(03)75009-7
47. von der Hocht I, Enderlein J (2007) Fluorescence correlation spectroscopy in cells: confinement and excluded volume effects. Exp Mol Pathol 82:299–304. https://doi.org/10.1016/j.yexmp.2006.12.009
48. Gennerich A, Schild D (2000) Fluorescence correlation spectroscopy in small cytosolic compartments depends critically on the diffusion model used. Biophys J 79:3294–3306. https://doi.org/10.1016/S0006-3495(00)76561-1
49. Chen JH, Kellner Y, Zagrebelsky M, Grunwald M, Korte M, Walla PJ (2015) Two-photon correlation spectroscopy in single dendritic spines reveals fast actin filament reorganization during activity-dependent growth. PLoS One 10:e0128241. https://doi.org/10.1371/journal.pone.0128241
50. Arellano JI, Benavides-Piccione R, Defelipe J, Yuste R (2007) Ultrastructure of dendritic spines: correlation between synaptic and spine morphologies. Front Neurosci 1:131–143. https://doi.org/10.3389/neuro.01.1.1.010.2007

51. Dunaevsky A, Tashiro A, Majewska A, Mason C, Yuste R (1999) Developmental regulation of spine motility in the mammalian central nervous system. Proc Natl Acad Sci U S A 96:13438–13443. https://doi.org/10.1073/PNAS.96.23.13438
52. Digman MA, Gratton E (2011) Lessons in fluctuation correlation spectroscopy. Annu Rev Phys Chem 62:645–668. https://doi.org/10.1146/annurev-physchem-032210-103424
53. Ringemann C, Harke B, Von Middendorff C, Medda R, Honigmann A, Wagner R, Leutenegger M, Schönle A, Hell SW, Eggeling C (2009) Exploring single-molecule dynamics with fluorescence nanoscopy. New J Phys 11:103054. https://doi.org/10.1088/1367-2630/11/10/103054
54. Gao P, Prunsche B, Zhou L, Nienhaus K, Nienhaus GU (2017) Background suppression in fluorescence nanoscopy with stimulated emission double depletion. Nat Photonics 11:299–304. https://doi.org/10.1038/nphoton.2016.279
55. Lanzanò L, Scipioni L, Di Bona M, Bianchini P, Bizzarri R, Cardarelli F, Diaspro A, Vicidomini G (2017) Measurement of nanoscale three-dimensional diffusion in the interior of living cells by STED-FCS. Nat Commun 8:1–9. https://doi.org/10.1038/s41467-017-00117-2
56. Kirshner H, Aguet F, Sage D, Unser M (2013) 3-D PSF fitting for fluorescence microscopy: implementation and localization application. J Microsc 249:13–25. https://doi.org/10.1111/j.1365-2818.2012.03675.x

INDEX

A

Active zones.......245–258
Anomalous diffusion.......29, 48, 51, 52, 123, 321–324
Astigmatism.......204, 248, 257, 263, 266, 287
Astrocytes.......139, 140, 150, 278
Axonal transport.......157, 158, 169
Axons.......1, 2, 5, 9, 60, 82, 92, 101, 103, 105, 108, 111, 157, 158, 165–169, 177, 287, 295–297, 300, 301, 303, 304, 307, 310, 322

B

Bessel beam.......263–265, 268–270, 287
Biocompatibility.......243
Bleaching.......5, 10, 51, 52, 54, 55, 68, 84, 91, 95, 96, 121, 182, 183, 188, 205, 209, 301, 304, 311, 314
Blinking.......27, 29, 120, 121, 124, 141, 145, 146, 173–175, 178, 181–183, 186, 188–190, 196, 197, 205, 207–211, 213, 215–220, 222–224, 251, 263, 264, 272, 275, 283, 287, 290, 302, 320

C

Caenorhabditis elegans.......11–34
CAMP response element (CRE).......65–67, 75
CAMP response element binding protein (CREB).......60, 62, 65–67, 69, 71, 74, 75
Channel rhodopsin-2 (ChR2).......66
Clustered regularly interspaced short palindromic repeats (CRISPR) genetic engineering.......15–17, 21, 33
Correlative light-electron microscopy (CLEM).......89, 90, 94, 104
Cortical neurons.......60, 66, 67, 70, 73–75, 149, 150

D

Dark-field.......157–169
Dendrites.......1, 2, 5, 177, 287, 295, 305, 310, 322
Density-based spatial clustering of applications with noise (DBSCAN).......192
Diffraction limits.......141, 173, 178, 203, 214, 229, 235, 241, 242, 262
Diffusions.......13, 14, 27–33, 42, 48–52, 55, 56, 88, 89, 100, 101, 116, 117, 121, 123–127, 132, 133, 144–151, 167, 179, 309–325
Dissociated cell culture.......134
Dorsal root ganglion (DRG) neuron.......159, 165
Drift correction.......89, 99, 174, 182, 184, 185, 191, 193, 257, 287

E

Effective rate constants.......116
Electron multiplying charge coupled device (EMCCD) camera.......212
Electroporation.......4, 6–8, 62, 67, 68, 296, 299, 300, 303
Endocytosis.......81, 84, 104, 109, 131, 145, 151, 158, 166
Endosomes.......82, 90, 104, 108, 157–169
Excitation polarization angle narrowing (ExPAN).......230, 231

F

Fluorescence correlation spectroscopy (FCS).......60, 310–318, 320–325
Fluorescent beads.......28, 53, 89, 99, 110, 182, 183, 239, 241, 252, 253, 257, 258, 266, 269, 270, 316
Fluorescent probes.......203, 204, 206, 207, 222, 224, 230
Fluorescent proteins.......2, 5, 12, 82, 91, 132, 140, 151, 175, 176, 204–206, 211, 230–233, 235, 262, 296, 301, 302, 304, 307, 311, 315, 318

G

Gaussian fitting.......13, 23, 29, 191
Gephyrin.......116, 118, 122, 123, 126, 127, 179, 181, 184, 189, 190, 192–195, 197
Glycine receptors (GlyR).......116, 121, 126, 179, 184, 189, 192–194, 196
Gold nanorods.......159, 160, 163, 166–169
Growth cones.......296, 297, 301–303, 305–307

Nobuhiko Yamamoto and Yasushi Okada (eds.), *Single Molecule Microscopy in Neurobiology*, Neuromethods, vol. 154, https://doi.org/10.1007/978-1-0716-0532-5,

H

Halo tag ... 5
Halo tag ligand ... 43, 54, 65, 68, 69, 71, 74, 77, 213
Highly inclined laminated optical sheet (HILO) illumination ... 16, 26, 27, 42, 46, 53–55, 60, 62–64, 69–71, 74, 75, 302
Hippocampal neurons ... 5, 6, 82–85, 87, 90, 97–99, 101, 102, 104, 105, 107, 136, 144, 146, 149, 150, 245–258, 261–290, 315, 321, 323, 324
HMSiR fluorophore ... 264
H-watershed segmentation ... 194, 195, 197

I

IGF2 mRNA-binding protein 1 (IMP1) ... 40, 41, 52
Image registration ... 182, 193
Imaging medium ... 136–138, 141–143, 151, 152, 214, 217, 314
Immunocytochemistry ... v, 184, 186, 247, 251, 256, 296
Inhibitory synapses ... 177, 179, 197
Insulin-like growth factor 2 (IGF2) ... 41
Intensity correlation analysis (ICA) ... 193, 195, 197
Intracellular transport ... 1–10, 157, 158
In vivo live animal imaging ... 12

K

Kinesin ... 1, 2, 5, 6, 9, 157, 158
Kohinoor ... 231, 232, 235–237, 242, 243

L

Lateral diffusion ... 23, 115, 131, 151
Lifetime determination ... 50
Liquid-liquid phase separation (LLPS) ... 39
Live-cell imaging ... 175, 213
Localization densities ... 85, 175
Localization microscopy technique ... 83, 111, 173, 229, 262, 290
Localization precisions ... 29, 30, 92, 111, 120, 121, 124–126, 128, 168, 174–176, 180, 182, 184, 190–192, 205, 223, 246, 254

M

Membrane diffusion ... 150
Membrane molecular dynamics ... 150
Microfluidics ... 109, 157–169
Microtubules ... 1, 2, 159, 177, 206, 213–218, 284, 285, 287, 290
mRNA translation ... 295, 298, 305, 307
Multi-colour SMLM ... 204, 207, 209, 216, 217
Multipolarization ... 157–169
Munc13 ... 245

N

Nanobodies ... 82, 84, 89–96, 99, 101, 103, 104, 106, 109, 180, 181
Nanocores ... 40, 42, 51
Neurotransmitter receptors ... 131, 132, 177–180, 186
Nuclear pore proteins ... 221, 222
Nyquist-Shannon criterion ... 175

O

On-state polarization angle narrowing (OnSPAN) ... 230
Optogenetics ... 59–76
Oxidative stress ... 39, 121

P

Packing coefficient ... 115–128
Particle tracking analysis (PTA) ... 70, 72–74, 77
PC12 cell ... 42, 44–46
PHluorin ... 82, 83, 91, 107, 108
Photo-activated localization microscopy (PALM) ... 175, 204, 207, 229, 262, 263
Phototoxicity ... 121, 222, 229, 230, 242, 303, 304, 313
Plasmid transfection ... 5–7
Pointillist SMLM images ... 191
Point spread function (PSF) ... 47, 55, 175, 191, 204, 234, 235, 240, 241, 266, 269, 270, 303–305, 317
Polarizations ... 159, 161, 162, 168, 230–237, 239
Postsynaptic density (PSD) ... 132, 179, 195
Presynapses ... 89, 99, 101, 103, 107, 112
Protein-DNA interaction ... 59
Protein synthesis ... 131, 296, 303, 307
Primary neurons ... 6, 8, 42, 75, 98, 110, 186

Q

Quantitative cell biology ... 88
Quantum dots (QDs) ... 119–122, 124, 126, 127, 132–134, 136, 140, 141, 143–146, 148, 149, 151, 152

R

Ras GTPase-activating protein-binding protein 1 (G3BP1) ... 40, 41, 48, 52, 54
Regularized maximum likelihood ... 235
Rendered SMLM images ... 191
Retinal ganglion cells ... 295–307

Reversible blinking 208, 209
Reversibly photoswitchable fluorescent proteins (RPFPs) 231, 232
RNA-binding proteins (RBP) 41, 42, 51
RNA protein (RNP) complex 16, 17, 20, 39, 40
RNA transfection 6
Rotations 159, 168, 232, 238–240, 258

S

Scientific complementary metal oxide semiconductor (sCMOS) camera 4
Selective plane illumination microscopy (SPIM) 263, 264
Single molecule coordinate 191
Single molecule imaging 42, 59–76, 81–112
Single molecule localization microscopy (SMLM) 173–175, 177–180, 188, 191, 192, 198, 203–210, 212–214, 216–221, 223, 224, 229, 242
Single molecule localizations 40, 48, 203, 205, 246, 253, 263, 301
Single molecules 2, 5, 6, 39–56, 60, 62, 65, 66, 69–72, 75, 77, 82, 83, 87, 93, 94, 100, 108, 109, 111, 120, 132, 158, 180, 204, 217, 222, 262, 263, 311
Single molecule tracking 11–34, 42, 52, 83, 84, 99, 103
Single particle tracking (SPT) 116–118, 121, 124, 126, 128, 132, 310, 311
Small organic fluorophores 204–206, 208, 216, 223
SNAP tag 5, 65, 206
Spatial resolution 26, 27, 175, 178, 181, 205, 222, 229, 230, 242, 243, 246, 258
Spatiotemporal resolution 111, 263, 287
Spinal cord neurons 179, 184, 194, 196
Spines 105, 177, 310, 323–325
Spinning-disk confocal microscope 213, 220, 221
Stochastic optical reconstruction microscopy (STORM) 175, 176, 186, 204, 207, 229, 246–249, 251–258, 262, 263
Stress granules (SGs) 39–42, 51, 52
Subdiffractional tracking of internalized molecules (sdTIM) 82–87, 89–95, 98, 99, 101, 103–112
Subsynaptic domains (SSDs) 177, 179, 194–197
Super-resolution optical microscopy 287
Super-resolution polarization demodulation (SPoD) 230, 232, 234
Synapses 16, 23, 33, 105, 116, 121, 131, 134, 144, 146, 148–150, 173, 174, 177–179, 186, 190, 193–195, 197, 198, 245, 246, 255, 256, 309, 310
Synaptic plasticity 60, 116, 178, 179, 195, 310
Synaptic proteins 141, 173–198, 245, 246
Synaptic transmission 82, 89, 107, 131, 245, 309
Synaptic vesicles 81–112, 150, 245

T

Tetracycline-inducible promoter (Tet-On) 65, 68
Tetramethyl rhodamine (TMR) 43, 45–47, 51, 54, 61, 68, 69
Three dimensional imaging 286, 290
Time-lapse SMLM 214–216
Total internal reflection fluorescence (TIRF) 3, 4, 12, 27, 42, 43, 46, 55, 60, 83, 93, 186–188, 212, 213, 220, 301, 302
Transcription factors 59–77, 296
Trans-synaptic nanocolumn 177
Two-photon microscopy 311, 314

V

Vesicle-associated membrane protein 2 (VAMP2) 82, 85, 89, 101, 105, 108
Voltage-gated calcium channels (VGCCs) 13, 14, 16, 17, 23, 25–34

X

Xenopus laevis 297, 299